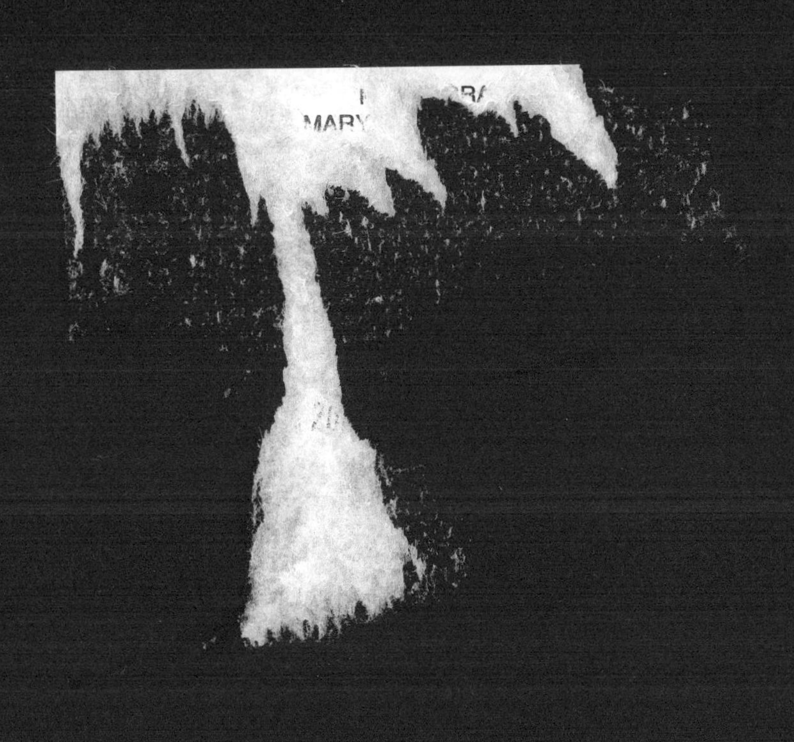

Work and Workers

SAGE LIBRARY IN BUSINESS AND MANAGEMENT

Work and Workers

VOLUME 3

edited by

Cary L. Cooper and
William H. Starbuck

SAGE Publications
London • Thousand Oaks • New Delhi

First published 2005

SAGE Publications Ltd
1 Oliver's Yard
55 City Road
London EC1Y 1SP

SAGE Publications Inc.
2455 Teller Road
Thousand Oaks, California 91320

SAGE Publications India Pvt Ltd
B-42, Panchsheel Enclave
Post Box 4109
New Delhi 110 017

British Library Cataloguing in Publication data

A catalogue record for this book is available from the British Library

ISBN 1-4129-0882-5 (set of three volumes)

Library of Congress Control Number: 2004096672

Production by Deer Park Productions, Tavistock, Devon
Typeset by TW Typesetting, Plymouth, Devon
Printed and bound in Great Britain by TJ International Ltd,
Padstow, Cornwall
Printed on paper from sustainable resources

Contents

VOLUME 3

PART EIGHT PERSONAL AND ORGANIZATIONAL IDENTITY

PART NINE TEAMWORK

Part Eight

Personal and Organizational Identity

44

Keeping an Eye on the Mirror: Image and Identity in Organizational Adaptation

Jane E. Dutton and Janet M. Dukerich

Source: *Academy of Management Journal* 34 (3) (1991): 517–554.

> The homelessness problem is perhaps a blight on that professionalism that we like to display, and that we are so proud of, and I think this is of great concern there. Again, there may be some conflicting issues on spending money to help solve the problem, but I think that's a value. We build beautiful facilities, we take pride in that, and the homelessness issue is something that obviously affects the perceptions of us (facility staff member, Port Authority of New York and New Jersey, 1989).

Theoretical Perspective

Models of how environments and organizations relate over time have typically assigned causal primacy to either environmental or organizational forces. Advocates of institutional theory, resource dependence, and population ecology have highlighted the environmental, and strategic choice theorists have emphasized the organizational. Still other theorists have assigned primacy to some combination of the two forces (e.g., Hambrick & Finkelstein, 1987; Hannan & Freeman, 1984; Singh, Tucker, & House, 1986; Tushman & Romanelli, 1985). None of these theories treat in depth the processes by which environments and organizations are related over time. Although the language theorists have used implies that a process determines how environments and organizations are connected — organizations chose strategies in response to environmental changes, or environmental selection mechanisms favor one structural form more than others — views of the process through which these relationships are accomplished are currently limited (Sandelands & Drazin, 1989).

In this research, we developed a framework for conceptualizing the process through which organizations adapt to and change their environments. Conceptually and empirically, we took seriously the assertion that organizations respond to their environments by interpreting and acting on issues (e.g., Daft & Weick, 1984; Dutton, 1988b; Dutton & Duncan, 1987; Milliken, 1990). Patterns of actions in response to issues over time create patterns of organizational action that in turn modify an organization's environment. Our claims were built from a case study of how the Port Authority of New York and New Jersey[1] has defined and responded to the issue of the rising number of homeless people present in the facilities it operates.

The case study was used to generate a framework for understanding how organizations and their environments interrelate over time. We employed the idea that organizations have identities (Albert & Whetten, 1985; Ashforth & Mael, 1989) that influence how individuals interpret issues as well as how they behave toward them. The assertion that organizational identity affects issue interpretations and actions has received some support from other studies of organizational adaptation (Meyer, 1982; Miles & Cameron, 1982). The present study also built on ideas from impression management (e.g., Tedeschi, 1981), suggesting that individuals seek to influence how others see and evaluate their organization. The article crosses between macro and micro organizational theory to explain how the Port Authority has dealt with the homelessness issue.

Issues as a Starting Point

Our perspective is that some organizational actions are tied to sets of concerns that we call issues. Issues are events, developments, and trends that an organization's members collectively recognize as having some consequence to the organization. Issues can arise from changes inside the organization, such as employees threatening to stage a strike or a new technology transforming a product or service, or changes originating externally, such as a demographic trend, a regulatory act, or a supply shortage.

The definition of an issue by a collectivity is a "social construction" (Hilgartner & Bosk, 1988). Issue definitions often emerge and evolve over time, and they can be contested (Dutton, 1988a; El Sawy & Pauchant, 1988; Feldman, 1989; Isabella, 1990; Weiss, 1989). Which issues gain attention and how they are interpreted are important concerns, as issues represent focal points that galvanize interest and direct attention in organizations because of the consequences associated with action or inaction. In some cases, issues activate decisions; in other cases, issues incite neglect or intentional inaction (Bachrach & Baratz, 1972).

A focus on issues as a starting point for interpretation and action in organizations charts a different course for seeing patterns of organizational action than a traditional decision-making view. Researchers who look at decisions as creators of patterns in organizational actions (e.g., Mintzberg, Raisinghini, & Théorêt, 1976; Nutt, 1984) have used the end point of a process — a choice or an absence of choice — as the defining referent and described who and what were involved in producing a certain pattern of action. Typically, researchers define a decision and trace backward from that point to find interpretations for it and actions relevant to it. In contrast, a focus on issues begins with an issue or a collective construction that some datum, fact, or event is of concern for an organization and then proceeds forward from this recognition point to find relevant actions and interpretations. Like the "garbage can model" of decision making (Cohen, March, & Olsen, 1972), an issue focus

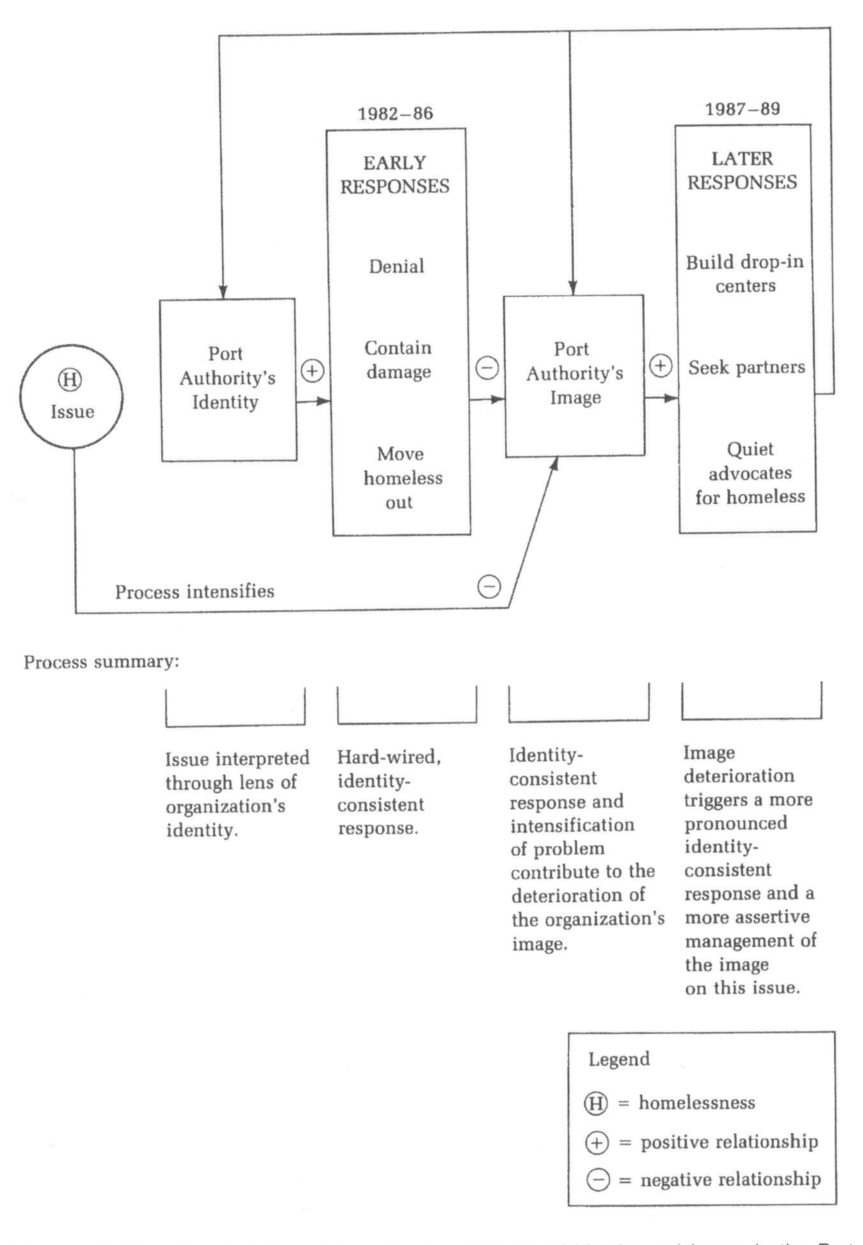

Figure 1: Simplified depiction of the role of organizational identity and image in the Port Authority's response to homelessness.

underlines the importance of attention allocation and sensitivity to context. Unlike the garbage can model, an issue focus is open to changes in issue interpretations over time. The present research adds to research on the temporal dimensions of interpretations (e.g., Dutton, 1988a; Isabella, 1990) by describing how organizational context contributes to how and when issue interpretive changes occur.

For organizations, some issues are routine and expected, and organizational members can easily classify them. The issues fit existing categories and, once classified, elicit a well-learned response (Starbuck, 1983; Starbuck & Milliken, 1988; Weick, 1988). The well-learned responses are types of organizational "recipes," or patterns of routinized behaviors that are easily available and rewarded in an organization (Weick, 1979). Other issues are not as easily interpreted or processed, however. Issues may be problematic because they are nontraditional: they have not been encountered in the past and thus do not easily fit well-used categorization schemes. Alternatively, issues may be problematic because of the feelings they evoke. Current models of issue diagnosis and organizational adaptation reveal very little about how the level of emotion an issue evokes affects individual and collective processes. Issues that are hot — those that evoke strong emotions — represent different types of stimuli and activate different responses from individuals and organizations than cooler, less affectively charged issues.

The Purpose of the Present Study

Our interest in how individuals and organizations make sense of and act on nontraditional and emotional strategic issues drew us to the case of the Port Authority of New York and New Jersey and its dealings with the issue of homelessness. The study was designed to generate new theory on how individual interpretations and organizational action on an issue are related over time.

In brief, our analysis revealed that an organization's identity and image are critical constructs for understanding the relationship between actions on and interpretations of an issue over time. Both constructs emerged clearly from a theme analysis of the data. An organization's identity, or what organizational members believe to be its central, enduring, and distinctive character (Albert & Whetten, 1985), filters and molds an organization's interpretation of and action on an issue. Organization members monitor and evaluate actions taken on issues because others outside the organization use these actions to make character judgments about it (Alvesson, 1990) and, by implication, its members. Organization members use an organization's image, which is the way they believe others see the organization, to gauge how outsiders are judging them. Deterioration of an organization's image is an important trigger to action as each individual's sense of self is tied in part to that image. Thus, individuals are motivated to take actions on issues that damage their organization's image (Ashforth & Mael, 1989; Cheney, 1983). At the same time, the organization's identity limits and directs issue interpretations and actions. These actions in turn may gradually modify the organization's future identity or make certain features of the identity more or less salient. Figure 1 presents a brief summary of the role of organizational identity and image in the Port Authority's response to homelessness.

Methods

A case study methodology was well suited to our goal of generating and building theory in an area where little data or theory existed (Yin, 1984), where we could study a process as it unfolded over time, and where we could use "controlled opportunism" to respond flexibly to new discoveries made collecting new data (Eisenhardt, 1989: 539).

We selected the case of how the Port Authority of New York and New Jersey has responded to the issue of homelessness because of the issue's social relevance and its visibility to both organization members and outside constituencies. In this sense, the case meets the criteria for an "extreme case," one in which the process of theoretical interest is more transparent than it would be in other cases (Eisenhardt, 1989).

Data Sources

The story of how the Port Authority and the issue of homelessness are related was built from five sources: (1) open-ended interviews with 25 employees of the Port Authority conducted from September 1988 to May 1989, (2) all reports, memos, and speeches prepared within the Port Authority on homelessness from November 1982 until March 1989, (3) articles from regional newspapers and magazines published from March 1986 through November 1988 that mentioned both the Port Authority and homelessness, (4) regular conversations with the head of the Homeless Project Team, a temporary task force of Port Authority employees charged with examining the corporation's response to the issue of homelessness, and (5) notes from an all-day training session with Port Authority facility staff members sponsored by the Homeless Project Team in May 1989. All informants were full-time employees of the Port Authority.

Informants

Individuals from four groups with different types of contact with and responsibility for the homelessness issue were informants. We interviewed the Port Authority's executive director and three top-level managers who were involved with the issue; all six members of the Homeless Project Team, line managers with responsibility for the facilities that were actively trying to deal with the issue; five staff members from the public affairs, corporate planning, and budget offices with responsibility for developing and analyzing ideas for a Port Authority response to the issue; and finally four people who dealt hands-on with the homeless in various Port Authority locations, including police officers and customer service managers.

Our initial research objective was to explore differences in how groups in the organization interpreted and responded to the issue. The objective was consistent with research on organizational culture (e.g., Martin & Meyerson, 1988) and the creation of meaning in organizations (e.g., Donnellon, Gray, & Bougon, 1986), which led us to expect a high degree of inconsistency,

Table 1: Interview Guide

Variable clusters	Illustrative questions
Issue interpretation	
Emotionality	As you think about the homelessness issue, what adjectives would you use to describe the issue?
Distinctiveness and similarity to other strategic issues	How do you see this issue as different from other strategic issues facing the Port Authority?
Perceived hotness	Imagine there was a thermometer for gauging how hot the homelessness issue was. Please indicate how hot you believe this issue is on a 7-point scale and explain the basis for your rating.
Interrelationships with other issues	What other issues inside or outside of the Port Authority is the homelessness issue related to?
Personal involvement in the issue	
Time spent on it Amount of direct contact with homeless people Change in involvement	Describe your involvement in the issue. When did you first get involved? How much of your time do you spend dealing with the issue? How has your involvement changed over time?
Organizational processing and actions on the issue	
When first noticed	Describe how and when the homelessness issue first became an issue at the Port Authority.
Major milestones	What have been the major milestones in the processing of the issue?
Major setbacks	What have been the major setbacks in the process?
Major successes	What have been the major points of success?
Perceived effectiveness of issue processing	
Costs and benefits of the Port Authority's involvement	What do you believe will be the major benefits and costs of the Port Authority's involvement in the homelessness issue?
Evaluation of the Homeless Project Team's handling of the issue	How has the Homeless Project Team affected you and how will you know if it's been a success?
Organizational context for the issue	
Shared values at the Port Authority	If you were to describe the values that people share at the Port Authority, what would they be?
Institutional mission	How would you describe the overall mission of the Port Authority?

disagreement, and ambiguity in how organization members interpret strategic issues. However, the data generated by the informants indicated a surprisingly consistent pattern of issue interpretations. Thus, the pattern of interpretations revealed in this study emphasizes the dominant logic (Prahalad & Bettis, 1986), collective beliefs (Walsh, Henderson, & Deighton, 1988), and consensual elements (Gioia & Sims, 1986) in how the homelessness issue was interpreted over time.

Interview Questions

The interview guide targeted data on five clusters of variables, which Table 1 describes. The average interview lasted two hours, with one researcher asking questions while the other took notes. More than half of the interviews were tape-recorded and transcribed verbatim.

Data Analysis

"Analyzing data is at the heart of building theory from case studies" (Eisenhardt, 1989: 11). Two analyses were critical for the purposes of this article: construction of the issue's history as depicted in interpretations, actions, and events from 1982 into 1989 and use of theme analysis to explain the pattern of interpretations and actions over time. Both analyses emerged from an identifiable set of steps.

Step 1: Devising and Coding Using a Contact Summary Form

Following the procedures Miles and Huberman (1984) recommended, we used a contact summary form for recording the main themes, issues, problems, and questions in each interview; one researcher originated each form and the other coded it. We defined themes as recurrent topics of discussion, action, or both on the part of the actors being studied (Bjorkegren, 1989). Like a recurring melody in music, a theme captures the central ideas or relationships in an interview (Bjorkegren, 1989).

Step 2: Developing a Complete Theme List

The contact summary forms for the 25 interviews generated 84 themes, which we collapsed into seven major groupings based on a very general classification of theme substance. For example, "organizational reactions to homelessness" and "the identity of the Port Authority" were broad theme categories. The first broad category included 14 different themes, each addressing unique ways that the Port Authority responded to the homelessness issue, such as denying being in the social service business or reacting negatively to other agencies' failures to take responsibility for the issue. We used the themes for two distinct purposes: to isolate commonalities in how Port Authority members interpreted homelessness and to suggest an explanation for the issue's history in terms of our dominant theme categories — the importance of organizational image and identity. Next, each theme was assigned a separate sheet on a coding form in preparation for step 3.

Step 3: Coding the Interview Data onto the Themes

Each interview was coded sentence by sentence onto a theme list in order to document and evaluate the degree and breadth of support for particular themes across informants. After completing the theme-based coding process, we were able to evaluate the degree of support for each theme indicated by the number of theme-related points mentioned both within and across interviews.

Step 4: Constructing an Issue History

We used questions on the meaning of the issue and on milestones in its processing to construct a history of how the Port Authority interpreted and responded to the issue over the period studied. Informants consistently identified 1982 as the year in which homelessness became an issue for the organization. Thus, we did not set the starting date but saw it emerge from informants' accounts of milestones in the issue's processing. Information from memos, speeches, and meeting minutes served as important supplements to interview data in constructing the issue history. We consulted members of the Homeless Project Team to validate the issue history once it was completed.

The Issue

The presence of homeless people has always been part of the scene at transportation facilities. Several informants noted the qualitative shift that took place in the early 1980s, when people previously referred to in the transportation trade as "bums, winos, and bag ladies" were transformed into "the homeless." During the last several years, the number of homeless people living and spending time at transportation facilities has dramatically increased. For the Port Authority, an agency that runs many diverse transportation-related facilities, the rising number of homeless people at its facilities caused increasing problems with the delivery of quality transportation service. One of our informants described the change this way:

> Well, a lot of it had to do with the change in the type of people. . . . And the bus terminal always had its share of down-and-out people, but you were able to move them along and get some kind of arrangement with them. But as the numbers increased, you couldn't do that. And the nature of the people began to change, and they began to get younger, and in some respects the people [the Port Authority's patrons] became more afraid of them because they were rowdier, they were more imposing.

In addition to the trend of rising numbers and change in type, three other issue characteristics were mentioned by more than ten informants as distinguishing homelessness from other strategic issues of importance to their organization. First, informants consistently mentioned the issue's broad scope and its linkages to other regional issues such as decreasing housing availability and changes in the skills represented in the region's labor market. Second, they emphasized the links between homelessness and other negative issues such as drugs and crime — links that magnified the fear and aversiveness that individuals expressed about the issue. Finally, close to two-thirds of the Port Authority informants mentioned the lack of control that they felt the organization had over the issue and possible solutions. One facility manager's description of his frustration with the issue captures that assessment well:

> I think with all of the building and fixing and all of those good, concrete, reassuring things that we did and still do, and the feeling, the good feeling that we got from being in control, I think this has been undermined in a way by the homeless problem. I think that it said to us, "Look, here is something that you really can't control, and you can't fix it, and you can't caulk it, you can't waterproof it, you can't dig it, and you can't make it go away."

This lack of control and other themes revealed in our analysis can be better understood in light of the distinctive features of the organizational context in which members of the Port Authority struggled to make sense of and respond to the homelessness issue. We describe the organizational context in two sections. First, we describe general features of the Port Authority.

Next, we discuss aspects of the organization's identity as perceived by its members. Those perceptions proved crucial for explaining the evolution of interpretations of the issue and actions on it over time. Although we did not originally intend to make the organization's identity so central to the explanation of how the organization adapted to this issue, individuals' senses of the organization's identity and image were metathemes that emerged from our data analysis, and we believe they organize the evolutionary story in a compelling way. Following descriptions of five phases into which we divided the history of the issue, we return to the substance of the Port Authority's identity and image to analyze how they give coherence to the evolution of interpretations, emotions, and actions and also to draw general inferences about the usefulness of these constructs for models of organizational adaptation.

The Site

General Features

The Port Authority of New York and New Jersey was established on April 30, 1921, the first interstate agency ever created under a clause of the Constitution permitting compacts between states with congressional consent. Its area of jurisdiction, the "port district," is a 17-county bistate region encompassing all points within a 25-mile radius of the Statue of Liberty. The mandate of the agency was to promote and protect the commerce of the bistate port and to undertake port and regional improvements that it was not likely private enterprise would invest in or that either state would attempt alone. The Port Authority provides wharfage for the harbor the two states share, improves tunnel and bridge connections between the states, and, in general, undertakes trade and transportation projects to improve the region.

Most public authorities in the United States were established to develop and operate a single public improvement project like a bridge or an airport; the Port Authority was the first multipurpose public authority (Caro, 1974). Today it

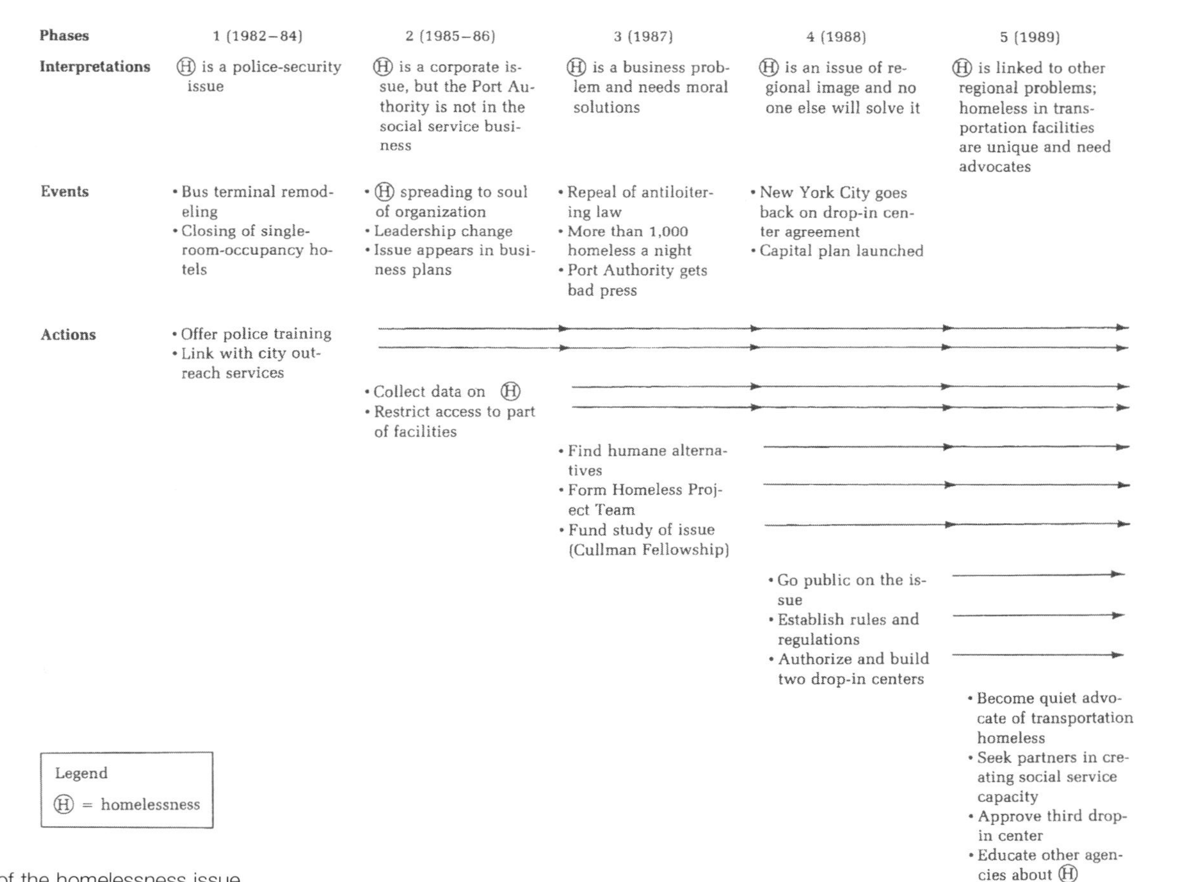

Figure 2: History of the homelessness issue.

owns and operates 35 facilities, including the World Trade Center; the Port Authority Bus Terminal at 42nd Street; Journal Square Path Center; Kennedy, LaGuardia, and Newark airports; PATH train service,[2] and many tunnels, bridges, and marine facilities. The mission of the Port Authority remains very broad — to protect the economic vitality of the New York–New Jersey Port District. The organization defines itself as being in the business of transportation.

The Port Authority is the largest public authority in the United States, employing 10,000 people and having total assets of approximately $5 billion and an annual budget of $1 billion. It supports itself through issuing bonds and collecting user fees and leasing revenues. An executive director and a board of commissioners selected by the governors of the two states run the organization.

The Identity of the Port Authority

Six attributes summarize the informants' views of the characteristics that distinguished their organization (Albert & Whetten, 1985). First, 100 percent of our informants called the Port Authority a professional organization with a uniquely technical expertise, ill-suited to social service activities. Second, informants (44%) referred to their organization as ethical, scandal-free, and altruistic. Third, 36 percent described it as a first-class, high-quality organization and a provider of superior service. Fourth, 36 percent of informants said the agency prided itself on its high commitment to the welfare of the region. Part of this dimension of the Port Authority's identity was a sense that the organization "spoke for the region" and symbolized its successes and shortcomings. Fifth, informants (32%) mentioned the loyalty of employees and their sense of the Port Authority as family. Finally, a fourth of our informants expressed a view of their organization as distinctive in terms of being a fixer, a "can-do" organization. As the story will reveal, the organization's identity was an important element of members' interpretations of the issue, acting both to prompt and constrain issue-related action and resulting in issue-related emotions.

Interpretations of and Actions on Homelessness

The Port Authority's struggle with the homelessness issue can be mapped onto five phases, each distinctive in terms of the interpretation of the issue current in the organization and its actions. Figure 2 presents a synopsis of the five issue phases as a timeline. The arrows indicate that once the actions so-designated were implemented, they continued over time. The arrows also show that the Port Authority's action repertoire expanded over the issue's history.

Although we present the five phases as though clear, identifiable signs separated one from another, they in fact shaded into each other. The path of understanding and responding to this issue can be thought of as an evolving history of interpretations, emotions, and actions. This history offers important insights into the organizational processes at work in creating patterns of action.

The five phases are described in terms of three components: key events, major interpretations, and major actions. The key events of each phase are the major developments and changes that informants identified as significant during a given phase of the issue's evolution. The events are crucial for comprehending how organization members interpreted the issue at each point in time and how and why the organization took certain actions. Although certain events appeared to have caused a certain action or interpretation, we refrain from making such causal inferences. Our purpose is to provide a relatively complete description of how interpretations and actions coevolved in the context of a series of unfolding events against the backdrop of this particular organization.

Phase 1: Homelessness is a Police-Security Issue (1982–84)

Homeless people have always been part of the landscape for transportation services. The features that are important for the delivery of effective service to transportation agency clients also attract the homeless. The facilities are warm in the winter and cool in the summer. They are clean, have toilets and running water, and guarantee people some degree of personal safety through the constant presence of police. Thus, for most transportation agencies and the police who patrol them, dealing with a certain number of homeless people has long been a normal part of business.

Key Events

In 1982, several factors converged to make homelessness a more prominent issue for the Port Authority, particularly at the bus terminal. First, organization members noted a marked rise in the number of homeless people present in their facilities. Second, a $226 million renovation that had just been completed at the bus terminal accentuated the visibility of the homeless. The renovation, which increased the building's square footage by about 40 percent, opened up new space for use by passengers and homeless people alike. At the same time, a large number of single-room-occupancy hotels in New York City closed. As one informant told us, "As the Manhattan real estate market picked up, these hotels were closed, and we had an increase in the number of homeless people, without many skills, without abilities, and without much money, all ending up out on the streets. A fair number of them ended up in the Port Authority bus terminal." The bus terminal's renovation accentuated the problem of the homeless by creating a strong contrast between the beautification of the facility, accomplished by adding space and expensive works of art, and the presence of homeless people who "smelled and looked dirty." To patrons and workers, homeless people marred the Port Authority's attempt to spruce up the bus terminal. For an organization that prided itself on being "the builder of beautiful structures," homeless people were a stain on its identity.

Major Interpretations

During 1982, organization members defined homelessness as a police or security issue: the presence of homeless people was problematic for Port Authority customers, and something had to be done. As one informant said, "The issue was 'How do we keep these people out of our facility?' Plain and simple, because they were interfering with our patrons in the sense that they felt that they were not safe because of their presence." The police were, and continue to be, a major source of organizational contact with the homeless at the bus terminal; police officers were also the organization members who carried out action on the issue. Customers confronted the police when they wanted someone from the Port Authority to "do something about this problem!" The organization employs 1,500 full-time officers, constituting the 26th largest police department in the United States, and 130 of them were assigned to the bus terminal. At this time, the police at the bus terminal and the facility's managers dealt with the issue; there was no coordinated corporate response.

Major Actions

The existence of an antiloitering law in New York City gave Port Authority police the option of insisting that homeless people leave the bus terminal. In 1982, bus terminal managers took two additional issue-related actions. First, they hired a consultant to train police officers on how to move people out of the facility in a manner that "acknowledged the difficult nature of the problem." Second, they established a relationship with the city's Human Resources Administration and the Manhattan Bowery Corporation[3] to develop an outreach program to "give the police some place to send these people." The officers helped workers from the Manhattan Bowery Corporation transport homeless persons from the Port Authority's facilities to shelters run by the Human Resources Administration.

Summary

Early Port Authority actions on homelessness were facility-based, limited in scope, and focused on the bus terminal. The organization framed the issue as primarily a police and security matter, an interpretation that, given the city's antiloitering law, helped contain the problem. Actions to engage the assistance of New York City's social service support system were also part of the facility-based solution at this time.

Phase 2: Homelessness is a Corporate Issue, but the Port Authority is not in the Social Service Business (1985–86)

Demarcations between phases in the relationship between the homelessness issue and the Port Authority are not clear-cut. However, in the 1985–86 period,

Port Authority members changed the way they talked about the issue. This change could be attributed to a number of different events and to the recognition that the problem extended beyond the bus terminal.

Key Events

Informants described having a growing awareness in 1985–86 that the homelessness issue was no longer confined to the bus terminal, where it was well understood and routines had been developed to deal with it. Now, the homeless were present in several Port Authority facilities. The appearance of homeless people at the World Trade Center and the airports — the organization's flagships — was the key to making the issue visible at the senior management level. Organization members did not expect to see the homeless in these facilities, and their presence conflicted with central components of the Port Authority's identity:

> It wasn't until homeless people started to show up at the World Trade Center . . . and the image of the World Trade Center as being a place where homeless people were began to raise its head, that people started to say, "Wait, geez, this is a problem. . . ." It [homelessness] started to show up finally in corporate documents as an issue. It never did before, because everybody knows the bus terminal is an aberration, but when it started to show up at the World Trade Center, and then ultimately, one or two people at the international arrivals building at Kennedy Airport and at LaGuardia Airport, then it began to touch upon the heart and soul of the organization.

The departure of the Port Authority's executive director and the appointment of a new director was another key event during this period. The leadership change was significant on several counts. First, facility managers and staff members assigned to work on homelessness argued that the momentum to recognize and deal with the issue at the bus terminal had come from the former director. That momentum dissolved with his departure, and advocates for the issue felt that they had to start over from the beginning. Second, the new executive director's vision for the organization was "returning to its basic businesses." The new director wanted to "[show others that] the Port Authority could run like a business." One implication of this change in vision was an emphasis on using business practices and business justifications as a basis for drawing attention to issues.

In 1986, for the first time the issue of homelessness appeared in business plans for several line departments. Simultaneously, the public affairs department became increasingly concerned about the issue as the rate and intensity of customer complaints increased. The new director openly expressed a strong personal aversion to straying from the main businesses of the Port Authority and "getting into the social service business."

Major Interpretations

In 1985–86, the interpretation of the issue shifted to a recognition that the problem was corporate-wide, not just a bus terminal police issue. The definition of homelessness as a corporate issue came about because Port Authority departments began to include the costs of dealing with the problem in their budgets. As one informant noted, "Corporate issues are identified theoretically through the business-planning process, which is both a strategic planning and a budgeting process." However, 85 percent of the informants mentioned that although they recognized at this time that homelessness was a corporate issue, they asserted they were not in the social service business. During this time, employees at all levels focused on how to minimize negative fallout from the issue by removing and restricting the problem as it presented itself at various facilities.

Major Actions

Three major actions distinguished the issue phase. First, the board and the executive director asked a group of staff members to collect data, analyze it, and make recommendations for a corporate policy on homelessness. Police and facility staff viewed this action as a sign that corporate attention was being directed at the issue. As one upper-level manager stated, the results from this analysis represented "the first time that it [homelessness] was explicitly recognized as a problem and put in writing." Second, actions at the facility level intensified: bus terminal managers (1) sought and obtained more extensive outreach services, with daytime as well as night-time assistance, through a contract with the Volunteers of America, a not-for-profit social service provider that sent volunteers to Port Authority facilities to assist homeless people and encourage them to go to shelters, and (2) closed or restricted access to areas of the bus terminal and removed patron benches from the waiting areas. The purpose of these actions was to make the bus terminal an undesirable place to be by "making it as unattractive and uncomfortable to the homeless as possible." As one informant told us, "I think some of it was motivated by aesthetics, that you didn't have the people sitting around and maybe they would find someplace else to go." The organization implemented similar types of outreach services and actions to make the facilities unattractive to the homeless at the two other Port Authority locations where the issue was visible, the World Trade Center and Journal Square Transportation Center.

The third action was an attempt by the bus terminal staff to manage patrons' understandings of and reactions to homeless people by issuing and posting a lengthy description of the types of homeless that patrons were observing at the bus terminal. This action was the first of many attempts to improve the image of the Port Authority using a well-learned recipe: "educating others or helping them get smart on the issue."

Summary

During this second issue phase, Port Authority members did not significantly change how they interpreted or acted in response to the issue. In fact, this phase can best be characterized as involving doing the same, but doing it harder. Although informants recognized a shift in corporate understanding of the issue, the organization maintained its fragmented, facility-based response with an overarching goal of "get[ting] the homeless out of here." Denial that the Port Authority was a social service agency accompanied the intense localized response. At the time, the staff at the bus terminal began to try to manage others' understanding of the issue of homelessness, an attempt that was to become more prominent as the staff became more involved with the issue and as the image of the bus terminal — and of the Port Authority through its affiliation with the bus terminal — deteriorated. This phase also marked the beginning of some serious soul-searching by employees and upper management focused in particular on what the role of the Port Authority should be with respect to this issue. As one informant put it, "And then we were saying to ourselves . . ., Can we get them out of there? Should we get them out of here? What are we supposed to do with them? Whose responsibility is this?" This type of concern ushered in the third issue phase.

Phase 3: Homelessness is a Business Problem and a Moral Issue (1987)

In 1987, several events contributed to changing the way the issue was framed and the level and type of the Port Authority's response to it.

Key Events

In late 1986, several events shifted the Port Authority's view of its responsibility for homelessness. First, informants indicated the nature of the homeless people spending time at transportation facilities abruptly changed, primarily because of the influx of crack, a derivative of cocaine that is easily obtained, relatively inexpensive, and very addicting. Links between homelessness, drugs, and crime accentuated the original problem. The increase in drug use and an associated increase in crime served to highlight the importance of police actions. However, at this same time the city's antiloitering law was repealed, significantly restricting the ability of facilities in the city to move the homeless out. For the police, the repeal of the antiloitering law "tied their hands," resulting in a real "blow to police morale." As one informant told us, "It's not that we ever arrested people for loitering. But the antiloitering law's existence allowed us, without as much hoopla, to ask people to move on or to leave."

The absence of a contract between the police officers' union and Port Authority management, dating from spring 1985, exacerbated the issue. There were tensions between the union and management, with the officers caught in the middle. "The individual police officers, in the middle of that issue, wondered who to take their direction from, management on the one hand reminding them of their oath to uphold the laws of the states of New York and New Jersey and the rules and regulations of the Port Authority. And on the other hand, the union advising them that they may end up losing their homes if they violate someone's civil rights."

The police union put pressure on the Port Authority to grant certain concessions by generating unfavorable press coverage about the organization. The union hired a public relations agency "to float stories about the Port Authority." The stories were intended to put pressure on the Port Authority to hire more police. "They [the public relations firm] generate publicity all the time, and the publicity is aimed at embarrassing the Port Authority and creating this climate of fear and stuff around its facilities to promote the police position, you know . . . that they need more cops and that sort of stuff." The bad press about the Port Authority peaked in late 1987 and early 1988, when 65 percent of the articles in the New York and New Jersey newspapers we reviewed were negative in tone. The Port Authority received negative press for its attempts to control homelessness through tightening regulations. A sample excerpt follows: "In its last board meeting before Christmas, the Port Authority of New York and New Jersey played Scrooge to Jersey City's poor by outlawing begging and sleeping at the Journal Square PATH Transportation Center" (*Jersey Journal*, December 11, 1987).

At the same time, in 1987 the number of homeless people congregating at Port Authority facilities surpassed 1,000 on some nights. This number represented an important threshold that, in the minds of organization members, made the issue no longer deniable for the organization.

Major Interpretations

The most significant change in the way the issue was defined during this period involved upper-level management's acceptance of some organizational responsibility for dealing with the issue and an acknowledgment that it was much more than a police problem. This interpretive shift represented an expanded concern for humane solutions and a heightened awareness of the issue's severity. An excerpt from an important internal memo from January 1987 illustrates this shift: "It is important to recognize that the agency is not in a position to solve the problems of the homeless. . . . The Port Authority's homeless policy is to encourage individuals to leave our facilities and find more appropriate shelter and services, and to minimize their return. . . . We seek to do this in a humane manner, through the assistance of social service agencies. . . ." The shift in the way that the issue was now being defined was subtle. There was still extensive denial of responsibility for solving the problem in any way beyond alleviating the burden on facility staffs, but there was new concern with choosing moral or humane solutions. Thirty-six percent of the informants noted the importance at this time of the Port Authority's acting and looking humane. In addition, there was a recognition that some of the social service mechanisms that were in place were having a positive effect and diminishing the burden on facility staffs.

Major Actions

The repeal of the antiloitering law provided a major impetus to the development (technically, an updating) of facility rules and regulations. The rules and regulations first appeared at the bus terminal, but the procedure spread rapidly to the other Port Authority locations. Police and facility staff viewed the regulations as important because they "gave us a mechanism to deal with certain types of personal conduct for anyone in our facilities." Nevertheless, the facility police viewed their options for dealing with the homeless as highly constrained, leaving many of them feeling "as if you're pumping out the ocean."

Informants at all levels acknowledged that space restrictions and closing off parts of the building were ineffective in minimizing the visibility of the homeless. Port Authority actions during this period indicated resignation to two facts: the problem could not be solved through outreach or restrictions alone, and the organization needed to take a stand.

> And then we kind of gave up, you know, we gave up some space. . . . They just sort of took over the waiting room. That was it. You know, we just didn't know what to do, you know, when you get 15 degree temperatures at night, and there's absolutely no place for them to go. And so, we said, well, how are we in good conscience going to throw them out of this facility? . . . And this was the first time that people really began to look at it and say, 'Wait a minute, you know, this is a real moral issue.' And this was when we decided to make the commitment. And while Grand Central and every place else was throwing them out, we weren't.

In 1987, top management reluctantly admitted the need to develop a coordinated corporate response to the issue. It was during late 1987 that the executive director decided to form a centralized project team, the Homeless Project Team, whose major responsibilities would include developing a Port Authority policy on homelessness, shifting the burden from the facility staffs, and reducing the amount of top management time spent on the issue. In many of our informants' minds, the formation of this team signaled that the Port Authority was ready to do something about this issue.

Another key symbol of top-level management's commitment to the issue was granting a one-year fellowship, the Cullman Fellowship, to a public affairs employee to study how the transportation industry was addressing the homelessness issue. The Port Authority established the Cullman Fellowship in 1962 to allow a staff member to undertake a one-year special project that was advantageous to both the individual's career and the agency. One informant described the significance of funding a fellowship that focused on this type of issue as follows: "It was a very risky thing for the Port Authority to do, because it is not typical of the transportation kind of issue or business or economic development issue that this kind of a conservative organization would generally grant."

Summary

In 1987, the level and type of attention being paid to the issue changed. Two important symbolic actions signaled internal and external constituencies that top management was now interested in the issue: the formation of the Homeless Project Team and the granting of the Cullman Fellowship. Early in 1987, the "batten down the hatches" response dominated, evidenced by the increase use of rules and regulations, restrictions on access to facilities and closings of parts of facilities. Although there was evidence that assistance from social service agencies and the use of rules and regulations were providing some relief, the problem worsened in terms of the numbers of homeless people. Several events transformed this early response into acceptance that the Port Authority needed to do something different and to do it in a way that did not violate the moral standards embedded in the organization's way of doing things. At this time, a rise in negative press coverage about the Port Authority severely damaged the organization's image. With the hands of facility police tied by the antiloitering law change, police-based solutions proved unsuccessful. In addition, the image of the authority as inhumane really bothered some of our informants and reaffirmed the importance of taking a more "humane stance" on the issue. Since the hotness of the homelessness issue increases with the coldness of the weather, a humane stance meant not endangering anyone "by throwing them out into the cold temperatures."

Phase 4: Homelessness is an Issue of Regional Image, and No One Else Will Deal with It (1988)

The year 1988 represents a period of significant action on homelessness for the Port Authority.

Key Events

Three events are important for understanding the unfolding of the interpretations, emotions, and actions concerning homelessness during this period. First, there was the launching of a $5.8 billion capital plan for the organization, aimed at updating facilities and improving the image of regional services to enhance the area's international competitiveness. This campaign introduced resource constraints and created expectations for positive press coverage and a corresponding positive image. As one informant said,

> We had embarked on this capital campaign at the airports and all of our facilities. We needed the resources to handle the program. It gave us the impetus . . . so we need to control other priorities as much as possible, particularly at the airports. From an organizational standpoint, we are focused on the major initiatives. We expected all of this positive press about the capital plan, and instead, all we have gotten is negative press about homelessness. It overshadows the positive.

The other two events were reactions to Port Authority actions on the issue during this phase. In order to do something "different," the organization decided to commit capital funds to establishing drop-in centers designed to provide social services to the homeless at two locations near its facilities. The two events related to this action were: (1) New York City informally agreed to take over the operation of the first center to be built but subsequently resisted doing so, and (2) there was organized opposition to the opening of a second drop-in center.

Major Interpretations

A speech given by the Port Authority's executive director in January to the Partnership for the Homeless in New York City publicized and structured the dominant interpretation of the homelessness issue and the organization's relationship to it for the first half of 1988. Many informants saw the speech as clear evidence that the Port Authority was publicly committed and was going to "do something" about the issue. This speech contained several critical points for understanding the actions and future interpretations of the Port Authority on this issue.

First, there was continued denial that the organization was "in the social service business." Second, the director described the homelessness problem as a regional responsibility, noting that the failure to solve it would have devastating consequences for the region. The speech symbolically associated homelessness with the fiscal crisis of New York City during the 1970s, an association that effectively communicated the seriousness of the issue for the entire region. The speech indicated that the issue's scope had broadened considerably and represented an attempt to involve others in the Port Authority's efforts to deal with the issue.

In the minds of organization members, positive actions could not overcome the damage to the Port Authority's image, and the stain from homelessness had spread to the entire region. As one top-level manager said, "The quality of life of the region is severely impacted by having as a kind of visible ornament, a large number of people who are described as homeless. . . . It creates an environment of extraordinary depression in a transportation mix which is already congested, difficult, and harassed. In some ways, like the graffiti on the subways, it is both a fact and a symbol that the environment is out of control." Some members believed that the Port Authority as an organization and the New York–New Jersey area as a region were unable to compete effectively in the international transportation market because of the image damage to the Port Authority.

At this time, the organization's leadership acknowledged that no one else would solve the issue, leaving them no choice but to get significantly involved:

> And so, once it became clear that we were really going to have to become more aggressive, I think at that point there was a kind of watershed which said, "We are going to have to do some things which clearly stretch our mandate, which commit both dollars and cents beyond what is appropriate,

> and what is probably on some level defensible, because the agencies that have this responsibility are just not prepared to act."

Informants were distinctly emotional when they described the realization that "the Port Authority was forced to get involved because no one else would." Anger, frustration, and disappointment that other organizations had shirked their responsibilities by not solving the problem were expressed by 56 percent of our informants.

Informants' descriptions of the Port Authority board's discomfort with the financial commitments to homelessness also revealed the negative emotions that accompanied heightened issue investment. One top-level manager expressed this feeling bluntly: "The board is very unhappy, and I think rightly so. They feel that we're spending money, which we are, which is money that is desperately needed for other things in terms of our mandate."

Emotional reactions, however, involved more than unease and anger at the organization's new role. Some informants described hurt and frustration brought on by accusations about their personal characters based, they believed, on outsiders' judgments of Port Authority actions on this issue. Many of the organization members felt good about what it was doing with the homeless but thought that others believed that the Port Authority was acting inhumanely. This discrepancy was distressing and hurtful for individuals. As one facility manager said,

> You know, the guy that's running the Lincoln Tunnel doesn't have a full perception of how the bus terminal or the homeless impact what he does on a day-to-day basis. But the minute he leaves and he goes to the cookout in his neighborhood and he meets somebody and this person says, "What do you do for a living?" "Oh, I work for the Port Authority." They say, "How can you stand that bus terminal, what can you do?" That's the name. That's the symbol of the Port Authority. It's the standard bearer. And you know, so personally everybody that's involved in any aspect of working for the Port Authority is identified with that place and with that issue.

Another facility manager described a case in which the press had "bashed" the Port Authority and made derogatory comments about the manager's personal character because of the Port Authority's refusal to set up tables in its facilities during Thanksgiving to serve the homeless. In fact, although the press did not report it, the Port Authority had paid for 400–500 Thanksgiving meals served at a local soup kitchen. The manager was deeply troubled because of the inaccuracy (in his mind) of the external portrait of the Port Authority and the misinterpretation of his actions: "When you see your name in print and they call you callous and you know that in your heart you are probably one of the more compassionate people about this issue, it's hard not to get angry."

During phases 3 and 4, the Port Authority's image suffered acutely from the association with homelessness. There was remarkable consensus from informants about the image's substance. Their view was that outsiders saw the Port

Authority as dirty (65 percent of informants used this term), dangerous (56%), ineffective (52%), and inhumane (24%) because of its association with homelessness.

At this time, the issue was clearly emotionally charged both individually and organizationally, and Port Authority actions heated up accordingly.

Major Actions

The most dramatic actions during this period involved financing and renovating facilities for two drop-in centers. In early 1988, the board approved expenditures for building and operating centers to service the bus terminal and the World Trade Center and was committed to opening them within a year. The total cost (initial operating and capital expenses) for these facilities was close to $2.5 million.

All our informants viewed the May 1988 completion and opening of the Open Door Drop-in Center, adjacent to the bus terminal, as a significant accomplishment, symbolizing the Port Authority's commitment to the issue. The center's opening reaffirmed members' views of the organization as able to "get things done." As an upper-level manager said, "There have been more major achievements than anybody would ever imagine because of the circumstances and the speed with which we have put this thing together."

In October 1988, New York City's Human Resources Administration went back on its informal agreement to take over the financing of the operation of the Open Door Drop-in Center, and the Port Authority altered its stance on the issue. First, some members of the Homeless Project Team and upper management expressed hesitancy about getting into building and managing drop-in centers. In their minds, the incident with the center taught them that they should not try to solve the problem of homelessness at that level because "we just get burned." As one informant told us, "Next time we will live with the problem much longer." Members of the task force and top management sensed that the process that had been used to get the center up and running created "expectations that the Port Authority would fund and operate facilities or created the impression that somehow the homeless at the bus terminal were the Port Authority's problem." Organization members became committed to eliminating this impression. Actions in the next issue phase were partly attempts to alter this false set of expectations.

Organization members also saw the financing and building of the second drop-in center as a significant milestone in processing the issue. This second drop-in center, the John Heuse House, officially opened in December to serve the homeless in lower Manhattan, near the World Trade Center. But the organized opposition of downtown business interests had made getting city approval for the facility a rocky process.

Summary

The year 1988 was a critical phase in the Port Authority's relationship to the homelessness issue. It marked a turning point in the sense that the organization now viewed the issue and justified action with a sense of resigned heroism — a sense that no one else would solve the problem, so the Port Authority would step in, in its usual, excellent way. The attachment of homelessness to concerns such as New York City's fiscal crisis and regional problems reframed the issue and broadened its boundaries (Feldman, 1989). The resigned admission that the organization had to take action on the issue was accompanied by a great deal of emotion about the unfavorable image the Port Authority had in the press, a sense of outrage that those responsible were not doing their job, and a sense of embarrassment and anger generated by negative press coverage of Port Authority actions on homelessness. The formation of the Homeless Project Team helped to congeal a set of initiatives that had already begun in earlier phases. Its members were important catalysts for establishing the two drop-in centers. Instrumental involvement in the issue significantly escalated during this period, evidenced by the expenditure of $2.5 million to fund the renovation for and initial operation of the Open Door Drop-in Center and the renovation for the John Heuse House.

Phase 5: Homelessness is an Issue of Regional Competitiveness, and the Port Authority is a Quiet Advocate (late 1988–early 1989)

Although the Port Authority's relationship to the issue of homelessness is still evolving, data collection for this study ended in May 1989.

Key Events

When active data collection was nearing an end, one event stood out in the minds of informants. In its February 27, 1989, issue, *Newsweek* published a particularly damaging article entitled "The Nightmare of 42nd Street." The article portrayed the bus terminal as a dangerous place for both commuters and the homeless, "a vortex of hopelessness, crime and despair." One day after this article was published, the Port Authority's board convened an emergency group to "try to do something dramatic to turn around the Port Authority image." The formation of this group signaled heightened frustration with the tarnishing of the organization's image through the equation of the Port Authority with the bus terminal and the strong association of the bus terminal and homelessness. The *Newsweek* article and information the organization collected during this period also led to the acknowledgment and articulation that the problem with the bus terminal was far broader than homelessness — it also involved the issues of loitering and drug abuse.

Major Interpretations

During the spring, informants indicated an increasing awareness that although there had been some significant victories, the homelessness problem was not going away. The press was still bashing the Port Authority although with less intensity than during the previous two years. Informants acknowledged that the previous winter had been mild, making the visibility of homeless people in Port Authority facilities unusually low. At the same time, several of the organization's initiatives, such as revising the rules and regulations and providing social service assistance, were producing some positive results. Top management claimed that the number of complaint letters received weekly was significantly lower than it had been the previous year, going from an average of seven letters a week at the bus terminal to an average of one letter a week.

Completion of the Port Authority–funded drop-in centers for the homeless signaled an increasing acknowledgment that the organization was getting more and more into the business of homelessness. As one informant put it, "Yeah, we're two feet deep into the business of homelessness, and we don't want to be." Another informant displayed the ambivalence that accompanied this change in level of involvement: "We may be throwing a lot of resources at this, but our heart just isn't in it."

A shift occurred in the Port Authority's definitions of its role in the homelessness issue. Members of the Homeless Project Team said that role was helping others "create capacity" for single men, the typical homeless people at transportation facilities. So, although management still adamantly denied that the organization was in the housing or social service business, they sought to accomplish some social service objectives "by increasing the capacity of other agencies that are better equipped to substantively address this issue."

Major Actions

The Port Authority continued to implement the formulas for dealing with the issue that it had developed over the previous six years. It established outreach services at the airports. It also financially backed a deal with Jersey City to set up a drop-in center and a single-room-occupancy hotel to be run by Let's Celebrate, originally a soup kitchen and pantry operator, near Port Authority facilities at Journal Square. The drop-in center concept was consciously modeled after the John Heuse House arrangement, which management viewed as a more successful and appropriate model than the Open Door Drop-in Center because it minimized the visibility of Port Authority involvement through turning operations over to a service group. The Port Authority encountered delays and resistance to these facility solutions but treated the resistance as "normal" and "part of the process." The sense of urgency and outrage that had accompanied previous setbacks with the first two drop-in centers were notably absent. As one informant told us, "You learn that those people who fight you the hardest, may turn around and be your biggest advocate."

Awareness of rising Port Authority involvement in the issue (spending more money, adding services at more facilities) coexisted with a conscious attempt to minimize the organization's public association with the issue. Management explicitly designed its policy to favor the role of "quiet advocate for the single homeless male." Consistent with this thrust was a desire to not take the credit for any action on or solutions to the problem. For example, one staff member who remarked that a local paper's coverage of an incident had been "balanced" and "good" explained that this meant the paper had not mentioned that the Port Authority had played any role in bringing about the successful solutions the article described. As a top manager explained, "I don't want any credit. Let them take the credit. Let the bastards who fought us six months earlier take the credit. It's easy to give the credit. I prefer to work behind the scenes."

Part of the quiet advocate role involved educating others about the special needs of homeless people at transportation facilities. The Port Authority began to actively seek connections with other transportation agencies on the issue. For example, members of the Homeless Project Team began to meet with their counterparts at the Metropolitan Transportation Authority. As one Homeless Project Team member explained, "We are trying to broaden the circle of people who participate, working with the business community as a team." The form of these partnerships and the sorts of solutions implied were not made explicit. However, the Homeless Project Team stated that the agency would offer its "special expertise and viewpoint on the issue to New York City and to businesses who needed it."

Publicity on the Cullman Fellowship and other efforts to manage outsiders' impressions of the Port Authority's stand on homelessness had an unintentional consequence. Increasingly, people both within and outside the organization viewed it as a leader on the issue. Informants described the Port Authority as "on the cutting edge of what a transportation agency can do on this issue" and as offering "the most creative solutions to this problem." However, some managers were quick to see that this reputation was a double-edged sword: "I think there is another temptation, which is a peculiar Port Authority temptation. There's a tendency in a lot of places around this organization that wants people to get involved in something, and they want to be leaders in it. I just want to deal with this problem, not become a leader on it."

Summary

The relationship of the Port Authority homelessness took a new turn in 1989. Although the organization's position was still not solidified (one informant said, "We are still like an amoeba with this issue"), its actions were increasingly deliberate and intentionally highlighted or down-played. During the part of 1989 in which we collected data, the Port Authority managed the context in which the issue was affecting it more actively than before. These efforts included searching for partners with whom to design new collective solutions to this regional crisis. Efforts involved presenting information about the issue and information about the Port Authority's actions on the issue in a way that would minimize image damage by disassociating the organization from the issue. The efforts took place within the constraints of taking actions consistent with the Port Authority's identity, actions that complemented its perceived expertise. At the same time, the organization was increasingly recognized as a leader on how to deal with homelessness in the transportation industry. Port Authority members expressed tremendous pride in the organization's method for dealing

Table 2: Organization's Identity and Issue-Related Behaviors

Characteristics of Port Authority's identity	Percentage of informants who mentioned characteristic	Examples of relationship to issue behaviors		
		Interpretations	Emotions	Actions
Professionalism, technical expertise, no social service expertise	100	Constrains what are considered legitimate versus illegitimate issues: Not in social service business (phase 2)	Evokes strong negative emotion if identity compromised: Engineers holding AIDS babies (phase 4)	Provides recipes for issue action: Getting selves and others ''smart'' on the issue (phase 2)
Ethicality, altruism, public service ethic	44	Activates salient issue categories: Moral and business issues (phase 3)	Negative emotion evoked if negative image assumed to be the identity: Anger at bad press (phase 3)	Sets parameters for acceptable and unacceptable action: No moving the homeless out into the cold (phase 3)
Commitment to quality	36	Reference point for assessing importance of the issue: Homeless spoil attempt at beautifying bus terminal (phase 1) and stain image of Port Authority flagships (phase 2)	Negative emotion evoked when not able to resolve the issue: Frustration in not being able to fix the problem (phase 2)	Provides guidelines for evaluating issue success: Speed of completion of drop-in centers (phase 4)
Commitment to region's welfare	36			
Employee loyalty and employees as family	32	Identity-inconsistent behaviors signal heightened issue commitment: Granting fellowship for study of nontraditional issues seen as risky (phase 3)		
Can-do mentality	25		Strong emotions expressed when identity reinforced in unusual situation: Port Authority–funded drop-in centers provide better service than New York City social services (phase 4)	

with the homeless. In their eyes, it was the "most humane approach" used by any transportation agency in the region.

The Role of Organizational Identity and Image

The story of the Port Authority's relationship to the issue of homelessness is still unfolding today. Despite the story's complexity, the evolution of interpretations, actions, and emotions is sufficiently suggestive to allow us to extract, examine, and build on several important themes.

Two central themes that emerged from our analysis of interviews, media coverage, and internal memos focus on the role that the organization's identity and image played in creating the pattern of how individuals in the organization interpreted and responded to the homelessness issue. Specifically, we found that the Port Authority's identity, or how organization members saw it, played a key role in constraining issue interpretations, emotions, and actions. At the same time, the organization's image — how organization members thought others saw it — served as a gauge against which they evaluated and justified action on the issue. In addition, the organization's image was an important mirror for interpretations that triggered and judged issue action because of a close link between insiders' views of the organization and insiders' and outsiders' inferences about the characters of organizational members.

Over time, actions taken on issues reposition an organization in its environment by modifying tasks, allocation of resources, and assignments of personnel. The pattern of action on issues can therefore reinforce or, potentially, transform the organization's identity and image through individuals' sense-making efforts, and the process of adaptation continues.

The Importance of Organizational Identity

The Port Authority's identity is a critical construct for understanding the evolution of issue interpretations, emotions, and actions over time. We discussed the consensual attributes of that identity earlier and present them again in Table 2, which also summarizes the relationship between the Port Authority's identity and issue interpretations, emotions, and actions by using examples from the phases described in the issue history. The elements in this table provide important material for the beginning of a theory of how organizational identity affects adaptation processes through its effect on issue interpretations, emotions, and actions.

Identity and Issue Interpretations

The Port Authority's identity shaped its members' interpretations of homelessness in at least three different ways. First, the organization's identity served as an important reference point that members used for assessing the importance of the issue. Perceptions of issue importance are in turn important predictors of willingness to invest in an issue (Dutton, Stumpf, & Wagner, 1990). The issue was important because it threatened key elements of identity. In particular, informants' sense of the Port Authority as a high-quality, first-class institution made the presence of homeless people problematic. The expanding scope of the issue over time can be seen as an indication that the issue was being seen as more important and urgent as it threatened central identity components. Although Port Authority members were uncomfortable with the stain on the organization's identity when the problem worsened at the bus terminal, they interpreted it as even more threatening when the presence of homeless people affected the quality of flagship facilities such as the World Trade Center and the airports. Further, the intractability of the issue and members' sense of not being able to control it were anathema in an organization that considered itself to be a "fixer" and "doer." Additionally, Port Authority members not only emphasized the importance of "looking humane" in their actions, but also focused on "being humane." Thus, the organization's identity defined what aspects of the issue were seen as a threat and helped to locate solutions that could transform the issue into an opportunity (Jackson & Dutton, 1988). For example, some informants described the use of partnering strategies in phase 5 as representing an opportunity for the Port Authority "to show its stuff" to other transportation agencies. As Meyer (1982) found in his study of hospital employees' interpretations of a doctors' strike, ideology — in this case, beliefs about identity — shaped the meanings given to the event and the set of legitimate solutions.

Port Authority members' sense of the issue's importance was also related to the occurrence of identity-inconsistent responses. When the organization took actions that members saw as inconsistent with its identity, they judged the issue as more important and the organization as more committed to it than they had previously. Informants' interpretations of the significance of the Port Authority's granting the fellowship to study homeless people at transportation facilities illustrates this connection. The grant was seen as risky and unconventional, and several informants viewed the nontraditional character of this action as a sign that top management saw the issue as serious and worthy of action commitments.

The Port Authority's identity also constrained what members saw as legitimate interpretations. In the early issue phases, the organization's identity was a critical force in defining homelessness as an issue to which the Port Authority should *not* respond. Organization members justified nonaction using the rationale that the Port Authority excelled in its technical skills but lacked the social service skills necessary to deal with homelessness.

The organization's identity affected the meanings members gave the issue. Two terms frequently applied were "moral issue" and "business issue." Each issue category had associated with it a set of routines and solutions for dealing with the issue (Dutton & Jackson, 1987). However, more important for the argument developed here, different aspects of the Port Authority's identity were associated with each category: homelessness as a business issue with the high-quality-organization identity component, and homelessness as a moral issue with the ethical and altruistic identity component. Thus, these two aspects drove the application of different categories to the issue, which engaged different interpretations of the issue's significance and activated different recipes for solving the problem over time.

Identity and Issue Emotions

The organization's identity was also significant in explaining the direction and level of emotional expression about the issue. This connection was most vivid in phase 4. Informants expressed negative emotion when inappropriate involvement of individuals or the organization in certain activities compromised the Port Authority's identity. For example, informants told us stories about architects holding babies with AIDs, engineers changing diapers, and sanitation engineers cleaning filthy bathrooms — all related to the issue of homelessness. Whether the substance of the stories was accurate is less important than the values that the stories conveyed, a great disdain about the inappropriate diversion of technical skills for the delivery of social services. This disdain was a strong defense for not responding to the homelessness issue, particularly in the 1982–86 period. The sense of not being able to control homelessness further delayed Port Authority involvement. However, these defenses were no longer sustainable when the problem worsened and the issue's visible appearance in Port Authority facilities other than the bus terminal severely damaged the organization's image.

At the same time, the Port Authority's identity also produced positive emotions when organizational actions were identity-consistent, especially when those actions were in arenas in which organization members did not expect action. For example, opening the two drop-in centers in the Port Authority's record-breaking style was a source of pride and a sense of accomplishment for informants at all levels of the organization.

Identity and Issue Actions

The Port Authority's identity also affected the pattern of issue-related actions. First, the identity affected action through the link to issue interpretations and emotions discussed above. However, it also affected action directly by providing guidelines for evaluating success, recipes for solutions, and parameters for acceptable ways of resolving the issue. An argument could be made that objective characteristics of the situation — the increase in the number of homeless people in Port Authority facilities and increased constraints on feasible actions as a result of the repeal of the antiloitering law — created the push for action. The present emphasis on organizational identity doesn't negate the influence of such other forces; rather, it is meant to enrich understanding of the particular responses this organization made. Thus, although a resource dependency perspective (Pfeffer & Salancik, 1978) could be used to explain the increase in the number of actions the Port Authority took, particularly after phase 3, the concept of identity is helpful in understanding how those actions were shaped.

The Port Authority's identity offered implicit guidelines for evaluating the effectiveness of its actions on the issue. Using the speed with which the two drop-in centers were completed as a criterion for the success of the Homeless Project Team and overall success in dealing with the issue typified this connection. Organization members used efficiency in task completion as an important barometer of the Port Authority's success with the issue even though they admitted that the actual problem, in terms of the number of homeless at facilities, had not changed.

Individuals' senses of the Port Authority's identity were associated with a set of routines, or standard procedures for dealing with the issue, whose activation engaged ways of doing things members identified as "typical of the Port Authority." In this sense, an organization's identity is closely tied to its culture because identity provides a set of skills and a way of using and evaluating those skills that produce characteristic ways of doing things (Nelson & Winter, 1982; Swidler, 1986). As Child and Smith (1987) pointed out, "cognitive maps" like identity are closely aligned with organizational traditions. An organization's identity is one of the vehicles through which "pre-conceptions determine appropriate action" (Weick, 1988: 306). For example, when the homelessness issue was no longer deniable, the Port Authority went to work to "get smart on the issue." The phrase describes the organization's ideal approach to a problem — investigating and analyzing it from all angles. Members learned a great deal about the unique attributes of homeless people at transportation facilities. Some informants saw this engagement of learning routines as typical of the Port Authority and indicative of its professionalism. Members also saw searching for partners for dealing with the issue and framing the issue as related to the region's future as actions that "typified the Port Authority's approach to things."

Finally, individuals' senses of the organization's identity did more than activate a set of familiar routines for dealing with the issue. That identity also constrained what were considered acceptable or legitimate solutions (Meyer, 1982). The frequent claims that throwing homeless people out in the cold was not the Port Authority's way of dealing with the issue well illustrate that link. Several informants directly compared the Port Authority's response to that of Grand Central Station, where police were moving homeless people out "into the cold," to illustrate the limits of what they saw as legitimate action for coping with the issue.

The Port Authority's upper-level managers were also concerned about doing too much on the issue, such as providing direct outreach or other social services to the homeless. Three considerations fueled this concern. First, these managers were adamant about not straying from their main business of transportation. Providing social services was perceived as a "deviation from our basic area of business" because it would have required hiring people trained in social services. Second, upper-level managers did not want to appear to be leaders on the issue, for they felt that taking such a role would "blur accountability" for the homeless, relieving city agencies of their responsibilities. Third, there was a continual concern over attracting more homeless to Port Authority facilities if services were provided. Thus, upper management sought to maintain a policy of moderation, focusing on actions consistent with the organization's identity.

In sum, a knowledge of individuals' beliefs about an organization's identity is crucial for discerning the importance of an issue, its meanings, and its emotionality. These interpretations, shaped by the organization's identity, move individuals' commitment, involvement, indifference, and resistance in particular directions and thereby direct and shape organizational actions.

The Importance of Organizational Image

An organization's identity describes what its members believe to be its character; an organization's image describes attributes members believe people outside the organization use to distinguish it. Organizational image is different from reputation: reputation describes the actual attributes outsiders ascribe to an organization (Fombrun & Shanley, 1990; Weigelt & Camerer, 1988), but image describes insiders' assessments of what outsiders think. Both organizational image and identity are constructs held in organization members' minds. They capture two of the key ways that an organization becomes meaningful to individuals and motivate individuals to action in particular ways and at particular times. In the case of the Port Authority and its dealings with homelessness, image changes triggered the organization's later, more substantive response to the issue, particularly in 1987. Active attempts to manage the organization's image on this issue also explain the changing issue-related actions.

Organizational Image and Individuals' Motivation

An organization's image matters greatly to its members because it represents members' best guesses at what characteristics others are likely to ascribe to them because of their organizational affiliation. An organization's image is directly related to the level of collective self-esteem derivable from organizational membership (Crocker & Luhtanen, 1990; Pierce, Gardner, Cummings, & Dunham, 1989). Individuals' self-concepts and personal identities are formed and modified in part by how they believe others view the organization for which they work.

Impetus to take action to improve the damaged image resulting from the Port Authority's association with homelessness was more than organizationally based. As the story revealed, the damage to the organization's image hurt individuals personally. Spoiled organizational images transfer to organization members (Sutton & Callahan, 1987), and this link tightens when actions that affect the organization's image are public and irrevocable. As Weick noted, in such situations actions "become harder to undo" and "harder to disown" (1988: 310). As a result, individuals are strongly motivated and committed to take actions that will restore their organization's image.

The close link between an individual's character and an organization's image implies that individuals are personally motivated to preserve a positive organizational image and repair a negative one through association and disassociation with actions on issues. This explanation complements Sutton and Callahan's (1987) description of how companies' bankruptcy filings caused their managers' efforts to restore their own self-images in the eyes of critical organizational audiences. Similarly, in the Port Authority's struggle with the issue of homelessness we observed defensive tactics designed to actively manage outsiders' impressions of the organization; however, the Port Authority's actions were subject to the constraint of doing things that were consistent with the organization's identity.

Organizational Image and Impression Management

Individuals in organizations actively monitor organizational actions on social issues because such actions can be especially character-enhancing or damning. Port Authority members became aware of their organization's image through personally distant media, like the press, and through close ones, like conversations with friends. Informants' accounts documented the triggers to personal and organizational action the negative press coverage set off. As the story suggested, press coverage of the Port Authority on this issue was particularly vivid and disturbing during phase 4. Most staff members working on this issue also mentioned friends and family as active sources of feedback on the organization's image and the pride or shame that this close feedback provided. The connection between individuals' senses of self and the Port Authority's image created incentives to manage the impression others had of the organization's actions.

As our history ended in 1989, the Port Authority members were continuing to try a variety of impression and image management tactics to see if they could transform the organization's image without violating attributes that defined its core identity. The evolution of actions was a continuous experimentation and learning process that became more deliberate over time. Although organization members denied responsibility for the problem throughout, when they saw no alternative, they took identity-consistent action in deliberate and significant ways. However, as the significance of actions on the issue increased — that is, as the human and monetary resources invested increased — the Port Authority began to plan which actions it wanted to highlight and which it wanted to conceal. When we stopped collecting data in mid-1989, the organization was acting as an advocate for the homeless, educating and sharing information with other transportation agencies on what could be done, but it was intentionally maintaining a low profile in the development of programs and services. In the minds of the members of Homeless Project Team and most of upper management, the costs of being associated with taking responsibility for homelessness far outweighed any gains from being seen as a builder of superior drop-in centers.

The evolution of actions that we observed over time was partially trial-and-error image management that became more assertive (designed to create a positive image) and less defensive (designed to mend a negative image) over time (Tedeschi & Melburg, 1984). The facility-based solutions were largely reactive, based on attempts to conceal, contain, and eliminate the problem. However, as the problem became more severe and image deterioration amplified emotional reactions to the issue, the organization went into high gear on homelessness in an instrumental sense and low gear in a public sense. In a way that was consistent with its technically expert, high-quality, ethical, and fixer-doer identity, the organization proposed and funded major outreach facilities for the homeless near three of its affected facilities.

In sum, deterioration in the Port Authority's image was an important trigger for and accelerator of issue-related action. Changes in the organization's image fueled investment in and motivation to work on the issue in two distinct ways. First, it prompted personal investment because of members' concerns about how the organization's image was affecting others' views of themselves. Second, it provided important political ammunition for justifying and legitimating further issue commitment (Pettigrew, 1987). The Port Authority's image became a direct target for action as management became more aggressive and deliberate in its actions on the issue.

Discussion and Implications

The ideas of image and identity and their links to patterns of issue interpretation, action, and emotion reinforce some well-known ideas about organizational adaptation and suggest important new directions for theory and research.

The story of the Port Authority and the role of identity and image in it suggest that organizational context matters in explaining patterns of change. Treatments of organizational adaptation and strategic change have argued and documented that claim well (e.g., Bartunek, 1984; Miles & Cameron, 1982; Pettigrew, 1987; Tushman & Romanelli, 1985). The Port Authority's struggle with the homelessness issue also supports adaptation researchers' assertions that organizational context affects patterns of change through its effect on how issues are interpreted (e.g., Dutton & Duncan, 1987; Milliken, 1990; Meyer, 1982; Normann, 1977). However, two persistent themes — that what people see as their organizations' distinctive attributes (its identity) and what they believe others see as distinctive about the organization (its image) constrain, mold, and fuel interpretations — help link individual cognitions and behaviors to organizational actions. Because image and identity are constructs that organization members hold in their minds, they actively screen and interpret issues like the Port Authority's homelessness problem and actions like building drop-in centers using these organizational reference points. In this way, organizational image and identity and their consistency or inconsistency help to explain when, where, and how individuals become motivated to push for or against organizational initiatives. As other change researchers have noted (Child & Smith, 1987; Hinings & Greenwood, 1988), it is inconsistency between various conditions in an organization and its context that precipitates action.

The relationship between individuals' senses of their organizational identity and image and their own sense of who they are and what they stand for suggests a very personal connection between organizational action and individual motivation. It suggests that individuals have a stake in directing organizational action in ways that are consistent with what they believe is the essence of their organization. Actions are also directed in ways that actively try to manage outsiders' impressions of the organizations' character (its image) to capture a positive reflection. This connection between organization, employees' self-concepts, and their motivation to invest in and act on issues in particular ways uncovers a new way of thinking about the organizational adaptation process, a perspective in which organizational impression management is an important driving force in adaptation.

Thinking about organizational adaptation processes as attempts at impression management raises several intriguing theoretical and research questions. First, what is the link between managing impressions of organizations and what and how issues are interpreted? Because an organization's association or disassociation with certain issues defined in particular ways has consequences for individuals' careers (Chatman, Bell, & Staw, 1986), impression management concerns are important in determining when and how issues are interpreted. Previous research has assumed these interpretations are important elements in the adaptation process (e.g., Dutton & Duncan, 1987); if that is so, impression management processes hold important clues for discovering how environments and organizations correlate over time. Second, how do impression management processes direct organizational actions? In the Port Authority's struggle with homelessness, we saw impression management concerns become more prominent over time as informants' senses of the organization's image deteriorated. Organization members cared how others judged Port Authority actions on this issue. They pushed for types of actions that reflected positively on the Port Authority and, by association, on themselves as well. Serious consideration of these questions reveals the role that impression management processes play in the adaptation process. By linking individual motivation to organizational action, we begin to see new links between microprocesses (individual motivations) and macro behaviors (patterns of organizational change).

Issue interpretations and actions by Port Authority members reflected changes in public awareness and attention to homelessness in the media and "other arenas of public discourse" (Hilgartner & Bosk, 1988: 53). The waxing and waning of the national attention given to this issue eased or accentuated internal difficulties in legitimating mobilization and investment in the issue. For adaptation researchers, this connection suggests that the rise and fall of issues in broad institutional environments affects issue interpretation and action within an organization. This viewpoint is consistent with population ecologists' and institutional theorists' claims that external context constrains organizational change patterns (e.g., Hannan & Freeman, 1984; Zucker, 1988). Other organizational theorists have linked external context to organizational change through the idea of industry recipes (e.g., Huff, 1982; Spender, 1989). The idea presented here is similar; we suggest that meanings in use and legitimated in a broad external context constrain what issues or ideas have currency in organizations. Such a view urges adaptation researchers to consider how changes occurring in a public issues arena mold and modify issue interpretations.

In conclusion, the story of the Port Authority's struggle with the homelessness issue provides fertile ground for unearthing new considerations for students of organizations. Consistent with the spirit of Glaser and Strauss (1967), the story reveals new ideas for theory building, particularly for the domain of organizational adaptation. The idea that an organization's identity and image are central to understanding how issues are interpreted, how reactions are generated, how and what types of emotions are evoked, and how these behaviors are related to one another in an organizational context is very simple. It suggests that individuals in organizations keep one eye on the organizational mirror when they interpret, react, and commit to organizational actions. Researchers in strategy, organization theory, and management might better understand how organizations behave by asking where individuals look, what they see, and whether or not they like the reflection in the mirror.

Notes

1. We may subsequently refer to the agency as the Port Authority.
2. PATH stands for Port Authority Trans-Hudson commuter line.
3. The Manhattan Bowery Corporation is a "community corporation," a neighborhood-based agency that administers social services where needed.

References

Albert, S., & Whetten, D. 1985. Organizational identity. In L. L. Cummings & B. M. Staw (Eds.), *Research in organizational behavior*, vol. 7: 263–295. Greenwich, CT: JAI Press.
Alvesson, M. 1990. Organization: From substance to image? *Organization Studies*, 11: 373–394.
Ashforth, B., & Mael, F. 1989. Social identity theory and the organization. *Academy of Management Review*, 14: 20–39.
Bachrach, P., & Baratz, B. 1962. The two faces of power. *American Political Science Review*, 56: 947–952.
Bartunek, J. 1984. Changing interpretive schemes and organizational restructuring. *Administrative Science Quarterly*, 29: 355–372.
Bjorkegren, D. 1989. *It doesn't have to be that way*. Paper presented at the Organizational Behavior Teaching Conference, Columbia, MO.
Caro, R. 1974. *The power broker: Robert Moses and the fall of New York*. New York: Vintage Books.
Chatman, J., Bell, N., & Staw, B. M. 1986. The managed thought. In H. P. Sims & D. A. Gioia (Eds.), *The thinking organization*: 191–214. San Francisco: Jossey-Bass.
Cheney, G. 1983. The rhetoric of identification and the study of organizational communication. *Quarterly Journal of Speech*, 69(2): 143–158.
Child, T., & Smith, C. 1987. The context and process of organizational transformation. *Journal of Management Studies*, 24: 565–593.
Cohen, M. D., March, J. G., & Olsen, J. P. 1972. A garbage can model of organizational choice. *Administrative Science Quarterly*, 17: 1–15.
Crocker, J., & Luhtanen, S. R. 1990. Collective self-esteem and ingroup bias. *Journal of Personality and Social Psychology*, 58: 60–67.
Daft, R., & Weick, K. 1984. Toward a model of organizations and interpretation systems. *Academy of Management Review*, 9: 284–296.
Donnellon, A., Gray, B., & Bougon, M. 1986. Communication, meaning and organized action. *Administrative Science Quarterly*, 31: 43–55.
Dutton, J. E. 1988a. Perspectives on strategic issue processing: Insights from a case study. In P. Shrivastava & R. Lamb (Eds.), *Advances in strategic management*, vol. 5: 223–244. Greenwich, CT: JAI Press.

Dutton, J. E. 1988b. Understanding strategic agenda building in organizations and its implications for managing change. In L. R. Pondy, R. J. Boland, & H. Thomas (Eds.), *Managing ambiguity and change*: 127–144. Chichester, England: John Wiley & Sons.

Dutton, J. E., & Duncan, R. B. 1987. Creation of momentum for change through the process of strategic issue diagnosis. *Strategic Management Journal*, 8: 279–295.

Dutton, J. E., & Jackson, S. B. 1987. Categorizing strategic issues: Links to organizational action. *Academy of Management Review*, 12: 76–90.

Dutton, J. E., Stumpf, S., & Wagner, D. 1990. Diagnosing strategic issues and managerial investment of resources. In P. Shrivastava & R. Lamb (Eds.), *Advances in strategic management*: 143–167. Greenwich, CT: JAI Press.

Eisenhardt, K. M. 1989. Building theory from case study research. *Academy of Management Review*, 14: 532–550.

El Sawy, O. A., & Pauchant, T. C. 1988. Triggers, templates and twitches in the tracking of emerging strategic issues. *Strategic Management Journal*, 9: 455–473.

Feldman, M. 1989. *Order without design*. Stanford, CA: Stanford University Press.

Fombrun, C., & Shanley, M. 1990. What's in a name? Reputation building and corporate strategy. *Academy of Management Journal*, 33: 233–258.

Gioia, D. A., & Sims, H. P. 1986. Introduction: Social cognition in organizations. In H. E. Sims & D. A. Gioia (Eds.), *The thinking organization*: 1–19. San Francisco: Jossey-Bass.

Glaser, B., & Strauss, A. 1967. *The discovery of grounded theory: Strategies for qualitative research*. London: Weidenfeld.

Hambrick, D., & Finkelstein, S. 1987. Managerial discretion: A bridge between polar views of organizational outcomes. In L. L. Cummings & B. M. Staw (Eds.), *New directions in organizational behavior*: 369–406. Greenwich, CT: JAI Press.

Hannan, M., & Freeman, J. 1984. Structural inertia and organizational change. *American Sociological Review*, 49: 149–164.

Hilgartner, S., & Bosk, C. L. 1988. The rise and fall of social problems: A public arenas model. *American Journal of Sociology*, 94: 53–78.

Hinings, C. R., & Greenwood, R. 1988. *The dynamics of strategic change*. New York: Basil Blackwell.

Huff, A. 1982. Industry influences on strategy reformulation: *Strategic Management Journal*, 3: 119–131.

Isabella, L. 1990. Evolving interpretations as a change unfolds: How managers construe key organizational events. *Academy of Management Journal*, 33: 17–41.

Jackson, S., & Dutton, J. 1988. Discerning threats and opportunities. *Administrative Science Quarterly*, 33: 370–387.

Martin, J., & Meyerson, D. 1988. Organizational culture and the denial, channeling, and acknowledgement of ambiguity. In L. R. Pondy, R. J. Boland, & H. Thomas (Eds.), *Managing ambiguity and change*: 93–126. Chichester, England: John Wiley & Sons.

Meyer, A. 1982. Adapting to environmental jolts. *Administrative Science Quarterly*, 27: 515–583.

Miles, M. B., & Huberman, A. M. 1984. *Qualitative data analysis*. Beverly Hills, CA: Sage Publications.

Miles, R. H., & Cameron, K. 1982. *Coffin nails and corporate strategies*. Englewood Cliffs, NJ: Prentice-Hall.

Milliken, F. 1990. Perceiving and interpreting environmental change: An examination of college administrators' interpretation of changing demographics. *Academy of Management Journal*, 33: 42–63.

Mintzberg, H., Raisinghini, A., & Theoret, L. 1976. The structure of unstructured decision processes. *Administrative Science Quarterly*, 21: 246–275.

Nelson, R., & Winter, S. G. 1982. *An evolutionary theory of economic change*. Cambridge, MA: Harvard University Press.

Normann, R. 1977. *Management for growth*. London: Wiley.

Nutt, P. 1984. Types of organizational decisions. *Administrative Science Quarterly*, 29: 414–450.

Pettigrew, A. 1987. Context and action in the transformation of the firm. *Journal of Management Studies*, 24: 649–670.

Pfeffer, J., & Salancik, G. R. 1978. *The external control of organizations.* New York: Harper & Row.
Pierce, J., Gardner, D., Cummings, L. L., & Dunham, R. B. 1989. Organization-based self-esteem: Construct definition measurement and validation. *Academy of Management Journal*, 32: 622–648.
Prahalad, C. K., & Bettis, R. 1986. The dominant logic: A new linkage between diversity and performance. *Strategic Management Journal*, 7: 485–501.
Sandelands, L. E., & Drazin, R. 1989. On the language of organization theory. *Organization Studies*, 10: 457–478.
Singh, H., Tucker, D., & House, R. 1986. Organizational legitimacy and the liability of newness. *Administrative Science Quarterly*, 31: 171–193.
Spender, J. C. 1989. *Industry recipes.* Cambridge, MA: Basil Blackwell.
Starbuck, W. H. 1983. Organizations as action generators. *American Sociological Review*, 48: 91–102.
Starbuck, W. H., & Milliken, F. 1988. Executives' perceptual filters: What they notice and how they make sense. In D. Hambrick (Ed.), *The executive effect: Concepts and methods of studying top managers*: 35–65. Greenwich, CT: JAI Press.
Sutton, R., & Callahan, A. L. 1987. The stigma of bankruptcy: Spoiled organizational image and its management. *Academy of Management Journal*, 30: 405–436.
Swidler, A. 1986. Culture in action: Symbols and strategies. *American Sociological Review*, 51: 273–286.
Tedeschi, J. T. (Ed.). 1981. *Impression management theory and social psychological research.* New York: Academic Press.
Tedeschi, J. T., & Melburg, V. 1984. Impression management and influence in the organization. In S. D. Bacharach & E. J. Lawler (Eds.), *Research in the sociology of organizations*, vol. 3: 31–58. Greenwich, CT: JAI Press.
Tushman, M., & Romanelli, E. 1985. Organizational evolutions: A metamorphosis model of convergence and reorientation. In L. L. Cummings & B. M. Staw (Eds.), *Research in organizational behavior*, vol. 7: 171–222. Greenwich, CT: JAI Press.
Walsh, J. P., Henderson, C. M., & Deighton, T. 1988. Negotiated belief structures and decision performance: An empirical investigation. *Organizational Behavior and Human Decision Processes*, 42: 194–216.
Weick, K. E. 1979. *The social psychology of organizing.* Reading, MA: Addison-Wesley.
Weick, K. E. 1988. Enacted sensemaking in crisis situations. *Journal of Management Studies*, 24: 305–317.
Weigelt, K., & Camerer, C. 1988. Reputation and corporate strategy: A review of recent theory and applications. *Strategic Management Journal*, 9: 443–454.
Weiss, J. A. 1989. The powers of problem definition: The case of government paperwork. *Policy Sciences*, 22(2): 97–121.
Yin, R. 1984. *Case study research.* Beverly Hills, CA: Sage Publications.
Zucker, L. (Ed.). 1988. *Institutional patterns and organizations.* Cambridge, MA: Ballinger Publishing Co.

45

Organizational Images and Member Identification

Jane E. Dutton, Janet M. Dukerich and Celia V. Harquail

Source: *Administrative Science Quarterly* 39 (2) (1994): 239–263.

Members vary in how much they identify with their work organization. When they identify strongly with the organization, the attributes they use to define the organization also define them. Organizations affect their members through this identification process, as shown by the comments of a 3M salesman, quoted in Garbett (1988: 2):

> I found out today that it is a lot easier being a salesman for 3M than for a little jobber no one has ever heard of. When you don't have to waste time justifying your existence or explaining why you are here, it gives you a certain amount of self-assurance. And, I discovered I came across warmer and friendlier. It made me feel good and enthusiastic to be "somebody" for a change.

This salesman attributes his new, more positive sense of self to his membership in 3M, a well-known company. What he thinks about his organization and what he suspects others think about his organization affects the way that he thinks about himself as a salesperson.

This paper explores the kind of connection that salesman had with 3M: a member's cognitive connection with his or her work organization derived from images that each member has of the organization. The first image, what the member believes is distinctive, central, and enduring about the organization, is defined as perceived organizational identity. The second image, what a member believes outsiders think about the organization, is called the construed external image (Dutton and Dukerich, 1991). Our model proposes that these two organizational images influence the cognitive connection that members create with their organization and the kinds of behaviors that follow.

When a person's self-concept contains the same attributes as those in the perceived organizational identity, we define this cognitive connection as organizational identification. Organizational identification is the degree to which a member defines him- or herself by the same attributes that he or she believes define the organization. The 3M salesman reflects his organizational identification when he describes himself as innovative and successful, just like the 3M organization. A person is strongly identified with an organization when (1) his or her identity as an organization member is more salient than alternative identities, and (2) his or her self-concept has many of the same characteristics he or she believes define the organization as a social group. We build our arguments on the core assumption that people's sense of membership in the social group "the organization" shapes their self-concepts (Tajfel and Turner, 1985; Ashforth and Mael, 1989; Kramer, 1991). Organizational scholars have explored how a person's self-concept is shaped by membership in occupational groups (Van Maanen and Barley, 1984) and work groups (Alderfer and Smith, 1982). Here we consider how a person's self-concept is shaped by the knowledge that she or he is a member of a specific organization.

The images that members hold of their work organizations are unique to each member. A person's beliefs therefore may or may not match a collective organizational identity that represents the members' shared beliefs about what is distinctive, central, and enduring about their organization (Albert and Whetten, 1985). In addition, each member's own construal of the organization's external image may or may not match the reputation of the organization in the minds of outsiders. We focus on the relationship between a member's individual images of his or her organization as a social group and the effects of those images on the strength of organizational identification and member behavior.

Linking Organizational Images to Members' Self-Concepts

A person's well-being and behavior are affected both by the attributes they ascribe to themselves and by those they believe others infer about them from their organizational membership. As the quote from the 3M salesman illustrates, organizational membership can confer positive attributes on its members, and people may feel proud to belong to an organization that is believed to have socially valued characteristics. When members believe that outsiders see the organization in a positive light, they "bask in the reflected glory" of the organization (Cialdini et al., 1976: 366). Strong organizational identification may translate into desirable outcomes such as intraorganizational cooperation or citizenship behaviors.

Organizational membership can also confer negative attributes on a member. If members interpret the external organizational image as unfavorable, they may experience negative personal outcomes, such as depression and stress. In turn, these personal outcomes could lead to undesirable organizational outcomes, such as increased competition among members or reduced effort on long-term tasks. Over time, members may either disengage themselves from prior organizational roles (Kahn, 1990) or exit the organization (Hirschman, 1970). Just such a negative external organizational image has created problems for members of the Port Authority of New York and New Jersey (PA). For more than a decade, the PA has struggled with what to do about the rising number of homeless people who seek shelter in its transportation facilities, including the Bus Terminal. As problems with the homeless became more severe, the media increasingly depicted the PA's facilities as dangerous and dirty and the PA organization as ineffective and inhumane. Negative press about the PA indirectly hurt the employees. When PA members began to construe the organization's external image in these negative and socially undesirable terms, they felt demeaned and hurt by the criticism that they inferred from outsiders. Through a story told about one PA member at a cookout, another member emphasized the connection between organizational actions, a negatively construed external image, and the member's self-concept:

> You know, the guy that's running the Lincoln Tunnel doesn't have a full perception of how the Bus Terminal or the homeless impact what he does on a day-to-day basis. But the minute he leaves and he goes to the cookout in his neighborhood and he meets somebody and the person says, "What do you do for a living?" "Oh, I work for the Port Authority." They say, "How can you stand that Bus Terminal, what can you do?" That's the name. That's the symbol of the Port Authority. It's the standard bearer. And you know, so personally everybody that's involved in any aspect of working for the Port Authority is identified with that place, and with that issue. (Dutton and Dukerich, 1991: 538)

An employee who worked far away from the homeless issue in the Bus Terminal still felt that actions in one segment of the organization affected how others saw him.

Similarly, Exxon employees felt that the public regarded them in a negative light after the Valdez oil spill. Because the press made Exxon's inadequate action so visible and public, Exxon employees found that they were expected to defend the company's actions in social situations. As described in Fanning (1990: 25):

> But the targets of the most scorn at the moment are probably oil-company executives. Take the case of Exxon corporation. It was only recently that executives at that company were able to admit their place of employment without the fear of being attacked by environmentalists infuriated by the company's handling of last year's oil spill at Prince William Sound. Slowly but surely, Exxon executives began to reappear at cocktail parties across the country, and occasionally, even had a good time.

Even employees working at Exxon's Credit Card Center did not escape the public's wrath: They received oil-soaked, cut-up credit cards from angry customers. As one newspaper explained: "Employees, confronted daily by criticisms of Exxon in the media and by friends and family members, are questioning their faith in the corporate giant" (*Star Ledger*, 1989: 3). When people identify strongly with their work organizations, they experience such threats personally (Schwartz, 1987).

As the Port Authority and Exxon examples suggest, outsiders actively judge employees by the characteristics attributed to the organization through its public reputation. Inside the organization, members interpret and infer the reputation of their organization and react to the external image they construe of their organization. As the media publicizes information about an organization, public impressions of the organization and of the organization's members become part of the currency through which members' self-concepts and identification are built or are eroded.

Organizational Identification

Members become attached to their organizations when they incorporate the characteristics they attribute to their organization into their self-concepts. The self-concept is an interpretive structure that mediates how people behave and feel in a social context (Gecas, 1982; Schenkler, 1985; Markus and Wurf, 1987) and refers to "the totality of self-descriptions and self-evaluations subjectively available to an individual" (Hogg and Abrams, 1988: 24). A person's self-concept may be composed of a variety of identities, each of which evolves from membership in different social groups, such as a social group based on race, gender, or tenure (Stryker and Serpe, 1982; Breakwell, 1986). But self-concepts are also influenced by memberships in social groups such as work organizations (Ashforth and Mael, 1989), through which a member may come to identify with the organization. While some researchers have focused on organizational identification as value congruence between a member and his or her organization (e.g., Hall, Schneider, and Nygren, 1970; Lee, 1971; Hall and Schneider, 1972), we focus on the cognitive connection between the definition of an organization and the definition a person applies to him- or herself, viewing identification as a process of self-definition (Brown, 1969). Defining organizational identification as a cognitive link between the definitions of the organization and the self is consistent with attitudinal approaches to commitment (e.g., Porter et al., 1974; Mowday, Porter, and Steers, 1982). As part of the commitment process, the level of organizational identification indicates the degree to which people come to see the organization as part of themselves. Organizational identification is one form of psychological attachment that occurs when members adopt the defining characteristics of the organization as defining characteristics for themselves.

Organizational identification can have both positive and negative effects on a member's sense of self. Organizational identification as defined here does not necessarily connote a pride in affiliation with the organization — a characteristic that is central in Kelman's (1958) view of identification and is used by O'Reilly and Chatman (1986) in their work on identification as a means for psychological attachment. As illustrated by the discomfort felt by Exxon executives and by PA members, identification with an organization can result in feelings of shame, disgrace, or embarrassment.

The strength of a member's organizational identification reflects the degree to which the content of the member's self-concept is tied to his or her organizational membership. When organizational identification is strong, a member's self-concept has incorporated a large part of what he or she believes is distinctive, central, and enduring about the organization into what he or she believes is distinctive, central, and enduring about him- or herself. When organizational identification is strong, the organization-based content of a member's self-concept is salient and central (Gergen, 1968; Stryker and Serpe,

1982), other identities in the self-concept have receded, and organizational membership is a central and frequently used basis for self-definition (Kramer, 1991).

Several researchers have described the formation of and change in a member's organizational identification. Borrowing from Tajfel and Turner (1985), Ashforth and Mael (1989) described organizational identification as a process of self-categorization. They proposed that organizational identification strengthens when members categorize themselves into a social group — in this case the organization — that has distinctive, central, and enduring attributes. The premise that identification is caused by self-categorization provides the foundation for our model of organizational identification. The model focuses on two key images that organizational members have of their work organization — perceived organizational identity and construed external image.

A Model of Member Identification

Several points support the idea that organizations have collective identities, consisting of the beliefs that members share as distinctive, central, and enduring. First, it is common practice for organizational leaders to articulate and claim what is distinctive, central, and enduring about their organization (Pfeffer, 1981; Albert and Whetten, 1985). Whether or not these claims of distinctiveness are empirically valid (e.g., Martin et al., 1983) is less important than the fact that powerful organizational members engage in communication and influence processes in an effort to create a collective identity for members. Organizations have a broad repertoire of cultural forms such as rituals, symbols, ceremonies, and stories that encode and reproduce shared organizational patterns of behavior and interpretation (Allaire and Firsirotu, 1984). Rituals, ceremonies, and stories objectify and communicate the collective organizational identity to organizational members.

Distinctive organizational attributes often remain hidden to members until the organization's collective identity is challenged (Albert and Whetten, 1985; Fiol, 1991) or until some precipitating event calls organizational actions or performance into question (Ginzel, Kramer, and Sutton, 1993). Sometimes major stakeholders' actions or changes in the organization's environment such as regulatory changes or competitive moves can cause the organization's collective identity to surface. The collective organizational identity also becomes more salient when members believe that the organization's actions are inconsistent with its collective identity (e.g., when a social service agency buys expensive office furniture) or when individual members act in ways that contradict the collective organizational identity (e.g., when professors in a teaching college consistently miss class). In these cases, organizational or individual actions interrupt the flow of normal organizational routines, prompting individuals to ask, "What is this organization really about?" These actions motivate members to review and acknowledge what they believe defines the organization, and this affects the strength of their connection to the organization.

Perceived Organizational Identity

Whereas an organization's collective identity represents the set of beliefs that members share, perceived organizational identity refers to the beliefs of a particular individual organizational member. Because organizations imperfectly socialize members to a collective view, perceived organizational identity may depart from the organization's collective identity. The perceived organizational identity — a member's beliefs about the distinctive, central, and enduring attributes of the organization — can serve as a powerful image influencing the degree to which the member identifies with the organization.

Perceived Organizational Identity and Identification

The degree to which the perceived organizational identity affects a person's identification level depends on the attractiveness of this image to the person, which requires a subjective evaluation. An attractive perceived organizational identity strengthens a member's identification. Stated in proposition form:

> Proposition 1 (P1): The greater the attractiveness of the perceived organizational identity, the stronger a person's organizational identification.

Three principles of self-definition — self-continuity, self-distinctiveness, and self-enhancement — account for the attractiveness of an organizational image and explain why it strengthens identification.

Self-Continuity and the Attractiveness of the Perceived Organizational Identity

People generally want to maintain the continuity of their self-concepts over time and across situations (Steele, 1988). A member's perception of his or her organization's identity adds or subtracts from the continuity that he or she experiences in his or her self-concept over time. Two arguments support the idea that similarity between the self-concept and perceived organizational identity enhances continuity and that continuity of self strengthens a member's identification by making the perceived organizational identity more attractive.

First, people find a perceived organizational identity more attractive when it matches their own sense of who they are (i.e., their self-concept) simply because this type of information is easy to process and understand. Social psychologists argue that people attend to and process "self-relevant" information differently than "self-irrelevant" information (Markus and Wurf, 1987). The general finding is that people more easily focus on, process, recognize, and retrieve self-relevant than self-irrelevant information. The ease of recognizing, processing, and retrieving self-relevant information makes organizational identities that match the self more attractive than organizational identities that do not match the self.

Second, when a person's self-concept and the perceived organizational identity are similar, a member is drawn to the organization because it provides easy opportunities for self-expression (Shamir, 1991). People are drawn to organizations that allow them to exhibit more of themselves and to enact a fuller range of characteristics and values in their self-concept. For example, a triathlete who values physical prowess and involvement in competitive athletics will be drawn more to a shoe manufacturer like Nike, which encourages its employees to work out on company time, than to another shoe manufacturer that is involved in community arts associations.

People value self-integrity and a sense that they are internally coherent (Steele, 1988). To maintain this integrity, people want to act authentically (Gecas, 1982), expressing the personality characteristics they think they have and which they value. People are drawn to organizations in which they can express themselves rather than hide the contents of their self-concept. This assertion is built on an assumption that "humans are not only pragmatic and goal oriented but also self-expressive" (Shamir, House, and Arthur, 1993: 580). For example, a vegetarian journalist will be drawn more to *Vegetarian Times* than to *Bow Hunter* magazine, since the former organization is more likely to provide the journalist with opportunities to express a sense of him- or herself as a supporter of animal rights. We see this relationship in Kunda's (1992) account of how engineers identified with Tech, where they could express more of the characteristics they valued about themselves. The engineers enjoyed Tech's technical orientation because it matched their own. As Kunda (1992: 177) explained: "Similarly, many engineers acknowledge attachment to Tech's technology, which they view as unique and through that, to the company. Says one: 'Once you've worked with Tech products in a Tech environment, it's hard to go to anything else. They just adjust so much better. It's an engineer's dream — if he's into technology'."

Chatman and her colleagues found similar results for behavioral outcomes. In a study of accounting recruits, Chatman (1991) found that a strong fit between the pattern of organizational values and members' values predicted members' satisfaction and intent to stay with the organization a year later. In addition, an increase in person-organization fit over the first year was significantly and positively related to members' satisfaction levels. In two different empirical contexts, O'Reilly and Chatman (1986) found that people who secured an attachment to the organization based on value-congruency (what they called attachment based on internalization) reported high intentions to stay with the organization (Study 1), and greater internalization was associated with lower actual turnover (Study 1 and Study 2). O'Reilly, Chatman, and Caldwell (1991) reported similar results in another series of studies. In all three studies, the greater the degree of fit between the person and the organization, the greater the degree of similarity between the perceived organizational identity and a person's self-concept. These results suggest that greater person-organization fit resulted in attitudes and behaviors consistent with stronger identification and lead to our second proposition:

> Proposition 2 (P2): The greater the consistency between the attributes members use to define themselves and the attributes used to define an

organizational image (e.g., perceived organizational identity), the stronger a member's organizational identification.

The link between continuity, the attractiveness of the perceived organizational identity, and levels of members' organizational identification helps members maintain a stable self-concept over time. They identify strongly with their organization when their prior sense of self resembles what they believe is central, enduring, and distinctive about their organization. This relationship makes organizational identification a reciprocal and recursive process. Members who believe the organization is similar to them strengthen the self-associations that were already in place before they became organizational members.

Self-Distinctiveness and the Attractiveness of the Perceived Organizational Identity

Theories of social identity assert that people seek to accentuate their own distinctiveness in interpersonal contexts (Tajfel and Turner, 1985). As a result, members will find organizations attractive when their social identities there provide them with a sense of distinctiveness. A salesperson working for 3M may feel that his or her identification with the organization is a basis for distinctiveness relative to other salespeople working for organizations lacking a clear identity. Kunda (1992: 177) described how engineers in Tech identified with the organization because of its uniquely honest business practices. He related the experience of one project manager:

> I worked for a while for a company that built on those contracts. I worked on the ABM radar. It's not so much that I mind what the products end up doing. No. But all the dishonesty — the excessive costs, the stupidity, the unnecessary work — it really got me down. The norm was: hide the basic specs, follow the letter of the law and produce garbage, then get another contract. Disgusting stuff. Like telling reliability engineers to cook figures. At Tech at least we give customers an honest product. They get what they pay for. Most of the time I feel good about that.

Borrowing from social identity theory, Ashforth and Mael (1989: 24) asserted that "the distinctiveness of the group's [in this case the organization's] values and practices in relation to those of comparable groups increase members' tendency to identify with the organization." Mael and Ashforth (1992) found that alumni of a religious college who perceived their university as distinctive in attitudes, values, and practices had high levels of organizational identification, in terms of a perception of oneness or belongingness to an organization. Ashforth and Mael (1989) argued that this form of organizational identification is associated with cognitive identification, when a person defines him- or herself in terms similar to the organization. Organizational members who believe their organization has a distinctive culture, strategy, structure, or some other configuration of distinctive characteristics are likely to experience strong levels of organizational identification. The above arguments lead to the following proposition:

> Proposition 3 (P3): The greater the distinctiveness of an organizational image (e.g., perceived organizational identity) relative to other organizations, the stronger a member's organizational identification.

Self-Enhancement and the Attractiveness of the Perceived Organizational Identity

When members associate with organizations that have an attractive perceived identity, it enhances their self-esteem as they acquire a more positive evaluation of self. Consistent with past theory, we define self-esteem in terms of the degree to which one likes oneself (Brockner, 1988). For example, if members believe their work organization is defined by qualities associated with competence, power, efficacy, virtue, or moral worth (Gecas, 1982), they are likely to see the perceived organizational identity as attractive. Association with an organization possessing these qualities enhances members' self-esteem because this affiliation provides them with an opportunity to see themselves with these positive qualities, strengthening the degree to which a member likes him- or herself. The link between the attractiveness of the perceived organizational identity and self-esteem implies that members personally experience any decreases or increases in the attractiveness of organizational images.

In the larger study of the Port Authority from which Dutton and Dukerich's (1991) research was drawn, one-third of the respondents noted that the organization was distinctive in terms of being a first-class, high-quality institution. This conception of perceived organizational identity was an important source of self-esteem for Port Authority members. The PA's failure to act on the issue of the homeless raised questions about the basis for the organization's distinctiveness, and members experienced these questions as threats to a positive evaluation of self. As one facility manager expressed it:

> But I've always felt that the Port Authority is . . . and part of our self-image is, as I put my fingers on it, that we do things a little better than other public agencies. There's a whole psyche that goes with that . . . and that's why, when there's time like now, when times get tough, people are nervous a bit, because that goes to their self-image, which is that the Port Authority and therefore, we do things first class.

This example points to how perceived organizational identity and the evaluation of self by members are linked to members' self-esteem. It leads to the fourth proposition:

> Proposition 4 (P4): The more an organizational image (e.g., perceived organizational identity) enhances a member's self-esteem, the stronger his or her organizational identification.

There is an important reciprocal quality to the relationships described in these propositions. As members increasingly define themselves with characteristics that distinguish the organization, and as organizational identification is strengthened, this identity looks increasingly attractive. The more an employee at Tech takes on an attribute of Tech's identity by defining him- or herself as a uniquely honest person, the more attractive Tech's perceived identity looks. As Ashforth and Mael suggested (1989), one consequence of strong identification with an organization is a strengthening of its antecedents, in this case, the attractiveness of the perceived organizational identity. Another factor that affects the attractiveness of perceived organizational identity is a person's involvement with the organization.

Level of Contact and the Attractiveness of the Perceived Organizational Identity

The attractiveness of the perceived organizational identity varies with a member's length of tenure and intensity of exposure to the organization. As members gain tenure in an organization, they increase the level and breadth of exposure to the collective organizational identity, making these organizational attributes more accessible in memory (Bruner, 1957). Through the passage of time, doing organizational work, and intense daily interactions, people come to know themselves as members of the organization (Foote, 1953). Greater contact with the organization increases a member's perceptual readiness (Bruner, 1957) to categorize and define him- or herself as a member of this social group. The longer one remains with an organization, the more salient this group membership is for self-categorization and the more primary is organizational membership as opposed to other group memberships (Kramer, 1991).

O'Reilly and Chatman's (1986) research provides indirect support for the relationship between levels of contact, attractiveness of the perceived organizational identity, and levels of identification. They found a significant positive correlation between tenure in a university and the degree of pride and ownership that people felt with respect to their employing organization. If we assume that pride and ownership in the organization are associated with the attractiveness of the perceived identity, their studies suggest that intense and long contact with an organization (as reflected by greater tenure) increases the level of attractiveness of the organization's identity, contributing to a greater degree of identification. Mael and Ashforth (1992) also found a significant positive correlation between length of time in school and a person's organizational identification, providing more direct support for this link.

Researchers studying the socialization process (Van Maanen, 1975; Feldman, 1976) have asserted that members of organizations incorporate the meaning of the organization into their self-concept. Over time, members are exposed more and more to the totems or symbols that remind them of their union with the organization (Stern, 1988). Members change their level of inclusion in an organization, moving from the periphery of the organization to the center of things (Van Maanen and Schein, 1979: 222) as they interact with other members. As members experience increasing inclusion and contact with the organization, the attractiveness of the perceived organizational identity increases, strengthening organizational identification. As was the case with the previous proposition, rising levels of identification, in turn, motivate members to increase their levels of contact with the organization:

> Proposition 5 (P5): The more contact a member has with an organization (in terms of intensity and duration), the greater the attractiveness of the perceived organizational identity and the stronger the organizational identification.

Construed External Image

Members' identification is also sensitive to how they think outsiders view the organization. While the perceived organizational identity is a member's assessment of the organization's character, construed external image refers to a member's beliefs about outsiders' perceptions of the organization. The construed external image provides more than just information about the probable social evaluation of the organization. For members, the construed external image answers the question, "How do outsiders think of me because of my association with this organization?" The construed external image acts as a potentially powerful mirror, reflecting back to the members how the organization and the behavior of its members are likely being seen by outsiders. This ties in to the concept of the corporate image.

Practitioners and academics use the term "corporate image" in a variety of ways. Consultants use the term "corporate image" to refer to the impression that an organization makes to outsiders and insiders (e.g., Selame and Selame, 1988). Researchers in marketing assert that corporate images matter to a firm's customers (e.g., Arora and Cavusgil, 1985). Human resource researchers study how information shapes the attractiveness of an organization's image during recruiting (e.g., Gatewood, Gowan, and Lautenschlager, 1993). Researchers interested in the processes of organizational impression management describe how various tactics for enhancing organizational image alter how outside parties view the organization and its actions (e.g., Elsbach and Sutton, 1992). Dutton and Dukerich (1991) used the term organizational image to refer to what organizational insiders believe outsiders think is distinctive, central, and enduring about the organization. Here, however, to clarify and distinguish whose beliefs are of interest, we use the term construed external image.

We distinguish between two different uses of the term organizational image: one focusing on the beliefs of outside members, the other focusing on the beliefs of inside members. Organizational reputation refers to outsiders' beliefs about what distinguishes an organization; construed external image captures internal members' own assessment of these beliefs (Dutton and Dukerich, 1991). This distinction between reputation and construed external image is important. Insiders and outsiders to an organization have access to different information about the organization and apply different values and goals in interpreting this information. Distinguishing between construed external image and reputation allows these two organizational images to differ from one another.

Sometimes an organization's reputation and insiders' construed external images are closely aligned. When an organization's reputation is widely disseminated through extensive press or media attention, for example, the organization's reputation is likely to be highly correlated with the external image of the organization construed by insiders. Despite their public media campaigns and the creation of pseudo-events that are planned for the explicit purpose of being reported (Boorstin, 1961), most organizations are unable to align fully outsiders' beliefs about an organization (i.e., reputations) and insiders' readings of these beliefs (i.e., construed external images). Organizational members sometimes have a distorted impression of what others believe, either believing their organization is perceived in a more positive or a more negative light than outsiders see it. Ginzel, Kramer, and Sutton (1993) described how top managers at Dow Corning Wright tried to control the reputational damage to their firm that occurred because of the continued production and sale of silicon breast implants. As their attempts to manage the impressions of external audiences created new reputational crises (e.g., as a production problem became a problem of organizational integrity and honesty), top managers seemed unaware of the amount of damage that their reputation had sustained. In this case, the construed external image (by top management and perhaps members more generally at Dow Corning) was more positive than the firm's actual reputation. This inconsistency between reputation and construed external image delayed the firm's crisis response and contributed to the interpretive conflict that top managers experienced as they tried to manage the firm's reputation in the minds of both sympathetic and antagonistic audiences.

Construed External Image and Identification

The earlier example of Exxon's executives' struggle with social contact at cocktail parties illustrates that organizational members use the construed external image of an organization to assess the social value of their affiliation with an organization. Because construed external image summarizes a member's beliefs about how people outside the organization are likely to view the member through his or her organizational affiliation, the construed external image is a powerful reflection of public opinion. When the construed external image of an organization is assessed as attractive (i.e., members believe the image contains attributes that distinguish the organization in positive, socially valued terms), the construed external image strengthens members' organizational identification.

People try to maintain a positive social identity (Tajfel and Turner, 1985) because positive social identities (1) create self-gratifying social opportunities (Brown, 1969), (2) heighten social prestige (Perrow, 1961; Cheney, 1983; Ashforth and Mael, 1989), (3) facilitate social interaction (Foote, 1953), and (4) create social credits. When members construe the external image as attractive — meaning that they believe this image has elements that others are likely to value — then organizational affiliation creates a positive social identity (Tajfel, 1982) that increases the level of overlap between how a member defines him- or herself and the organization. Empirical research supports this claim. Vardi, Wiener, and Poppa (1989) found that members in an organization that produced a product for the military market in Israel more strongly identified with their organization than members in a matched firm producing a similar product for a commercial market. The firm's positive social role as a manufacturer of products for the military market — a market that is socially valued in Israel — could explain these findings: Members working in the first organization viewed the construed external image of the firm as attractive, which thereby strengthened their identification. This logic and example provide the basis for proposition 6:

> Proposition 6 (P6): The greater the attractiveness of an organization's construed external image, the stronger a member's organizational identification.

The same principles of self-definition account for why members are likely to see the construed external image of their work organization as attractive or not: (1) self-continuity, (2) self-distinctiveness, and (3) self-enhancement.

Self-Continuity and the Attractiveness of the Construed External Image

Proposition 2 asserted that members will find organizational images more attractive if they contribute to a consistent sense of self. This proposition is as valid for how members value construed external images as it is for the perceived organizational identity. When members believe outsiders view the organization in terms that are close to how they see themselves, then membership provides an opportunity to maintain a coherent and consistent sense of self.

Self-Distinctiveness and the Attractiveness of the Construed External Image

A construed external image that enhances a member's distinctiveness in interpersonal contexts will also be seen as more attractive, following the logic of proposition 3, which derives from social identity theory. Members gain distinctiveness from their own sense of what uniquely characterizes their work organization (perceived organizational identity) and from what they believe outsiders think about the organization (construed external image). This may be the route by which advertising and public relations efforts affect people's attachments to their work organizations. When these advertising efforts and public image campaigns make insiders believe that outsiders think the organization is unique in some way, these efforts may yield benefits not only in attracting customers, they may also enhance members' commitment by strengthening identification.

Self-Enhancement and the Attractiveness of the Construed External Image

According to proposition 4, the construed external image of an organization can become attractive to a person because it enhances his or her self-evaluation by providing important information about how others are likely to appraise a member's character based on his or her organizational affiliation (Dutton and Dukerich, 1991). People want to maintain a positive self-concept (Turner, 1978; Brockner, 1988). They use an organization's construed external image to estimate the reflected appraisal of outsiders and are drawn to images that portray them as competent and morally virtuous (Gecas, 1982). Thus, if a member believes outsiders are likely to view the organization favorably, the image is attractive. An attractive image encourages further alignment between a member's self-concept and organizational definition. Greenberg (1990) argued, for example, that organizations that are construed as "fair" have members who more strongly identify with it. This may occur because possibilities for self-enhancement are afforded by one's affiliation with such a positively construed external image.

Visibility of Affiliation and Identification

The relationship between the attractiveness of these organizational images and the strength of identification depends on people's visible affiliation with their work organization. Two sets of arguments support this assertion: one is based on a simple self-perception logic, the other is tied to the logic of impression management.

When people are visibly associated with an organization, they are more frequently reminded of their organizational membership. Visible affiliations, such as those made through public organizational roles, serve as vivid reminders of organizational membership (Charters and Newcomb, 1952) and increase the potency of the organization as a source of self-definition (Brown, 1969). These reminders make people's membership in the organization accessible and salient to them (Turner, 1982). When a person is visibly affiliated with an organization, self-perception processes heighten his or her own awareness of the attractiveness of the organization. For perceived organizational identity, the attractiveness of this image will have a greater effect on the strength of a member's identification if he or she is visibly affiliated with the organization:

> Proposition 7 (P7): The greater the visibility of a member's affiliation with the organization, the stronger the relationship between the attractiveness of the perceived organizational identity and his or her organizational identification.

The visibility of a member's organizational affiliation can have an even greater moderating effect on the relationship between the attractiveness of the construed external image and member identification because of the motivation to manage impressions. Public knowledge that a person is affiliated with an organization

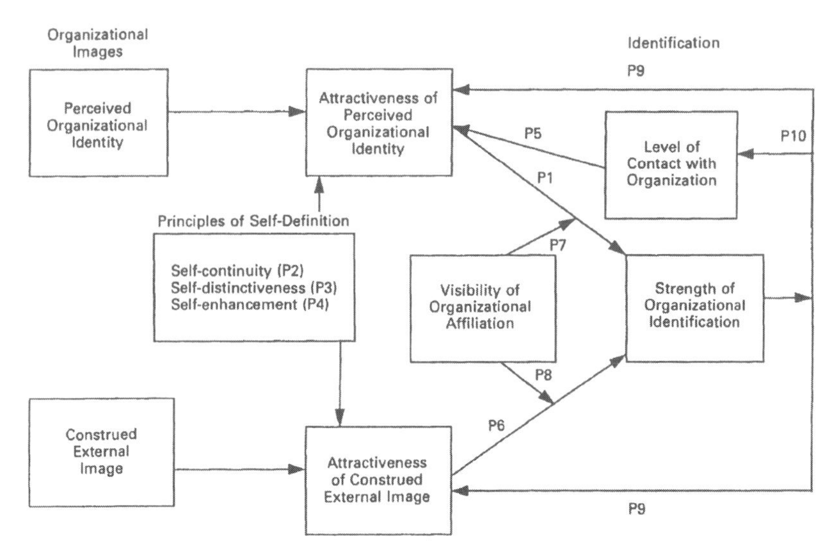

Figure 1: Linking perceived organizational identity and construed external image to strength of organizational identification.

creates expectations about how he or she is likely to behave and the types of attitudes he or she is likely to hold (Tetlock and Manstead, 1985). People expect a member who is visibly affiliated with the Rotary Club to behave in ways and to hold attitudes that are appropriate to Rotarians, whereas people not affiliated with the Rotary Club would not be subject to these expectations. These expectations, and members' awareness of them, encourage members to take on the qualities embodied in the perceived organizational identity. If one is visibly associated with the Rotary Club and this organization's perceived organizational identity includes the attributes of community service, this quality will more likely become part of the member's own self-concept, thus strengthening identification, if the member is visibly affiliated with the organization.

When people have organizational affiliations that are visible, either through physical displays, such as Rotarians' use of lapel pins, or their organizational locations (e.g., having leadership or boundary spanning roles), they are in the position of having to explain and justify their role and standpoint frequently (Turner, 1978: 15). This strengthens the correlation between the attractiveness of the image and the strength of identification. The desire to create an impression for others that is consistent with the construed external image is also more intense when one's organization affiliation is visible. This strengthens the correlation between the attractiveness of the construed external image and the strength of identification. Both arguments support the eighth proposition:

> Proposition 8 (P8): The greater the visibility of a member's affiliation with the organization, the stronger the relationship between the attractiveness of the construed external image and a member's organizational identification.

Figure 1 summarizes our model of how perceived organizational identity and the construed external image are linked to the strength of a member's identification with the organization.

Consequences of Organizational Identification

Organizational identification has several consequences for individuals' beliefs and behaviors. Using the model of member identification developed thus far, we will discuss these consequences to explain more fully how the process we model affects the organization, its members, and even the world outside it.

Strengthening the Antecedents of Organizational Identification

Research on self-affirmation processes (Steele, 1988) and self-justification processes (Staw, 1980) indicates that people attempt to preserve a sense of integrity and self-worth. These beliefs about the self are sustained by positively evaluating the groups with which one identifies, including the organization. As mentioned earlier, one consequence of organizational identification is the strengthening of its antecedents. As members identify more strongly with the organization, their beliefs about the organization are likely to become more positive. Members who strongly identify with an organization are likely, for example, to believe that the organization is producing valuable outputs. A particular consequence of this tendency to evaluate the organization more positively is that the attractiveness of the perceived organizational identity and construed external image are likely to increase.

> Proposition 9 (P9): The greater the strength of organizational identification, the more members will evaluate the perceived organizational identity and construed external image as attractive.

The feedback loops in Figure 1 which connect strength of organizational identification to attractiveness of the perceived organizational identity and the construed external image, depict the effect of organizational identification on its antecedents.

In addition to affecting beliefs, strong organizational identification affects behaviors. These behaviors also strengthen the antecedents of organizational identification. The logic of self-perception helps to explain this reinforcing cycle. People are motivated to maintain consistency between their self-perceptions and behavior (Festinger, 1957), and those who strongly identify with an organization will seek more contact with the organization. More contact with the organization enhances the sense of continuity of one's self-concept that people value. Kunda's (1992) account of the behaviors of the strongly identified engineers at Tech described how they worked to increase their contact with and submersion in the organization by depicting themselves as Tech members. One memorable example is Mary:

Mary is unmarried. Over the desk, where others keep family pictures there is a glossy picture of her at a trade show with colleagues. A row of ribbons and name tags from various such events are pinned to the wall next to it. Above it is an 'I love Tech' bumper sticker. On a shelf there is a golf section with a few trophies. 'Most Improved Golfer' from *Golfer's Digest* and a Tech trophy. Next to it a color print of a sailing boat with a large Tech logo in the billowing sail. An orderly row of beer bottles and mugs with a Tech logo, all with their handles facing left. (p. 194)

These arguments suggest another proposition that links the strength of organizational identification to one of its antecedents:

Proposition 10 (P10): The greater the strength of organizational identification, the more a member will seek contact with the organization.

Patterns of Social Interaction

Strong identification with the organization keeps members attuned to the future viability of the organization. When people strongly identify with their work organization their sense of survival is tied to the organization's survival. This link has at least two effects. One effect involves interpersonal dynamics: Strong identification prompts increased cooperation with organizational members as part of the organizational group and increased competition with nonmembers. Second, members direct additional effort toward tasks that contribute to coworkers and to the organization.

When a member's level of organizational identification is strong, his or her sense of self as an organizational member helps to form a pattern of in-group and out-group dynamics. While these dynamics play out between units within an organization (Ashforth and Mael, 1989), they can also explain differences in members' behavior across organizations. In early studies of intergroup discrimination, Tajfel et al. (1971) designed an experiment in which subjects were separated into groups by random criteria and were told only about their group affiliation. Even under this stripped-down condition, now known as the minimal group paradigm, subjects discriminated in favor of in-group members and against out-group members. Social identification theory explains that the perception of a shared categorical identity creates an in-group bias, which leads to intragroup cohesion (Turner, 1978; Kramer, 1991), through the accentuation of perceived similarities with other group members (Hogg and Abrams, 1988) and the resulting positive attitudes toward these in-group members (Turner, 1978: 28). Strong identification with an organization makes cooperative behavior toward other organizational members likely because of a heightened sense of in-group (organizational) trust and reciprocity, heightened social attraction toward in-group members, and presentation of a favorable image of the organization to self and others (Kramer, 1991):

Proposition 11 (P11): The stronger the organizational identification, the greater a member's cooperation with other members of the organization (in-group cooperation).

When people identify themselves as organizational members they rely on organizational-level categories to determine the relevant in-groups and out-groups. Along with increasing the cooperativeness among people perceived as being part of the organizational in-group, identification increases the competitiveness that organizational members perceive between themselves (as the in-group) and other out-groups.

Ample empirical evidence links intergroup-level categorization with the emergence of competitive behaviors between groups (see Hogg and Abrams, 1988; Kramer, 1991, for reviews). Although intergroup differentiation does not necessarily cause competition, in our culture, competition between groups is relatively easy to trigger (Hogg and Abrams, 1988). Within the minimal group paradigm, simply categorizing people into groups has been shown to be sufficient to produce competitive behavior between people in different groups (Tajfel et al., 1971). Further, in interdependent social situations in which one group "wins" something valued only when other groups "lose," members who identify with the organization are more aware of the collective consequences of winning and are thus more competitive with out-group members than are members who don't identify with the organization. This logic supports the twelfth proposition:

> Proposition 12 (P12): The greater the strength of organizational identification, the greater a member's competitive behavior directed toward out-group members.

Along with cooperation between members, other patterns of interaction also change. People who strongly identify with the organization are likely to focus on tasks that benefit the whole organization rather than on purely self-interested ones. This is organizational citizenship behavior. Organizational citizenship behaviors are organizationally functional behaviors that extend beyond role requirements and are not contractually guaranteed (Smith, Organ, and Near, 1983; Organ, 1990). O'Reilly and Chatman's (1986) studies demonstrated a positive correlation between attachment based on internalization and identification and levels of extrarole behavior, which indirectly supports this assertion even though they defined and measured the two forms of attachment (identification and internalization) in ways that differ from the strength of identification. As members become more psychologically attached to an organization, their relationship to the organization changes, resulting in systematically different behavioral displays of psychological involvement. Mael and Ashforth (1992) found some evidence of this relationship in their study of alumni's identification with their alma mater. They found that more organizational identification correlated with the ranked priority of contributions and advising one's son and others to attend an all-male, religious college. We suggest that effort directed toward preserving, supporting, and improving the organization proceeds naturally from the congruence between a member's self-definition and the organization's definition. Organizational identification aligns individual interests and behaviors with interest and behaviors that benefit the organization. It means exertion on behalf of the organization is also exertion on behalf of the self (Shamir, House, and Arthur, 1993).

Members' efforts to benefit the organization result in behaviors that are acts of obedience, loyalty, and participation, such as spending time helping newcomers, working on long-term organizational projects, pushing superiors to perform to higher standards, and providing ideas for improving the organization (Van Dyne, Graham, and Dienesch, 1994). Strong organizational identification enables members to contribute more frequently and more freely to the organization. When strong overlap exists between what defines the person and what is thought to define the organization, this overlap enables a member to contribute simultaneously to him- or herself and to the organization:

> Proposition 13 (P13): The stronger the organizational identification, the more often a member exhibits organizational citizenship behaviors.

Discussion

In this paper we assert that images of organizations shape how members define themselves. When members define themselves with attributes that overlap with the attributes they use to define the organization, they are strongly identified with the organization. Strong identification with an organization is also apparent when the social identity of an organization is more available and salient than other social identities.

The psychology of social identity theory is powerful because it implies that members may change their behavior by merely thinking differently about their employing organization. If members believe that the perceived organizational identity has been altered either in content (e.g., in what attributes distinguish this organization) or in its evaluation (making it more or less attractive), members are likely to modify their behavior. This change in members' behavior does not require interacting with others, altering employees' jobs and rewards, or changing bosses. Rather, if members think of their employing organization differently (by changes in the perceived organizational identity or construed external image), we argue they will behave differently.

Organizational images shape the strength of members' identification with the organization, serving as important cognitive reference points that either connect or disconnect a member from the organization. When the images are attractive, they increase the degree to which self-definitions approximate the organizational definition. Members' images of their employing organization are vital sources of their self-construction. By providing members with images of the social group to which they belong that specify the content of what it means to be a member, organizations provide vital input for members' self-definition.

The model shows how the construed external organizational image affects the level of connection between individuals and their organizations. By doing so, it brings the insights of symbolic interactionists into depictions of individual attachment to organizations. Symbolic interactionists assert that organizational members come to know themselves through the impressions of others and that these anticipated impressions shape people's everyday behavior. This assertion implies that there are attachment consequences to believing that outsiders see the organization in a particular way. By examining the relationships between construed external image and members' identification, we recognize that individual-organizational attachment is more than an intrapersonal phenomenon. Members' degree of cognitive attachment (e.g., strength of identification) to the organization links to the anticipated reflected appraisal by others, making cognitive attachment a social and interpersonal process as well.

Our model is relevant to three approaches in research on organizational impression management. One approach uses attribution theory to look at organizational accounts for success and failure (Salancik and Meindl, 1984). Another focuses on how organizations construe actions and events to maintain a positive image in the minds of key stakeholders (Sutton and Kramer, 1990; Elsbach and Sutton, 1992; Ginzel, Kramer, and Sutton, 1993). And a third concentrates on the content and effectiveness of organizational image management (Elsbach, 1994). All three approaches emphasize leaders' and organizational efforts to create organizational images and acknowledge that these images affect outside stakeholders' impressions of an organization's legitimacy. This literature ignores how these images affect organizational insiders — the members who are associated with these images as part of their everyday work behavior. Our model suggests that researchers interested in the social psychology of organizational impression management should consider how the images created for outsiders shape the experience, attachments, and behaviors of insiders.

Future Research

Extending and testing the model presented in this paper requires operationalizing strength of organizational identification. There are at least three different approaches to measuring this variable. The first strategy involves directly assessing one's organizational identification. Mael and Ashforth (1988) developed a scale-based measure of strength of organizational identification that is reliable and empirically distinguishable from concepts such as organizational commitment and involvement, but, like O'Reilly and Chatman's (1986) operationalization, their scale includes assessments of how individuals feel about the organization. By involving more than the cognitive connection between a member and an organization, these scales tap into a broader concept of psychological attachment than what we intend by organization identification. A possible alternative might be to modify Jackson's (1981) index of commitment to an identity index that uses an array of items to assess the importance of a given social identity to a member's self-definition.

A second strategy would be to ask members to evaluate a set of social identities and indicate the relative degree to which these identities accurately describe them as individuals, either by rating each identity or ranking them in a hierarchy (e.g., Hoelter, 1985; Harquail, 1994). Using this approach, strong organizational identification is indicated when a person ranks or rates the organizational identity higher than other social identities.

A third approach involves directly assessing the level of overlap between the characteristics that a member believes typify him or her as an individual (i.e., are enduring, central, and distinctive) and the characteristics that typify the organization. High levels of overlap between the two lists of central, distinctive, and enduring attributes would indicate strong organizational identification.

Research assessing the strength of organizational identification should also address the desirable and undesirable outcomes associated with strong organizational identification. For example, Ashforth and Mael (1992) presented a study in which they found that strong organizational identification was associated with tyrannical behavior of managers toward their subordinates, such as belittling of subordinates and increased use of noncontingent punishment. Their results are unsettling and add credence to warnings about the dark side of organizational identification (Schwartz, 1987; Kunda, 1992).

Future research should consider the array of organizational images that may affect members' attachments to an organization. We have not treated all organizational images that have currency in an organization as equal but, rather, have singled out perceived organizational identity and construed external image as particularly important and worthy of empirical study. Also, while we have considered only members' present images of the organization, psychologists and organizational researchers have shown that future-based images of the organization — what Gioia and Thomas (1991) and Reger et al. (1994) called the ideal organization or what Whetten, Lewis, and Mischel (1992) called the desired organization image — also shape members' behaviors in organizations. The concept of "possible selves" discussed by Markus and Nurius (1986) suggests one avenue for exploring how these future images of organizations shape members' behavior by affecting the content and evaluation of possible selves. This concept describes individuals' ideas about what they might become and what they are afraid of becoming and can provide a conceptual link between self-schema and members' motivations (Markus and Nurius, 1986: 954). Future research should elaborate how future and past images of the organization connect to a member's self-concept and direct his or her behavior.

We also need research on how changing conditions affect members' images of their work organization and the behaviors that result. Changes in structure, culture, organizational performance, organizational boundaries, or an organization's competitive strategy may induce members to revise their perceived organizational identity and construed external image. These redefinitions could have significant psychological effects. A decline in organizational performance, for example, may lower the perceived attractiveness of the organization's construed external image, weakening organizational identification and creating less cooperation and fewer organizational citizenship behaviors.

Mergers and acquisitions represent changes in both structure and culture and may alter members' organizational images. These strategic changes revise both the boundaries and the content of a member's perceived organizational identity. When the retail discount giant, KMart, bought the upscale and highbrow

Borders bookstores, employees' sense of the perceived organizational identity changed, as did the construed external image (Bridgeforth, 1992). Some employees believed that the basis for the distinctiveness of Borders — its ability to attract "readers with discriminating tastes" — would be compromised by its association with a large discount store (Bridgeforth, 1992: C1). Based on our model, we would predict that some Borders employees would weaken their level of organizational identification, resulting in fewer displays of affiliation, less cooperative behavior with inside members, fewer organization citizenship behaviors, and a host of other outcomes. Alternatively, the model suggests that if the perceived organizational identity does change, and Borders members spend a large amount of time seeing themselves as part of this changed social group (a KMart-Borders), then, over time, members may alter how they see themselves.

Future researchers should consider the relationship between members' perceived organizational identity and their expectations about organizational actions, which would help us understand how members respond when the organization's actions exceed or fail to meet their expectations. Members expect the organization's actions to reflect what is distinctive, central, and enduring about the organization's identity, and they discern their organization's identity in part by interpreting the organization's actions (Dutton and Dukerich, 1991). When members who identify with an organization observe a consistency between expected and actual organizational actions, their organizational identification strengthens because the organizational identity proves to be a reliable source of self-definition.

When members perceive major inconsistencies between expected and actual organizational actions, a different set of responses is likely. As suggested by balance theory (Heider, 1958), self-perception theory (Bem, 1967), and cognitive dissonance theory (Festinger, 1957), members will work to resolve this inconsistency (Turner and Oakes, 1986). One response involves downplaying the importance of these inconsistencies by offering excuses or justifications (Bies and Sitkin, 1991), with no major change in the strength of their organizational identification. Another response involves revising one's perceptions of the organization's identity. If this revision enhances the attractiveness of the perceived identity, member identification will be strengthened. Conversely, if the revision erodes the attractiveness of the identity, members' identification will weaken. A third possible response implies that members experience these inconsistencies as threats to their own identities and respond by changing their self-definitions (Breakwell, 1986).

Finally, we must consider the generalizability of this model of organizational identification to different national or societal contexts. Our model is built on a conceptualization of the self that assumes that people want to become "independent from others and to discover and express their own uniqueness" (Markus and Kitayama, 1991: 226). This assumption about how a person's self is organized underlies the hypothesized links between perceived organizational identity, construed external image, strength of organizational identification, and individual-level outcomes, but this assumption may be limited in its cultural generalizability (Erez and Earley, 1993). Markus and Kitayama (1991: 227) explained how an alternative view of self — one that is built on the "fundamental connectedness of human beings to each other" and one that is typically associated with non-Western cultures — may create very different connections between an organization and an individual's beliefs and actions. When Japan Air Lines (JAL) experienced the worst single aircraft accident ever (520 people were killed), the president, Yasumoto Takagi, "followed Japanese custom and took responsibility for the crash by resigning" (*Los Angeles Times*, 1985: 2). Organizational members in non-Western cultures may feel accountable for the images that are created of their organization, thus implying an even stronger connection between these images and subsequent behaviors.

This paper shows that it is more than economic transactions that connect members to their work organizations. Members' attachments to an organization are fundamentally tied both to the images that they have of what the organization means to them and what they think it means to others. If the images provide them with continuity, distinctiveness, or positive evaluations, then their attachments strengthen through organizational identification. Economic depictions of organizational attachments ignore these images and their dynamic qualities. Our model suggests that these images should be at center stage if we are to understand what makes the 3M salesman get up eagerly for work in the morning and see new possibilities and meaning in his life as an organizational member.

References

Albert, Stuart, and David A. Whetten 1985 "Organizational identity." In L. L. Cummings and Barry M. Staw (eds.), *Research in Organizational Behavior*, 7: 263–295. Greenwich, CT: JAI Press.

Alderfer, Clayton P., and Kenwyn K. Smith 1982 "Studying intergroup relations embedded in organizations." *Administrative Science Quarterly*, 27: 35–65.

Allaire, Yvan, and Michael E. Firsirotu 1984 "Theories of organizational culture." *Organization Studies*, 5: 193–226.

Arora, R. G., and S. T. Cavusgil 1985 "Image and cost factors in the choice of mental health-care organizations: A causal model." *Journal of Academy of Marketing Science*, 13: 119–129.

Ashforth, Blake E., and Fred Mael 1989 "Social identity theory and the organization." *Academy of Management Review*, 14: 20–39.

Ashforth, Blake E., and Fred Mael 1992 "The dark side of organizational identification." Paper presented at the Academy of Management Meeting, Las Vegas.

Bem, Daryl J. 1967 "Self-perception: The dependent variable of human performance." *Organizational Behavior and Human Performance*, 2: 105–121.

Bies, Robert J., and Sim B. Sitkin 1991 "Explanation as legitimation: Excuse-making in organizations." In M. McLaughlin, M. Cody, and S. Read (eds.), *Explaining One's Self to Others: Reason-giving in a Social Context*: 183–198. Hillsdale, NJ: Erlbaum.

Boorstin, Daniel 1961 *The Image: A Guide to Pseudo-events in America*. New York: Atheneum,

Breakwell, Glynis M. 1986 *Coping with Threatened Identities*. New York: Metheun.

Bridgeforth, Arthur, Jr. 1992 "Bordering on change: Borders bookstore chain turns a page in its history, but won't close the book on providing good reading." *Ann Arbor News*, October 25, C1.

Brockner, Joel 1988 *Self-esteem at Work: Research, Theory and Practice*. Lexington, MA: D.C. Heath.

Brown, Michael E. 1969 "Identification and some conditions of organizational involvement." *Administrative Science Quarterly*, 14: 346–355.

Bruner, Jerome S. 1957 "Going beyond the information given." In H. Gruber, K. Hammond, and R. Jessor (eds.), *Contemporary Approaches to Cognition*: 41–69. Cambridge, MA: Harvard University Press.

Charters, W. W., and Theodore M. Newcomb 1952 "Some attitudinal effects of experimentally increased salience of group membership." In G. E. Swanson et al. (eds.), *Readings in Social Psychology*: 415–420. New York: Holt, Rinehart and Winston.

Chatman, Jennifer A. 1991 "Matching people and organizations: Selection and socialization in public accounting firms." *Administrative Science Quarterly*, 36: 459–484.

Cheney, George 1983 "On the various and changing meanings of organized membership: A field study of organizational identification." *Communication Monographs*, 50: 342–362.

Cialdini, Robert B., Richard J. Borden, Avril Thorne, Marcus Randall Walker, Stephen Freeman, and Lloyd Reynolds Sloan 1976 "Basking in reflected glory: Three (football) field studies." *Journal of Personality and Social Psychology*, 34: 366–375.

Dutton, Jane E., and Janet M. Dukerich 1991 "Keeping an eye on the mirror: The role of image and identity in organizational adaptation." *Academy of Management Journal*, 34: 517–554.

Elsbach, Kimberly D. 1994 "Managing organizational legitimacy in the California cattle industry: The construction and effectiveness of verbal accounts." *Administrative Science Quarterly*, 39: 57–88.

Elsbach, Kimberly D., and Robert I. Sutton 1992 "Acquiring organizational legitimacy through illegitimate actions: A marriage of institutional and impression management theories." *Academy of Management Journal*, 35: 699–738.

Erez, Miriam, and P. Christopher Earley 1993 *Culture: Self-Identity and Work*. New York: Oxford University Press.

Fanning, D. 1990 "Coping in industries that the public hates." *New York Times*, August 19: F 25.

Feldman, Daniel C. 1976 "A contingency theory of socialization." *Administrative Science Quarterly*, 21: 433–452.

Festinger, Leon 1957 *A Theory of Cognitive Dissonance*. New York: Harper and Row.

Fiol, Marlene 1991 "Managing culture as a competitive resource: An identity-based view of sustainable competitive advantage." *Journal of Management*, 17: 191–206.

Foote, Nelson N. 1953 "Identification as a basis for a theory of motivation." *American Sociological Review*, 16: 14–21.

Garbett, Thomas 1988 *How to Build a Corporation's Identity and Project Its Image*. Lexington, MA: D.C. Heath.

Gatewood, Robert D., Mary A. Gowan, and G. J. Lautenschlager 1993 "Corporate image, recruitment image, and initial job choice decisions." *Academy of Management Journal*, 36: 414–427.

Gecas, Viktor 1982 "The self-concept." In R. H. Turner and J. F. Short, Jr. (eds.), *Annual Review of Sociology*, 8: 1–33. Palo Alto, CA: Annual Reviews.

Gergen, Kenneth J. 1968 "Personal consistency and the presentation of self." In Chad Gordon and K. J. Gergen (eds.), *The Self in Social Interaction*, 1: 299–308. New York: Wiley.

Ginzel, Linda E., Roderick M. Kramer, and Robert I. Sutton 1993 "Organizational impression management as a reciprocal influence process: The neglected role of the organizational audience." In L. L. Cummings and Barry M. Staw (eds.), *Research in Organizational Behavior*, 15: 227–266. Greenwich, CT: JAI Press.

Gioia, Dennis A., and James B. Thomas 1991 "Sensemaking, sensegiving and action taking in a university: Toward a model of strategic interpretation." Paper presented at the Academy of Management Meeting, Miami.

Greenberg, Jerald 1990 "Looking fair versus being fair: Managing impressions of organizational justice." In Barry M. Staw and L. L. Cummings (eds.), *Research in Organizational Behavior*, 12: 111–157. Greenwich, CT: JAI Press.

Hall, Douglas T., and Benjamin Schneider 1972 "Correlations of organizational identification as a function of career pattern and organizational type." *Administrative Science Quarterly*, 17: 340–350.
Hall, Douglas T., Benjamin Schneider, and Harold T. Nygren 1970 "Personal factors in organizational identification." *Administrative Science Quarterly*, 17: 340–350.
Harquail, Celia V. 1994 "Is sisterhood powerful? Social identification and women's advocacy in organizations." Unpublished Ph.D. dissertation proposal, University of Michigan.
Heider, Fritz 1958 *The Psychology of Interpersonal Relations*. New York: Wiley.
Hirschman, Albert O. 1970 *Exit, Voice and Loyalty: Responses to Decline in Firms, Organizations and States*. Cambridge, MA: Harvard University Press.
Hoelter, Jon W. 1985 "The structure of self-conception: Conceptualization and measurement." *Journal of Personality and Social Psychology*, 48: 1392–1407.
Hogg, Michael A., and Dominic Abrams 1988 *Social Identifications: A Social Psychology of Intergroup Relations and Group Processes*. London: Routledge.
Jackson, Susan E. 1981 "Measurement of commitment to role identities." *Journal of Personality and Social Psychology*, 40: 138–146.
Kahn, William A. 1990 "Psychological conditions of personal engagement and disengagement at work." *Academy of Management Journal*, 33: 692–724.
Kelman, Herbert C. 1958 "Compliance, identification and internalization." *Journal of Conflict Resolution*, 2: 51–60.
Kramer, Roderick M. 1991 "Intergroup relations and organizational dilemmas: The role of categorization processes." In L. L. Cummings and Barry M. Staw (eds.), *Research in Organizational Behavior*, 13: 191–228. Greenwich, CT: JAI Press.
Kunda, Gideon 1992 *Engineering Culture*. Philadelphia: Temple University Press.
Lee, Sang M. 1971 "An empirical analysis of organizational identification." *Academy of Management Journal*, 14: 213–226.
Los Angeles Times 1985 "JAL replaces top officers in shake-up over crash." December 19, part 4: 2.
Mael, Fred, and Blake E. Ashforth 1988 "A reconceptualization of organizational identification." In T. L. Keon and A. C. Bluedorn (eds.), *Proceedings of the Midwest Academy of Management Meetings*: 127–129.
Mael, Fred, and Blake E. Ashforth 1992 "Alumni and their alma mater: A partial test of the reformulated model of organizational identification." *Journal of Organization Behavior*, 13: 103–123.
Markus, Hazel R., and Shimbu Kitiyama 1991 "Culture and the self: Implications for cognition, emotion and motivation." *Psychological Review*, 98: 224–253.
Markus, Hazel R., and Paula Nurius 1986 "Possible selves." *American Psychologist*, 41: 95469.
Markus, Hazel R., and Elissa Wurf 1987 "The dynamic self-concept: A social-psychological perspective." In Mark R. Rosenzweig and Lyman W. Porter (eds.), *Annual Review of Psychology*, 38: 299–337. Palo Alto, CA: Annual Reviews.
Martin, Joanne, Martha S. Feldman, Mary Jo Hatch, and Sim B. Sitkin 1983 "The uniqueness paradox in organizational stories." *Administrative Science Quarterly*, 28 : 438–453.
Mowday, Richard, Lyman M. Porter, and Richard M. Steers 1982 *E-O Linkages: The Psychology of Commitment, Absenteeism and Turnover*. New York: Academic Press.
O'Reilly, Charles, and Jennifer Chatman 1986 "Organizational commitment and psychological attachment: The effects of compliance, identification and internalization on prosocial behavior." *Journal of Applied Psychology*, 71: 492–499.
O'Reilly, Charles, Jennifer Chatman, and David F. Caldwell 1991 "People and organizational culture: A profile comparison approach to assessing person-organization fit." *Academy of Management Journal*, 34: 487–516.
Organ, Dennis W. 1990 "The motivational basis of citizenship behavior." In Barry M. Staw and L. L. Cummings (ads), *Research in Organizational Behavior*, 12: 43–72. Greenwich, CT: JAI Press.
Perrow, Charles 1961 "Organizational prestige: Some functions and dysfunctions." *American Journal of Sociology*, 61: 373–391.

Pfeffer, Jeffrey 1981 "Management as symbolic action." In L. L. Cummings and Barry M. Staw (eds.), *Research in Organizational Behavior*, 3: 1–52. Greenwich, CT: JAI Press.

Porter, Lyman W., Richard M. Steers, Richard T. Mowday, and Paul V. Boulian 1974 "Organizational commitment, job satisfaction and turnover among psychiatric technicians." *Journal of Applied Psychology*, 59: 603–609.

Reger, Rhonda K., Loren T. Gustafson, Sam M. DeMarie, and John V. Mullane 1994 "Reframing the organization: Why implementing Total Quality is easier said than done." *Academy of Management Review*, vol. 19 (in press).

Salancik, Gerald R., and James R. Meindl 1984 "Corporate attributions as strategic illusions of management control." *Administrative Science Quarterly*, 29: 238–254.

Schlenker, Barry R. 1985 "Self-identification: Toward an integration of the private and public self." In R. Baumiester (ed.), *Public Self and Private Self*: 21–62. New York: Springer-Verlag.

Schwartz, Howard S. 1987 "Anti-social actions of committed organizational participants: An existential psychoanalytic perspective." Organization Studies, 8: 327–340.

Selame, Elinor, and Joe Selame 1988 *The Company Image: Building Your Identity and Influence in the Marketplace*. New York: Wiley.

Shamir, Boas 1991 "Meaning, self and motivation in organizations." *Organization Studies*, 12: 405–424.

Shamir, Boas, Robert J. House, and Michael B. Arthur 1993 "The motivational effects of charismatic leadership: A self-concept based theory." *Organization Science*, 4: 577–594.

Smith, Ann C., Dennis W. Organ, and Janet P. Near 1983 "Organizational citizenship behavior: Its nature and antecedents." *Journal of Applied Psychology*, 68: 653–663.

Star Ledger (Newark) 1989 "Exxon workers assailed." May 22: 3.

Staw, Barry M. 1980 "Rationality and justification in organizational life." In Barry M. Staw and L. L. Cummings (eds.), *Research in Organizational Behavior*, 2: 45–80. Greenwich, CT: JAI Press.

Steele, Claude M. 1988 "The psychology of self-affirmation: Sustaining the integrity of the self." In Leonard Berkowitz (ed.), *Advances in Experimental Social Psychology*, 21: 261–302. New York: Academic Press.

Stern, S. 1988 "A symbolic representation of organizational identity." In Michael O. Jones, Michael D. Moore, and Richard C. Snyder (eds.), *Inside Organizations*: 281–295. Newbury, CA: Sage.

Stryker, Sheldon, and Richard T. Serpe 1982 "Commitment, identity salience and role behavior: Theory and research example." In W. Ickers and E. Knowles (eds.), *Personality, Roles and Social Behavior*: 199–219. New York: Springer-Verlag.

Sutton, Robert I., and Roderick M. Kramer 1990 "Transforming failure into success: Impression management, the Reagan Administration, and the Iceland Arms Control Talks." In Robert L. Kahn and Mayer Zald (eds.), *International Cooperation and Conflict Perspectives from Organization Theory*: 221–245. San Francisco: Jossey-Bass.

Tajfel, Henri 1982 "Social psychology of intergroup relations." In Mark R. Rosenzweig and Lyman W. Porter (eds.), *Annual Review of Psychology*, 33: 1–39. Palo Alto, CA: Annual Reviews.

Tajfel, Henri, C. Flament, M. G. Billig, and R. F. Bundy 1971 "Social categorization and intergroup behavior." *European Journal of Social Psychology*, 1: 149–177.

Tajfel, Henri, and John C. Turner 1985 "The social identity theory of intergroup behavior." In Steven Worchel and William G. Austin (eds.), *Psychology of Intergroup Relations*, 2: 7–24. Chicago: Nelson-Hall.

Tetlock, Philip E., and Anthony S. Manstead 1985 "Impression management versus intrapsychic explanations in social psychology: A useful dichotomy?" *Psychological Review*, 92: 59–77.

Turner, John C. 1982 "Towards a cognitive redefinition of the social group." In Henri Tajfel (ed.), *Social Identity and Intergroup Relations*: 15–40. Cambridge: Cambridge University Press.

Turner, John C., and Penelope J. Oakes 1986 "The significance of the social identity concept for social psychology, with reference to individualism, interactionism and social influence." *British Journal of Social Psychology*, 25: 237–252.

Turner, Ralph H. 1978 "The role and the person." *American Journal of Sociology*, 84: 1–23.
Van Dyne, Linn, Jill W. Graham, and Richard M. Dienesch 1994 "Organizational citizenship behavior: Construct redefinition, operationalization and validation." *Academy of Management Journal*, vol. 37 (in press).
Van Maanen, John 1975 "Police socialization: A longitudinal examination of job attitudes in an urban police department." *Administrative Science Quarterly*, 20: 207–228.
Van Maanen, John, and Stephen R. Barley 1984 "Occupational communities: Culture and control in organizations." In Barry M. Staw and L. L. Cummings (eds.), *Research in Organizational Behavior*, 6: 287–365. Greenwich, CT: JAI Press.
Van Maanen, John, and Edgar H. Schein 1979 "Toward a theory of organizational socialization." In Barry M. Staw and L. L. Cummings (eds.), *Research in Organizational Behavior*, 1: 209–264. Greenwich, CT: JAI Press.
Vardi, Yoav, Yoash Wiener, and Micha Poppa 1989 "The value content of organizational mission as a factor in the commitment of members." *Psychological Reports*, 65: 27–34.
Whetten, David A., Debra Lewis, and Leann Mischel 1992 "Towards an integrated model of organizational identity and member commitment." Paper presented at Academy of Management Meeting, Las Vegas.

46

Social Identity and Self-Categorization Processes in Organizational Contexts

Michael A. Hogg and Deborah J. Terry

Source: *Academy of Management Review* 25 (1) (2000): 121–140.

Organizations are internally structured groups that are located in complex networks of intergroup relations characterized by power, status, and prestige differentials. To varying degrees, people derive part of their identity and sense of self from the organizations or workgroups to which they belong. Indeed, for many people their professional and/or organizational identity may be more pervasive and important than ascribed identities based on gender, age, ethnicity, race, or nationality. It is perhaps not surprising that social psychologists who study groups often peek over the interdisciplinary fence at what their colleagues in organizational psychology are up to. Some, disillusioned with social cognition as the dominant paradigm in mainstream social psychology, vault the fence, thus fueling recent and not so recent laments within social psychology that the study of groups may be alive and well, but not in social psychology (e.g., Levine & Moreland, 1990; Steiner, 1974).

Over the past 10 or 15 years, however, there has been a marked revival of interest among social psychologists in the study of groups and group processes (e.g., Abrams & Hogg, 1998; Hogg & Abrams, 1999; Hogg & Moreland, 1995; Moreland, Hogg, & Hains, 1994), even spawning two new journals: *Group Dynamics* in 1996 and *Group Processes and Intergroup Relations* in 1998. The *new* interest in groups is different. There is less emphasis on interactive small groups, group structure, and interpersonal relations within groups, and there is more emphasis on the self concept: how the self is defined by group membership and how social cognitive processes associated with group membership-based self-definition produce characteristically "groupy" behavior. This revival of interest in group processes and identity has been influenced significantly by the development within social psychology of social identity theory and self-categorization theory. A search of *PsychLit* in mid 1997 for the key terms *social identity* and *self-categorization* resulted in a list of almost 550 publications since 1991.

In this article we introduce social identity theory as a platform from which to describe in detail how social categorization and prototype-based depersonalization actually produce social identity phenomena. We explain how these processes, which are the conceptual core of self-categorization theory, relate to the original and more familiar intergroup and self-enhancement motivational perspective of social identity theory. We show how recent conceptual advances based largely, although not exclusively, on self-categorization theory have great but as yet largely unexplored potential for our understanding of social behaviors in organizational contexts. We have tried to energize this potential by describing various speculations, hypotheses, and propositions that can act as a framework for empirical research.

Some of the key theoretical innovations we promote are based on the ideas that (1) social identity processes are motivated by subjective uncertainty reduction, (2) prototype-based depersonalization lies at the heart of social identity processes, and (3) groups are internally structured in terms of perceived or actual group prototypicality of members. After introducing social identity theory and describing self-categorization mechanisms, we discuss cohesion and deviance, leadership, group structure, subgroups, sociodemographic groups, and mergers and acquisitions. We have chosen these group phenomena because they particularly benefit from the self-categorization-based extension of social identity theory. They capture the interplay of intergroup and intragroup relations and the conceptual importance of prototypicality, depersonalization, and uncertainty. They are also particularly organizationally relevant phenomena, where social identity theory can make a contribution.

Before we begin, we underscore two caveats. First, consistent with social identity theory's group level of analysis and cognitive definition of the social group (e.g., Turner, 1982; Turner, Hogg, Oakes, Reicher, & Wetherell, 1987), we consider organizations to be groups, units or divisions within organizations to be groups, professions or sociodemographic categories that are distributed across organizations to be groups, and so forth — all with different social identities and group prototypes. Thus, intergroup relations can exist between organizations, between units or divisions within an organization, between professions that are within but transcend organizations, and so forth. Salience mechanisms, described below, determine which group and, therefore, intergroup relationship is psychologically salient as a basis for self-conceptualization in a given context.

Second, social identity theory is not entirely new to organizational psychologists. Although already adopted to some extent by organizational researchers, Ashforth and Mael (1989) first systematically introduced the theory to organizational psychology (also see Ashforth & Humphrey, 1993, and Nkomo & Cox, 1996) and subsequently published some related empirical work (e.g., Mael & Ashforth, 1992, 1995). Others have also applied it to organizational settings (e.g., Dutton, Dukerich, & Harquail, 1994; Pratt, in press; Riordan & Shore, 1997; Tsui, Egan, & O'Reilly, 1992).

This work, however, often touches only the surface of social identity theory. It focuses on some aspects but does not systematically incorporate significant theoretical developments made since 1987 that focus on self-categorization, group prototypicality, contextual salience, and depersonalization processes (see Pratt, in press). These developments have enabled social identity theorists to extend the theory's conceptual and empirical focus on intergroup phenomena to incorporate a focus on what happens within groups; it has become what could be called an *extended social identity theory*. For example, in recent work on social psychology, researchers have explored social influence and norms (e.g., Turner, 1991); solidarity and cohesion (e.g., Hogg, 1992); attitudes, behavior, and norms (e.g., Terry & Hogg, 1999); small groups (e.g., Hogg, 1996a); group motivation (Hogg, in press a,b; Hogg & Abrams, 1993a; Hogg & Mullin, 1999); and group structure and leadership (e.g., Hogg, 1996b, 1999).[1]

Social Identity and Self-Categorization

Tajfel first introduced the concept of social identity — "the individual's knowledge that he belongs to certain social groups together with some emotional and value significance to him of this group membership" (1972: 292) — to move from his earlier consideration of social, largely intergroup, perception (i.e., stereotyping and prejudice) to consideration of how self is conceptualized in intergroup contexts: how a system of social categorizations "creates and defines an individual's *own* place in society" (Tajfel, 1972: 293). Social identity rests on intergroup social comparisons that seek to confirm or to establish ingroup-favoring evaluative distinctiveness between ingroup and out-group, motivated by an underlying need for self-esteem (Turner, 1975).

Tajfel (1974a,b) quickly developed the theory to specify how beliefs about the nature of relations between groups (status, stability, permeability, legitimacy) influence the way that individuals or groups pursue positive social identity. Tajfel and Turner (1979) retained this emphasis in their classic statement of social identity theory. The emphasis on social identity as part of the self-concept was explored more fully by Turner (1982). In a comprehensive coverage of relevant research, Hogg and Abrams (1988) then integrated and grounded intergroup, self-conceptual, and motivational emphases. At about the same time, Turner and his colleagues (Turner, 1985; Turner et al., 1987) extended social identity theory through the development of self-categorization theory, which specified in detail how social categorization produces prototype-based depersonalization of self and others and, thus, generates social identity phenomena.

Social identity theory and/or self-categorization theory has been described by social identity theorists in detail elsewhere (e.g., Hogg, 1992, 1993, 1996b; Hogg & Abrams, 1988; Hogg, Terry, & White, 1995; Tajfel & Turner, 1986; for historical accounts see Hogg, in press c; Hogg & Abrams, 1999; Turner, 1996). Because the original form of social identity theory is familiar to organizational psychologists, we do not redescribe it here. Instead, we focus on self-categorization theory, which is less familiar to and less accessible for organizational psychologists in terms of its processes, its relationship to social identity theory, and its potential for explicating organizational processes (Pratt, in press).

Self-categorization theory clearly evolves from Tajfel's and Turner's earlier ideas on social identity. We view it as a development of social identity theory or, more accurately, as that component of an extended social identity theory of the relationship between self-concept and group behavior that details the social cognitive processes that generate social identity effects (e.g., Abrams & Hogg, in press; Hogg, 1996a, in press c; Hogg & Abrams, 1988, 1999, in press; Hogg & McGarty, 1990; Hogg, Terry, & White, 1995). We see no incompatibility between self-categorization theory and the original form of social identity theory but view self-categorization theory, rather, as an important and powerful new conceptual component of an extended social identity theory. The self-categorization component of social identity theory has been very influential in recent developments within social psychology (e.g., Abrams & Hogg, 1999; Oakes et al., 1994; Spears et al., 1997) but has hitherto attracted little attention in organizational psychology (Pratt, in press).

Self-Categorization Theory

Self-categorization theory specifies the operation of the social categorization process as the cognitive basis of group behavior. Social categorization of self and others into ingroup and outgroup accentuates the perceived similarity of the target to the relevant ingroup or outgroup prototype (cognitive representation of features that describe and prescribe attributes of the group). Targets are no longer represented as unique individuals but, rather, as embodiments of the relevant prototype a process of *depersonalization*. Social categorization of self — self-categorization — cognitively assimilates self to the ingroup prototype and, thus, depersonalizes self-conception. This transformation of self is the process underlying group phenomena, because it brings self-perception and behavior in line with the contextually relevant ingroup prototype. It produces, for instance, normative behavior, stereotyping, ethnocentrism, positive ingroup attitudes and cohesion, cooperation and altruism, emotional contagion and empathy, collective behavior, shared norms, and mutual influence. Depersonalization refers simply to a change in self-conceptualization and the basis of perception of others; it does not have the negative connotations of such terms as *deindividuation* or *dehumanization* (cf. Reicher, Spears, & Postmes, 1995).

Representation of Groups As Prototypes

The notion of prototypes, which is not part of the earlier intergroup focus of social identity theory, is absolutely central to self-categorization theory. People cognitively represent the defining and stereotypical attributes of groups in the form of prototypes. Prototypes are typically not checklists of attributes but, rather, fuzzy sets that capture the context-dependent features of group membership, often in the form of representations of exemplary members (actual group members who best embody the group) or ideal types (an abstraction of group features). Prototypes embody all attributes that characterize groups and distinguish them from other groups, including beliefs, attitudes, feelings, and behaviors. A critical feature of prototypes is that they maximize similarities within and differences between groups, thus defining groups as distinct entities. Prototypes form according to the principle of metacontrast: maximization of the ratio of intergroup differences to intragroup differences. Because members of the same group are exposed to similar social information, their prototypes usually will be similar and, thus, shared.

Prototypes are stored in memory but are constructed, maintained, and modified by features of the immediate or more enduring social interactive context (e.g., Fiske & Taylor, 1991). They are highly context dependent and are particularly influenced by what outgroup is contextually salient. Enduring changes in prototypes and, therefore, self-conception can arise if the relevant comparison outgroup changes over time — for instance, if Catholics gradually define themselves in contradistinction to Muslims rather than to Protestants, or if a car manufacturer compares itself to a computer software manufacturer rather than to another car manufacturer. Such changes are also transitory in that they are tied to whatever outgroup is salient in the immediate social context. For instance, a psychology department may experience a contextual change in self-definition if it compares itself with a management school rather than with a history department. Thus, social identity is dynamic. It is responsive, in type and content, to intergroup dimensions of immediate comparative contexts.

> Proposition 1: Changes in the interorganizational comparative context affect the content of organizational prototypes.

As we will see, the content of prototypes strongly influences the group phenomena discussed later in the article.

Self-Enhancement and Uncertainty Reduction Motivations

According to social identity theory, social identity and intergroup behavior are guided by the pursuit of evaluatively positive social identity, through positive intergroup distinctiveness, which, in turn, is motivated by the need for positive self-esteem — the self-esteem hypothesis (e.g., Abrams & Hogg, 1988; see also Hogg & Abrams, 1990, 1993b; Hogg & Mullin, 1999; Long & Spears, 1997; Rubin & Hewstone, 1998). Self-categorization theory's focus on the categorization process hints at an additional (perhaps more fundamental), epistemic, motivation for social identity, which has only recently been described — the uncertainty reduction hypothesis (Hogg, in press a,b; Hogg & Abrams, 1993b; Hogg & Mullin, 1999). In addition to being motivated by self-enhancement, social identity processes are also motivated by a need to reduce subjective uncertainty about one's perceptions, attitudes, feelings, and behaviors and, ultimately, one's self-concept and place within the social world. Uncertainty reduction, particularly about subjectively important matters that are generally self-conceptually relevant, is a core human motivation. Certainty renders existence meaningful and confers confidence in how to behave and what to expect from the physical and social environment within which one finds oneself. Self-categorization reduces uncertainty by transforming self-conception and assimilating self to a prototype that describes and prescribes perceptions, attitudes, feelings, and behaviors.

Because prototypes are relatively consensual, they also furnish moral support and consensual validation for one's self-concept and attendant cognitions and behaviors. It is the prototype that actually reduces uncertainty. Hence, uncertainty is better reduced by prototypes that are simple, clear, highly focused, and consensual, and that, thus, describe groups that have pronounced entitativity (Campbell, 1958; also see Brewer & Harasty, 1996; Hamilton & Sherman, 1996; Hamilton, Sherman, & Lickel, 1998; Sherman, Hamilton, & Lewis, 1999), are very cohesive (Hogg, 1992, 1993), and provide a powerful social identity. Such groups and prototypes will be attractive to individuals who are contextually or more enduringly highly uncertain, or during times of or in situations characterized by great uncertainty.

> Proposition 2: Subjective uncertainty may produce a prototypically homogenous and cohesive organization or work unit with which members identify strongly.

Uncertainty reduction and self-enhancement are probably independent motivations for social identity processes, and in some circumstances it may be more urgent to reduce uncertainty than to pursue self-enhancement (e.g., when group entitativity is threatened), whereas in others it may be the opposite (e.g., when group prestige is threatened). However, uncertainty reduction may be more fundamentally adaptive because it constructs a self-concept that defines who we are and prescribes what we should perceive, think, feel, and do.

The uncertainty reduction hypothesis has clear relevance for organizational contexts. Indeed, the hypothesis is not inconsistent with Lester's (1987) uncertainty reduction theory that plays an important role in Saks and Ashforth's (1997) multilevel process model of organizational socialization. Although in both cases uncertainty motivates group socialization behaviors, the uncertainty reduction hypothesis specifies self-categorization as the social cognitive process that resolves uncertainty through prototype-based self-depersonalization.

Salience of Social Identity

The responsiveness of social identity to immediate social contexts is a central feature of social identity theory — and self-categorization theory within it. The cognitive system, governed by uncertainty reduction and self-enhancement motives, matches social categories to properties of the social context and brings into active use (i.e., makes salient) that category rendering the social context and one's place within it subjectively most meaningful. Specifically, there is an interaction between category accessibility and category fit so that people draw on accessible categories and investigate how well they fit the social field. The category that best fits the field becomes salient in that context (e.g., Oakes et al., 1994; Oakes & Turner, 1990).

Categories can be accessible because they are valued, important, and frequently employed aspects of the self-concept (i.e., chronic accessibility) and/or because they are perceptually salient (i.e., situational accessibility).

Categories fit the social field because they account for situationally relevant similarities and differences among people (i.e., structural fit) and/or because category specifications account for context-specific behaviors (i.e., normative fit). Once fully activated (as opposed to merely "tried on") on the basis of optimal fit, category specifications organize themselves as contextually relevant prototypes and are used as a basis for the perceptual accentuation of intragroup similarities and intergroup differences, thereby maximizing separateness and clarity. Self-categorization in terms of the activated ingroup category then depersonalizes behavior in terms of the ingroup prototype.

Salience is not, however, a mechanical product of accessibility and fit (Hogg, 1996b; Hogg & Mullin, 1999). Social interaction involves the motivated manipulation of symbols (e.g., through speech, appearance, and behavior) by people who are strategically competing with one another to influence the frame of reference within which accessibility and fit interact. People are not content to have their identity determined by the social cognitive context. On the contrary, they say and do things to try to change the parameters so that a subjectively more meaningful and self-favoring identity becomes salient. For instance, a mixed-sex conversation about the communication of feelings is likely to make sex salient, because the chronically accessible category "sex" is situationally accessible and has good structural and normative fit. Male interactants who find the self-evaluative implications of gender stereotypes about feelings unfavorable might change the topic of conversation to politics so that sex becomes situationally less accessible and now has poor structural and normative fit. In this way a different, and self-evaluatively more favorable, identity may become salient.

This dynamic perspective on identity and self-conceptual salience has clear implications for organizational contexts. Manipulation of the intergroup social comparative context can be a powerful way to change organizational identity (self-conception as a member of a particular organization) and, thus, attitudes, motives, goals, and practices. Organizations or divisions within organizations that have poor work practices or organizational attitudes can be helped to reconstruct themselves, through surreptitious or overt changes in the salience of relevant intergroup comparative contexts (different levels of categorization or different outgroups at the same level of categorization). Such changes affect contextual self-categorization and, therefore, people's internalized attitudes and behaviors (e.g., Terry & Hogg, 1996).

One way in which organizations may deliberately manipulate the intergroup social comparative context is by "benchmarking." An organization selects specific other organizations as a legitimate comparison set, which threatens the group's prestige. This motivates upward redefinition of organizational identity and work practices, to make the group evaluatively more competitive.

Self-categorization theory's focus on prototypes allows some important conceptual developments in social identity theory, which have direct implications for organizational contexts. When group membership is salient, cognition is attuned to and guided by prototypicality. Thus, within groups people are able to distinguish among themselves and others in terms of how well they match the prototype. An intragroup prototypicality gradient exists — some people are or are perceived to be more prototypical than others (Hogg, 1996a,b, 1999). This idea allows social identity theorists to now explicate social identity-based intragroup processes, such as cohesion and social attraction, deviance and overachievement, and leadership and intragroup structural differentiation.

Cohesion and Deviance

A development of social identity theory made possible by focusing on how social categorization produces prototype-based depersonalization is the social attraction hypothesis, which approaches group solidarity and cohesion as a reflection of depersonalized, prototype-based interindividual attitudes (Hogg, 1987, 1992, 1993). A distinction is drawn between interindividual evaluations, attitudes, and feelings that are based on and generated by being members of the same group or members of different groups (depersonalized *social attraction*) and those that are based on and generated by the idiosyncrasies and complementarities of close and enduring interpersonal relationships (*personal attraction*).

When a group is salient, ingroup members are liked more if they embody the ingroup prototype. Where the prototype is consensual, certain people are consensually liked, and where all members are highly prototypical, there is a tight network of social attraction. Of course, outgroup members generally are liked less than ingroup members. When a group is not salient, liking is based on personal relationships and idiosyncratic preferences. The prediction is that patterns of liking in an aggregate, and the bases of that liking, can change dramatically when an aggregate becomes a salient group (e.g., when uncertainty or entitativity are high, or when the group is under threat or is engaged in intergroup competition over a valued scarce resource). Social and personal attraction are not isomorphic (see Mullen & Copper, 1994). These predictions have been supported repeatedly by a program of research with laboratory, quasi naturalistic, sports, and organizational groups (Hogg, Cooper-Shaw, & Holzworth, 1993; Hogg & Hains, 1996; Hogg & Hardie, 1991, 1992, 1997; Hogg, Hardie, & Reynolds, 1995; see overviews by Hogg, 1992, 1993).

One practical implication of the idea of depersonalized social attraction in an organizational setting is that organizational or workgroup solidarity and, thus, adherence to group norms are unlikely to be strengthened by activities that strengthen only personal relationships or friendships. Indeed, such activities may compromise solidarity and norm adherence by fragmenting the group into friendship pairs or cliques that show interpersonal dislike for other pairs or cliques. To increase social attraction and solidarity within an organization, managers might, among other things, create uncertainty (this motivates identification), focus on interorganizational competition (this makes the group salient), and emphasize desirable attributes of the organization (this provides positive distinctiveness).

> Proposition 3: Social attraction may foster organizational cohesion, and thereby identification and adherence to organizational norms; conversely,

interpersonal attraction may fragment the organization and disrupt identification and adherence to norms.

Cohesion and solidarity, and the feelings people have for one another within a group, hinge on the perceived group prototypicality of others. We now discuss two organizationally relevant implications of this idea: (1) the dynamic interplay of group and demographic prototypes that affects cohesion within an organization (relational demography) and (2) the perception and treatment of non-prototypical group members (negative outliers and high flyers).

Relational Demography

The social attraction analysis of cohesion has relevance for recent organizational research on relational demography (e.g., Mowday & Sutton, 1993; Riordan & Shore, 1997; Wesolowski & Mossholder, 1997). Relational demography theorists propose that people in organizations or work units compare their own demographic characteristics (e.g., race, gender, ethnicity) with those of individual other members or the group as a whole, and that perceived similarity enhances work-related attitudes and behavior. The organization or unit provides the context within which similarity comparisons are made. Social identity theory and the social attraction hypothesis provide a much more textured analysis, based on the relative salience of the demographic or organizational group and on the correspondence between demographic and organizational norms/prototypes (see the discussion of group structure below).

Demographic homogeneity may strengthen organizational ingroup prototypes, social attraction, and identification and, thus, adherence to norms, particularly if group norms are not inconsistent with demographic category norms. If the organizational group's norms clash with those of the wider demographic category, then demographic homogeneity may make the wider category and its norms salient and, thus, weaken adherence to the organization's norms. Demographic diversity may weaken the impact of demographic group membership, make the organizational group itself more salient, and, thus, strengthen adherence to organizational norms, particularly if societal relations between demographic groups are harmonious. If, however, relations between demographic groups are conflictual and are emotionally charged, diversity will highlight intergroup relations outside the organization or unit, thus making demographic membership salient and strengthening adherence to demographic — not organizational — norms. This analysis revolves around the contextual salience of demographic or organizational identity — not just the degree of perceived demographic similarity.

Proposition 4: Intraorganizational demographic similarity/diversity will impact organizational behavior via organizational or demographic identity salience; organizational salience and behavior are enhanced by demographic similarity, if organizational and demographic norms are consistent, and by demographic diversity, if there is societal harmony among demographic categories.

Negative Outliers and High Flyers

A further implication of the social attraction hypothesis is that prototypically marginal ingroup members will be liked less than prototypically central members and that this process will be accentuated under high salience so that marginal members may be entirely rejected as "deviants." A program of laboratory research by Marques and his associates provides good evidence for this process (Marques, 1990; Marques & Paez, 1994; Marques & Yzerbyt, 1988; Marques, Yzerbyt, & Leyens, 1988). By being aprototypical, particularly in a direction that leans toward a salient outgroup, a marginal ingrouper jeopardizes the distinctiveness and prototypical clarity and integrity of the ingroup. This may introduce the threat of uncertainty. Thus, fellow ingroupers, especially those for whom uncertainty is particularly threatening, will strongly reject the deviant in order to consolidate a clear prototype to which they can strongly assimilate themselves through self-categorization.

So-called black sheep studies focus on "negative" deviants: ingroup members who are inclined toward the outgroup prototype. But what about "positive" deviants? These are group members who are aprototypical, but in evaluatively favorable ways — for example, overachievers or high flyers. On the one hand, overachievers should be socially unattractive because they are aprototypical, but, on the other, they should be socially attractive because the group can bask in their reflected glory (cf. Burger, 1985; Cialdini et al., 1976; Cialdini & de Nicholas, 1989; Sigelman, 1986; Snyder, Lassegard, & Ford, 1986; Wann, Hamlet, Wilson, & Hodges, 1995). There is some evidence that people are evaluatively particularly harsh on overachievers who suffer a setback or experience a fall (e.g., Feather, 1994), but this research does not differentiate between overachievers who are members of a salient ingroup and those who are not.

To investigate this, researchers are conducting a series of laboratory experiments (Fielding & Hogg, 1998). From social identity theory we predict that the immediate and intergroup social context of overachievement determines the evaluation of positive ingroup deviants. There are two dimensions to the model:

(1) A functional dimension. Where solidarity and consensual prototypicality are important to the group, perhaps owing to uncertainty concerns, positive deviants are dysfunctional for the group; they will be evaluatively downgraded, much like negative deviants. Where solidarity is less critical and prototypicality less consensual but self-enhancement is important, positive deviants are functional for the group; they will be upgraded as they contribute to a favorable redefinition of ingroup identity.

(2) A social attribution dimension. Where positively deviant behavior can be "owned" by the group, the deviant will be favorably evaluated; this would be likely if the deviant modestly attributed the behavior to the support of the group rather than to personal ability and if the deviant had little personal history of overachievement (i.e., was a "new" deviant). Where positively deviant behavior cannot readily be "owned" by the group, the deviant will be unfavorably

evaluated; this would be likely if the deviant took full personal credit for the behavior without acknowledging the group's support (i.e., "boasted") and if the deviant had a long personal history of overachievement (i.e., was an enduring deviant).

> Proposition 5: Organizations will reject negative organizational deviants. Positive deviants will be accepted where organizational prestige is important but will be rejected where organizational solidarity and distinctiveness are important.

Leadership

In contrast to deviants, prototypical group members are reliably and consensually favorably evaluated when group membership is salient. This idea has recently been extended in order to develop a social identity model of leadership processes in groups (Hogg, 1996b, 1999; also, see Fielding & Hogg, 1997; Hains, Hogg, & Duck, 1997; Hogg, 1996a; Hogg, Hains, & Mason, 1998). From this perspective, leadership — the focus is largely on emergent leaders — is a structural feature of ingroups (i.e., leaders and followers), which is produced by the processes of self-categorization and prototype-based depersonalization. As group membership becomes more salient, being a prototypical group member may be at least as important for leadership as having characteristics that are widely believed to be associated with a particular type of leader (i.e., being stereotypical of a nominal leader category; see leader categorization theory: Lord, Foti, & De Vader, 1984; Nye & Forsyth, 1991; Nye & Simonetta, 1996; Rush & Russell, 1988). There are three aspects of the process:

(1) Self-categorization constructs a gradient of actual or perceived prototypicality within the group so that some people are more prototypical than others, and they act as a focus for attitudinal and behavioral depersonalization. The person who occupies the contextually most prototypical position embodies the behaviors that others conform to and, thus, *appears* to have exercised influence over other group members. If the social context remains stable, the prototype remains stable, and the same individual *appears* to have enduring influence. However, the process is automatic. The "leader" merely embodies the aspirations, attitudes, and behaviors of the group but does not actively exercise leadership.

(2) Social attraction ensures that more prototypical members are liked more than less prototypical members; if the prototype is consensual, more prototypical members are consensually liked. There are a number of important implications of this. First, being socially attractive furnishes the leader with the capacity to actively gain compliance with his or her requests — people tend to agree and comply with people they like. Second, this empowers the leader and publicly confirms his or her ability to exercise influence. Third, the prototypical leader is likely to identify strongly with the group and, thus, exercise influence in empathic and collectively beneficial ways, which strengthens his or her perceived

prototypicality and consensual social attractiveness. Fourth, consensual attractiveness confirms differential popularity and public endorsement of the leader, imbues the leader with prestige and status, and instantiates an intragroup status differential between leader(s) and followers.

(3) The final process is an attribution one, in which members make the fundamental attribution error (Ross, 1977) or show correspondence bias (Gilbert & Jones, 1986; see also Gilbert & Malone, 1995, and Trope & Liberman, 1993). Members overattribute or misattribute the leader's behavior to personality rather than to his or her prototypical position in the group. Because the behavior being attributed, particularly over an enduring period, includes the appearance or actuality of being influential over others' attitudes and behaviors, being consensually socially attractive, and gaining compliance and agreement from others, this constructs a charismatic leadership personality for the leader.

A number of factors accentuate this process. First, because prototypicality is the yardstick of group life, it attracts attention and renders highly prototypical members figural against the background of the group, thus enhancing the fundamental attribution error (Taylor & Fiske, 1978). Second, the emerging status-based structural differentiation between leader(s) and followers further enhances the distinctiveness of the leader(s) against the background of the rest of the group. Third, to redress their own perceived lack of power and control, followers seek individualizing information about the leader, because they believe that such information is most predictive of how the leader will behave in many situations (Fiske, 1993; Fiske & Depret, 1996). Fourth, cultural theories of causes of leadership behavior (e.g., the "great person" theory of leadership) may accentuate the fundamental error (e.g., Morris & Peng, 1994). And fifth, the correspondence bias may be strengthened because followers perceive the leader's behavior to be relatively extreme and distinctive and because they then fail to properly consider situational causes of the behavior (e.g., Gilbert & Malone, 1995; Trope & Liberman, 1993).

Together, these three processes transform prototypical group members into leaders who are able to be proactive and innovative in exercising influence. This also equips leaders to maintain their tenure. They can simply exercise power (more of this below), but they can also manipulate circumstances to enhance their perceived prototypicality: they can exercise self-serving ideological control over the content of the prototype, they can pillory ingroup deviants who threaten the self-serving prototype, they can demonize outgroups that clearly highlight the self-serving ingroup prototype, and they can elevate uncertainty to ensure that members are motivated to identify strongly with a group that is defined as the leader wishes (uncertainty can be managed as a resource by people in power; e.g., Morris, 1996).

The most basic prediction from this model is that as group salience increases, perceived leadership effectiveness becomes more determined by group prototypicality and less determined by possession of general leadership qualities. This prediction has been confirmed in a series of three laboratory studies of emergent leadership (Hains et al., 1997; Hogg et al., 1998, Experiments 1 and 2) and replicated in a field study of outward-bound groups (Fielding & Hogg, 1997).

The social attraction and attribution aspects of the model remain to be investigated, as do the many implications described in this section.

We now suggest three organizationally relevant leadership consequences of excessively high group cohesiveness. Such groups may (1) produce leaders who are prototypical but do not possess task-appropriate leadership skills (cf. groupthink); (2) consolidate organizational prototypes that reflect dominant rather than minority cultural attributes and, thus, exclude minorities from top leadership positions; and (3) produce an environment that is conducive to the exercise, and perhaps abuse, of power by leaders.

Prototypical Leadership and Groupthink

This research may help cast light on groupthink: suboptimal decision-making procedures in highly cohesive groups, leading to poor decisions with potentially damaging consequences (e.g., Janis, 1982). There is now some evidence that the critical component of "cohesiveness" associated with groupthink is social attraction, rather than interpersonal attraction (Hogg & Hains, 1998; see also Turner, Pratkanis, Probasco, & Leve, 1992). If we assume that group prototypes do not necessarily embody optimal procedures for group decision making, then group prototypical leaders are quite likely to be less effective leaders of decision-making groups than are leadership-stereotypical leaders (i.e., leaders who, in this case, possess qualities that most people believe are appropriate for group decision making). This suggests that groupthink may arise because overly cohesive groups "choose" highly prototypical and, thus, perhaps, task-inappropriate members as leaders.

> Proposition 6: Strong organizational identification may hinder endorsement of effective leaders, because leadership is based on group prototypicality, and group prototypes may not embody effective leadership properties.

Minorities as Organizational Leaders

Another implication of this analysis of leadership relates to evidence that minorities (e.g., women and people of color) can find it difficult to attain top leadership positions in organizations (e.g., Eagly, Karau, & Makhijani, 1995). If organizational prototypes (e.g., of speech, dress, attitudes, and interaction styles) are societally cast so that minorities do not match them well, minorities are unlikely to be endorsed as leaders under conditions where organizational prototypicality is more important than leadership stereotypicality — that is, when organizational identification and cohesion are very high. This might arise under conditions of uncertainty when, for example, organizations are under threat from competitors or when there is an economic crisis — situations where leaders, rather than managers, may be badly needed.

> Proposition 7: Minorities may find it difficult to attain top leadership positions in organizations because they do not fit culturally prescribed organizational prototypes.

An important feature of the model is that the processes of social attraction and prototypical attribution decouple the leader from the group; they create a status-based structural differentiation of leaders(s) and followers, which is endorsed by both leader(s) and followers. This has implications for the role of power in leadership (Hogg, 1998, 1999; Hogg & Reid, in press). Traditionally, social identity theorists have said little about power, preferring to talk of influence.

Where leaders are merely prototypical, they have influence over followers by virtue of being prototypical; followers automatically comply through self-categorization. It is unnecessary to exercise power to gain influence, and there are strong mutual bonds of liking and empathy between prototypically united leaders and followers that would inhibit the exercise of power in ways that might harm members of the group.

However, once charisma and status-based structural differentiation gather pace, the leader becomes increasingly psychologically and materially separated from the group. This severs the empathic and social attraction bonds that previously guarded against abuse of power. A consensually endorsed, status-based intergroup relationship between leader(s) (probably in the form of a power elite) and followers has effectively come into existence; thus, typical intergroup behaviors are made possible. The leader can discriminate against followers, favor self and the leadership elite, and express negative social attitudes against and develop negative stereotypes of followers (e.g., Goodwin & Fiske, 1996; Goodwin, Gubin, Fiske, & Yzerbyt, in press). Under these conditions leaders are likely to exercise power (in Yukl & Falbe's, 1991, sense of personal power or in Raven's, 1965, sense of reward power, coercive power, or legitimate power) and are able to abuse power — for example, when they feel their position is under threat.

This rigidly hierarchical leadership scenario is most likely to emerge when conditions encourage groups to be cohesive and homogenous, with extremitized and clearly delimited prototypes that are tightly consensual. In an organizational context, extreme societal or organizational uncertainty might produce these conditions (Hogg, 1999; see also Pratto, Sidanius, Stallworth, & Malle, 1994, and Tyler, 1990).

> Proposition 8: Subjective uncertainty may produce a prototypically and demographically homogenous organization or work unit that has a hierarchical leadership structure with a powerful leader and that has rigid, entrenched, and "extremist" attitudes and practices.

The progression from benign influence to the possibly destructive wielding of power may not be inevitable. Conditions that inhibit the attribution of charisma and the process of structural differentiation, and that reground leadership in prototypicality, may curb the exercise of power. For example, if a group becomes less cohesive, more diverse, and less consensual about its prototype, followers are less likely to agree on and endorse the same person as the leader. The incumbent leader's power base is fragmented, and numerous new "contenders" emerge. This limits the leader's ability to abuse power and renders the exercise of power less effective. Paradoxically, a rapid increase in cohesiveness, caused, for example, by imminent external threat to the group, may, through a different process, have a similar outcome. Cohesion may make the group so consensual that leader and group become temporarily re-fused. The empathic bond is re-established so that the leader does not need to exercise power to gain influence, and any abuse of power would be akin to abuse of self.

> Proposition 9: Emergent leaders may tend to abuse their power unless the organization is highly diverse or highly cohesive.

Group Structure

Leadership is only one way in which groups can be internally structured. Groups, such as organizations, are also structured, in various ways, into functional or demographic subgroups. In this section we discuss the relevance of social identity theory to the analysis of relations among subgroups within organizations — in particular, sociodemographic subgroups based on gender, race, ethnicity, and so forth and organizational subgroups within a superordinate organization formed by a merger or acquisition.

Subgroup Structure

Almost all groups are vertically organized to contain subgroups, while they themselves are nested within larger groups. Sometimes subgroups are wholly nested within a superordinate group (e.g., a sales department within an organization), and sometimes subgroups are crosscut by the superordinate group (e.g., pilots within an airline). Social identity theorists and those with more general social categorization perspectives make predictions about the nature of relations between subgroups as a function of the nature of the subgroups' relationship to the superordinate group. Much of these scholars' work is framed by the "contact hypothesis," to investigate the conditions under which contact between members of different groups might improve enduring relations between the groups (e.g., Brown, 1996; Gaertner, Dovidio, Anastasio, Bachman, & Rust, 1993; Gaertner, Dovidio, & Bachman, 1996; Gaertner, Rust, Dovidio, Bachman, & Anastasio, 1995; Hewstone, 1994, 1996; Pettigrew, 1998).

Subgroups often resist attempts by a superordinate group to dissolve subgroup boundaries and merge them into one large group. This can be quite marked where the superordinate group is very large, amorphous, and impersonal. Thus, assimilationist strategies within nations, or large organizations, can produce fierce subgroup loyalty and intersubgroup competition. Subgroup members derive social identity from their groups and, thus, view externally imposed assimilation as an identity threat. The threat may be stronger in large superordinate groups because of optimal distinctiveness considerations (Brewer, 1991, 1993). People strive for a balance between conflicting motives for inclusion/sameness (satisfied by group membership) and for distinctiveness/uniqueness (satisfied by individuality). So, in very large organizations, people feel overincluded and strive for distinctiveness, often by identifying with distinctive subunits or departments.

Some research suggests that an effective strategy for managing intersubgroup relations within a larger group is to make subgroup and superordinate group identity simultaneously salient. For example, Hornsey and Hogg (1999, in press a,b) conducted a series of experiments in which they found intersubgroup relations to be more harmonious when the subgroups were salient within the context of a salient superordinate group than when the superordinate group alone or the subgroups alone were salient. This may recreate, in the laboratory, the policy of multiculturalism, adopted by some countries to manage ethnic diversity at a national level (cf. Prentice & Miller, 1999).

The implication for organizations is clear. To secure harmonious and cooperative relations among departments or divisions within a large organization, it may be best to balance loyalty to and identification with the subunit with loyalty to and identification with the superordinate organization, and not overemphasize either one to the detriment of the other. From a social identity perspective, managers might achieve this balance by having a distinct departmental or divisional structure, involving, for example, departmental activities and friendly interdepartmental rivalry, carefully balanced against a clear interorganization orientation and organization-wide activities that emphasize positive distinctiveness and positive organizational identity.

> Proposition 10: Harmonious relations among subgroups within an organization are often best achieved by simultaneous recognition of subgroup and organizational identity.

Sociodemographic Structure

Intragroup dynamics and structure also are influenced by the sociodemographic structure of society. Most groups, including organizational groups, have a membership that is diverse in terms of race, ethnicity, gender, (dis)ability, and so forth (e.g., Chung, 1997; Cox, 1991; Ibarra, 1995; Kandola, 1995). Organizations are a crucible in which wider intergroup relations, often evaluatively polarized and emotionally charged, are played out; conflict, disadvantage, marginalization, and minority victimization can arise (e.g., Williams & Sommer, 1997; cf. the expectation states theory notion of diffuse status characteristics [de Gilder & Wilke, 1994] and our earlier discussion of relational demography).

As a theory of intergroup relations, social identity theory has direct relevance for the study of sociodemographic diversity within organizations (Brewer, 1996; Brewer, von Hippel, & Gooden, 1999; see also Alderfer & Thomas, 1988; Brewer & Miller, 1996; Kramer, 1991; Oakes et al., 1994). Intraorganizational minority status rests on the dominant composition of the organization —

for example, gender may be a minority status in some organizations but not others. Because of the salience of their minority status in the organizational context, members of such groups are likely to be classified and perceived in terms of this status, thus occasioning stereotypical expectations and treatment from members of the dominant group. The likelihood of stereotyped responses increases if the demographic minority categorization (e.g., gender or ethnicity) converges with a role or employment classification within the organization — for instance, if there are relatively few female employees and they are all employed in secretarial or clerical positions. In such circumstances, categorization in terms of the employment classification is facilitated, because it covaries with a salient demographic categorization.

According to Brewer (1996; Brewer et al., 1999; see also Brewer & Miller, 1996), differentiations within categories are more likely to be made when minority status does not correlate with employment classification. If minority group status is not diagnostic of employment categorization, employees will find it necessary, in order to function within the organization, to acknowledge differences within both the minority group and the employment classification. One way in which convergence between minority group status and employment classification can be avoided, and hence stereotyped responses to the minority group can be reduced, is to crosscut organizational roles and social group membership. In a crosscutting structure, minority group memberships and employment classifications are independent of each other; knowing a person's group membership is undiagnostic of employment role or classification. Marcus-Newhall, Miller, Holtz, and Brewer (1993) found that when category membership and role assignment were not convergent (i.e., they were crosscut), category members were less likely to favor their own category on post-test ratings, and they were less likely to differentiate among the categories than in a convergent role structure.

From a social identity perspective, a crosscutting structure is one way to manage diversity effectively in organizations. Another strategy is to create a pluralistic or multicultural normative environment within the organization (Cox, 1991; Kandola, 1995). As discussed above, this involves minority members' balancing subgroup (i.e., demographic minority) and superordinate group (i.e., demographic majority or organization) identification, and majority members' exhibiting normative acceptance and support for cultural diversity within the organization.

To summarize, a crosscutting structure will assist the development of a pluralistic organizational environment, as will reduced marginalization of minority group members, through co-operative intergroup contact (Hewstone & Brown, 1986; see also Deschamps & Brown, 1983) and through intergroup contact that changes members' cognitive representation of the intergroup structure from the perception of separate groups to one that acknowledges plural identities or a common ingroup identity (e.g., Gaertner et al., 1996).

> Proposition 11: Conflict arising from sociodemographic diversity within an organization can be moderated by crosscutting demography with role assignments or by encouraging a strategy of cultural pluralism.

Finally, drawing on the uncertainty reduction hypothesis, we would expect organizational uncertainty to generally work against diversity. Organizations facing uncertainty would strive for homogeneity and consensual prototypicality that might marginalize sociodemographic minorities within the organization. The effect would be amplified under conditions of wider societal uncertainty that encourages ethnic, racial, religious, and national identification, and concomitant xenophobia and intolerance.

Mergers and Acquisitions

A special case of group structure is the merging of two organizations or the acquisition of one organization by another. Mergers and acquisitions pose special problems of intragroup relations for organizations (e.g., Hakansson & Sharma, 1996; Hogan & Overmyer-Day, 1994). When two organizations merge or, more commonly, one acquires the other, the postmerger entity embraces premerger intergroup relations between the merger "partners." These relations are often competitive and sometimes bitter and antagonistic. Indeed, negative responses and feelings toward the employees of the other organization may jeopardize the success of the merger.

Case studies of mergers confirm this. There are many examples of mergers failing because of "us" versus "them" dynamics that prevail if employees do not relinquish their old identities (e.g., Blake & Mouton, 1985; Buono & Bowditch, 1989). In a laboratory study Haunschild, Moreland, and Murrell (1994) found similar results. People who had worked on a task together in a dyad showed stronger interdyad biases when different dyads were subsequently required to merge than did people who had not previously worked together in their own dyad and, hence, were only nominal groups.

Social identity theorists make clear predictions about the success of a merger. The behaviors that group members adopt to pursue self-enhancement through positive social identity are influenced by *subjective belief structures*: beliefs about the nature of relations between the ingroup and relevant outgroups (Tajfel & Turner, 1979; see also Ellemers, 1993; Ellemers, Doojse, van Knippenberg, & Wilke, 1992; Ellemers, van Knippenberg, de Vries, & Wilke, 1988; Tajfel, 1975; Taylor & McKirnan, 1984; van Knippenberg & Ellemers, 1993). These beliefs concern (1) the stability and legitimacy of intergroup status relations (i.e., whether one's group deserves its status, and the likelihood of a change in status) and (2) the possibility of social mobility (psychologically passing from one group to another) or social change (changing the ingroup's evaluation). Social change can involve direct conflict but also socially creative behavior, such as ingroup bias on dimensions that are not related directly to the basis for the status differentiation (e.g., Lalonde, 1992; Terry & Callan, 1998).

At the interorganizational level, an organization that believes its lower-status position is legitimate and stable and believes that it is possible for members to pass psychologically into the more prestigious organization (i.e., acquire a social identity as a member of the prestigious organization) will be unlikely to show organizational solidarity or engage in interorganizational competition. Instead, members will attempt, as individuals, to disidentify and gain psychological entry to the other organization. This will increase their support for the merger and their commitment to and identification with the new, merged organization.

In contrast, an organization that believes its lower-status position is illegitimate and unstable, that passing is not viable, and that a different interorganizational status relation is achievable will show marked solidarity, engage in direct interorganizational competition, and actively attempt to undermine the success of the merger. Although members of low-status organizations are likely to respond favorably to conditions of high permeability (see Zuckerman, 1979), an opposite effect is likely for employees of the higher-status premerger organization (see Vaughan, 1978). Permeable boundaries pose a threat to the status they enjoy as members of a higher-status premerger organization, so they are likely to respond negatively to permeable intergroup boundaries.

> Proposition 12: Lower-status merger partners will respond favorably to a merger, if they believe their status is legitimate and that the boundary between the premerger partners is permeable, and unfavorably, if they believe their status is illegitimate and boundaries are impermeable. Higher-status merger partners will respond unfavorably to permeable boundaries.

In a recent study of employees involved in a merger between two airlines, Terry, Carey, and Callan (in press) found some support for these predictions. Perception of permeable intergroup boundaries in the new organization was associated positively with identification with the new organization and both job-related (organizational commitment and job satisfaction) and person-related (emotional well-being and self-esteem) outcomes among employees of the low-status premerger organization, but negatively with the person-related outcomes among employees of the high-status premerger organization. Analyses showed that these effects were significant after controlling for the type of individual-level constructs that have been considered in previous merger research (e.g., perceived positiveness of the change process and the use of both problem- and emotion-focused coping responses) — a pattern of results that reflects the importance of considering group-level variables in merger research.

Gaertner and colleagues (Anastasio, Bachman, Gaertner, & Dovidio, 1997; Gaertner et at., 1996) also found support for a social categorization approach, in the context of a bank merger. Perception of successful contact between the premerger organizations (e.g., contact between equal-status partners, positive interdependence between the groups, and many opportunities for interaction) reduced intergroup bias, through employees' cognitive representations of the merged group and through low intergroup anxiety. For intergroup evaluative bias on both work-related and sociability dimensions, the belief that the merged organization felt like one group (see van Knippenberg, 1997) was related negatively to intergroup anxiety (and, through reduced anxiety, to low intergroup bias), whereas the perception that the organization felt like two subgroups was related positively to work-related bias. Thus, in contrast to the optimization of subgroup relations in an organizational context (and the management of sociodemographic diversity; see above), a dual identity model does not appear to be useful in the context of a merger, presumably because heightened salience of premerger group identities may threaten the success of this type of organizational change.

One lacuna in social identity research on mergers is the temporal dimension (which is absent from most social psychological research on social identity processes). Mergers take time and move through stages, during which different social identity processes may operate. It would be valuable to track social identity and self-categorization processes in mergers over time, perhaps within the framework of Levine and Moreland's (1994; Moreland & Levine, 1997, in press) diachronic group socialization model.

Another lacuna is uncertainty. Mergers and takeovers often produce enormous uncertainty, which can instantiate precisely the conditions that work against a successful merger. To reduce self-conceptual uncertainty, merger partners resist change and may polarize and consolidate interorganizational attitudes around narrowly prescriptive norms and fierce pre-merger organizational identification.

Summary, Conclusions, and Prospects

The aim of this article has been to describe recent theoretical developments within social identity theory that focus, via self-categorization theory, on how social categorization produces prototype-based depersonalization, which is responsible for social identity phenomena. These developments extend social identity theory. They advance our understanding of social identity processes in intergroup contexts and the way in which people may internalize group norms and align their behavior with these norms. They also have produced a new conceptualization of motivation associated with social identity, a better understanding of salience processes, and a new focus on intragroup processes that is now producing social identity models of, for example, cohesion, deviance, group structure, and leadership.

We introduce these new developments to an organizational readership that is familiar with some aspects of social identity theory, but less familiar with more recent self-categorization theory-based developments, in order to show how these developments are relevant to understanding a range of social behaviors in organizational contexts. One of our main aims has been to derive, from these developments, a variety of more or less specific, but testable, speculations and propositions, in order to help frame future research directions in the study of social identity processes and social behavior in organizational contexts.

The challenge for the future is to integrate new social identity mechanisms centrally into theories of organizational behavior. To date, such mechanisms have played a relatively small role in the literature on organizational behavior. Thus, the important role that identifications with the workgroup, organization, and profession, as well as those that emanate from people's sociodemographic background, may play in organizational behavior has yet to be articulated fully.

We suggest that identity-related constructs and processes have the potential to inform our understanding of organizational behavior. Combined with multilevel approaches to organizational research, the use of both individual-level and group-level constructs in models of organizational phenomena could mark the beginning of a new phase of research in organizational behavior. By acknowledging the importance of work-related identities to people's sense of self, a social identity perspective adds to our understanding of organizational attitudes and behavior by drawing on the important link between such identities and the person's sense of self. Such a perspective should improve explanation and understanding of intergroup relations, both within and between organizations.

To maximize the usefulness of the social identity perspective in the organizational arena, there must be significant interchange between social and organizational psychologists. When one is deriving predictions from a social identity perspective on organizational behavior, one finds the results from laboratory-based social psychological research invaluable, as are the insights that have been gained from field research in the organizational context. It is this type of interchange that will further our understanding of how identity-related constructs and processes impact organizational phenomena, and it should lead, in turn, to extensions of and refinements to social identity theory itself. We are confident that the extent of theoretical interchange between organizational and social psychologists studying social identity mechanisms can increase (e.g., Hogg & Terry, in press), and we are extremely optimistic about the potential that this interchange has for the development of both disciplines.

Notes

1. For general developments, see books by Abrams and Hogg (1990, 1999); Hogg and Abrams (1988); Hogg and Terry (in press); Oakes, Haslam, and Turner (1994); Robinson (1996); Spears, Oakes, Ellemers, and Haslam (1997); Terry and Hogg (1999); Turner et al. (1987); and Worchel, Morales, Paez, and Deschamps (1998).

References

Abrams, D., & Hogg, M. A. 1988. Comments on the motivational status of self-esteem in social identity and inter-group discrimination. *European Journal of Social Psychology*, 18: 317–334.

Abrams, D., & Hogg, M. A. (Eds.). 1990. *Social identity theory: Constructive and critical advances.* London: Harvester Wheatsheaf.

Abrams, D., & Hogg, M. A. 1998. Prospects for research in group processes and intergroup relations. *Group Processes and Intergroup Relations*, 1: 7–20.

Abrams, D., & Hogg, M. A. (Eds.). 1999. *Social identity and social cognition.* Oxford: Blackwell.

Abrams, D., & Hogg, M. A. In press. Self, group and identity: A dynamic model. In M. A. Hogg & R. S. Tindale (Eds.), *Blackwell handbook of social psychology: Group processes.* Oxford: Blackwell.

Alderfer, C. P., & Thomas, D. A. 1988. The significance of race and ethnicity for understanding organizational behavior. In C. L. Cooper & I. Robertson (Eds.), *International review of industrial and organizational psychology*: 1–41. New York: Wiley.

Anastasio, P. A., Bachman, B. A., Gaertner, S. L., & Dovidio, J. F. 1997. Categorization, recategorization, and common ingroup identity. In R. Spears, P. J. Oakes, N. Ellemers, & S. A. Haslam (Eds.), *The social psychology of stereotyping and group life*: 236–256. Oxford: Blackwell.

Ashforth, B. E., & Humphrey, R. H. 1993. Emotional labor in service roles: The influence of identity. *Academy of Management Review*, 18: 88–115.

Ashforth, B. E., & Mael, F. A. 1989. Social identity theory and the organization. *Academy of Management Review*, 14: 20–39.

Blake, R. R., & Mouton, J. S. 1985. How to achieve integration on the human side of the merger. *Organizational Dynamics*, 13: 41–56.

Brewer, M. B. 1991. The social self: On being the same and different at the same time. *Personality and Social Psychology Bulletin*, 17: 475–482.

Brewer, M. B. 1993. The role of distinctiveness in social identity and group behaviour. In M. A. Hogg & D. Abrams (Eds.), *Group motivation: Social psychological perspectives*: 1–16. London: Harvester Wheatsheaf.

Brewer, M. B. 1996. Managing diversity: The role of social identities. In S. Jackson & M. Ruderman (Eds.), *Diversity in work teams*: 47–68. Washington, DC: American Psychological Association.

Brewer, M. B., & Harasty, A. S. 1996. Seeing groups as entities: The role of perceiver motivation. In E. T. Higgins & R. M. Sorrentino (Eds.), *Handbook of motivation and cognition. Volume 3: The interpersonal context*: 347–370. New York: Guilford.

Brewer, M. B., & Miller, N. 1996. *Intergroup relations*. Milton Keynes, UK: Open University Press.

Brewer, M. B., von Hippel, W., & Gooden, M. P. 1999. Diversity and organizational identity: The problem of entree after entry. In D. A. Prentice & D. T. Miller (Eds.). *Cultural divides: Understanding and overcoming group conflict*: 337–363. New York: Russell Sage Foundation.

Brown, R. J. 1996. Tajfel's contribution to the reduction of intergroup conflict. In W. P. Robinson (Ed.), *Social groups and identities: Developing the legacy of Henri Tajfel*: 169–189. Oxford: Butterworth-Heinemann.

Buono, A. F., & Bowditch, J. L. 1989. *The human side of mergers and acquisitions: Managing collisions between people, cultures, and organizations*. San Francisco: Jossey-Bass.

Burger, J. M. 1985. Temporal effects on attributions for academic performances and reflected-glory basking. *Social Psychology Quarterly*, 48: 330–336.

Campbell, D. T. 1958. Common fate, similarity, and other indices of the status of aggregates of persons as social entities. *Behavioral Science*, 3: 14–25.

Chung, W. V. L. 1997. *Ethnicity and organizational diversity*. New York: University Press of America.

Cialdini, R. B., Borden, R. J., Thorne, A., Walker, M. R., Freeman, S., & Sloan, L. R. 1976. Basking in reflected glory: Three (football) field studies. *Journal of Personality and Social Psychology*, 34: 366–375.

Cialdini, R. B., & de Nicholas, M. E. 1989. Self-presentation by association. *Journal of Personality and Social Psychology*, 57: 626–631.

Cox, T. 1991. The multicultural organization. *Academy of Management Executive*, 5: 34–47.

de Gilder, D., & Wilke, H. A. M. 1994. Expectation states theory and the motivational determinants of social influence. *European Review of Social Psychology*, 5: 243–269.

Deschamps, J.-C., & Brown, R. J. 1983. Superordinate goals and intergroup conflict. *British Journal of Social Psychology*, 22: 189–195.

Dutton, J. E., Dukerich, J. M., & Harquail, C. V. 1994. Organizational images and member identification. *Administrative Science Quarterly*, 39: 239–263.

Eagly, A. H., Koran, S. J., & Makhijani, M. G. 1995. Gender and the effectiveness of leaders: A meta-analysis. *Psychological Bulletin*, 117: 125–145.

Ellemers, N. 1993. The influence of socio-structural variables on identity management strategies. *European Review of Social Psychology*, 4: 27–57.

Ellemers, N., Doojse, B. J., van Knippenberg, A., & Wilke, H. 1992. Status protection in high status minority groups. *European Journal of Social Psychology*, 22: 123–140.

Ellemers, N., van Knippenberg, A., de Vries, N., & Wilke, H. 1988. Social identification and permeability of group boundaries. *European Journal of Social Psychology*, 18: 497–513.
Feather, N. T. 1994. Attitudes towards high achievers and reactions to their fall: Theory and research concerning tall poppies. *Advances in Experimental Social Psychology*, 26: 1–73.
Fielding, K. S., & Hogg, M. A. 1997. Social identity, self-categorization, and leadership: A field study of small interactive groups. *Group Dynamics: Theory, Research, and Practice*, 1: 39–51.
Fielding, K. S., & Hogg, M. A. 1998. *Positive and negative deviance in groups: A social identity analysis*. Unpublished manuscript, University of Queensland.
Fiske, S. T. 1993. Controlling other people: The impact of power on stereotyping. *American Psychologist*, 48: 621–628.
Fiske, S. T., & Depret, E. 1996. Control, interdependence and power: Understanding social cognition in its social context. *European Review of Social Psychology*, 7: 31–61.
Fiske, S. T., & Taylor, S. E. 1991. *Social cognition* (2nd ed.). New York: McGraw-Hill.
Gaertner, S. L., Dovidio, J. F., Anastasio, P. A., Bachman, B. A., & Rust, M. C. 1993. Reducing intergroup bias: The common ingroup identity model. *European Review of Social Psychology*, 4: 1–26.
Gaertner, S. L., Dovidio, J. F., & Bachman, B. A. 1996. Revisiting the contact hypothesis: The induction of a common ingroup identity. *International Journal of Intercultural Relations*, 20: 271–290.
Gaertner, S. L., Rust, M. C., Dovidio, J. F., Bachman, B. A., & Anastasio, P. A. 1995. The contact hypothesis: The role of a common ingroup identity in reducing intergroup bias among majority and minority group members. In J. L. Nye & A. Brower (Eds.), *What's social about social cognition*: 230–260. Newbury Park, CA: Sage.
Gilbert, D. T., & Jones, E. E. 1986. Perceiver-induced constraint: Interpretations of self-generated reality. *Journal of Personality and Social Psychology*, 50: 269–280.
Gilbert, D. T., & Malone, P. S. 1995. The correspondence bias. *Psychological Bulletin*, 117: 21–38.
Goodwin, S. A., & Fiske, S. T. 1996. Judge not, lest . . .: The ethics of power holders' decision making and standards for social judgment. In D. M. Messick & A. E. Tenbrunsel (Eds.), *Codes of conduct: Behavioral research into business ethics*: 117–142. New York: Russell Sage Foundation.
Goodwin, S. A., Gubin, A., Fiske, S. T., & Yzerbyt, V. Y. In press. Power can bias impression formation: Stereotyping subordinates by default and by design. *Group Processes and Intergroup Relations*.
Hains, S. C., Hogg, M. A., & Duck, J. M. 1997. Self-categorization and leadership: Effects of group prototypicality and leader stereotypicality. *Personality and Social Psychology Bulletin*, 23: 1087–1100.
Hakansson, H., & Sharma, D. D. 1996. Strategic alliances in a network perspective. In D. Iacobucci (Ed.), *Networks in marketing*: 108–124. Thousand Oaks, CA: Sage.
Hamilton, D. L., & Sherman, S. J. 1996. Perceiving persons and groups. *Psychological Review*, 103: 336–355.
Hamilton, D. L., Sherman, S. J., & Lickel, B. 1998. Perceiving social groups: The importance of the entitativity continuum. In C. Sedikides, J. Schopler, & C. A. Insko (Eds.), *Intergroup cognition and intergroup behavior*: 47–74. Mahwah, NJ: Lawrence Erlbaum Associates.
Haunschild, P. R., Moreland, R. L., & Murrell, A. J. 1994. Sources of resistance to mergers between groups. *Journal of Applied Social Psychology*, 24: 1150–1178.
Hewstone, M. R. C. 1994. Revision and change of stereotypic beliefs: In search of the illusive subtyping model. *European Review of Social Psychology*, 5: 69–109.
Hewstone, M. R. C. 1996. Contact and categorization: Social psychological interventions to change intergroup relations. In C. N. Macrae, C. Stangor, & M. R. C. Hewstone (Eds.), *Stereotypes and stereotyping*: 323–368. London: Guilford.
Hewstone, M. R. C., & Brown, R. J. 1986. Contact is not enough: An intergroup perspective on the "contact hypothesis." In M. R. C. Hewstone & R. J. Brown (Eds.), *Contact and conflict in intergroup encounters*: 1–44. Oxford: Blackwell.
Hogan, E. A., & Overmyer-Day, L. 1994. The psychology of mergers and acquisitions. *International Review of Industrial and Organizational Psychology*, 9: 247–282.

Hogg, M. A. 1987. Social identity and group cohesiveness. In J. C. Turner, M. A. Hogg, P. J. Oakes, S. D. Reicher, & M. S. Wetherell, *Rediscovering the social group: A self-categorization theory*: 89–116. Oxford: Blackwell.

Hogg, M. A. 1992. *The social psychology of group cohesiveness: From attraction to social identity*. London: Harvester Wheatsheaf.

Hogg, M. A. 1993. Group cohesiveness: A critical review and some new directions. *European Review of Social Psychology*, 4: 85–111.

Hogg, M. A. 1996a. Social identity, self-categorization, and the small group. In E. H. Witte & J. H. Davis (Eds.), *Understanding group behavior. Volume 2: Small group processes and interpersonal relations*: 227–253. Mahwah, NJ: Lawrence Erlbaum Associates.

Hogg, M. A. 1996b. Intragroup processes, group structure and social identity. In W. P. Robinson (Ed.), *Social groups and identities: Developing the legacy of Henri Tajfel*: 65–93. Oxford: Butterworth-Heinemann.

Hogg, M. A. 1998. *Group identification, leadership, and the exercise of power*. Paper presented at the 1998 convention of the Society for the Psychological Study of Social Issues, Ann Arbor, MI.

Hogg, M. A. 1999. *A social identity theory of leadership*. Unpublished manuscript (submitted for publication), University of Queensland.

Hogg, M. A. In press a. Subjective uncertainty reduction through self-categorization: A motivational theory of social identity processes. *European Review of Social Psychology*.

Hogg, M. A. In press b. Self-categorization and subjective uncertainty resolution: Cognitive and motivational facets of social identity and group membership. In J. P. Forgas, K. D. Williams, & L. Wheeler (Eds.), *The social mind: Cognitive and motivational aspects of interpersonal behavior*. New York: Cambridge University Press.

Hogg, M. A. In press c. Social identity and social comparison. In J. Suls & L. Wheeler (Eds.), *Handbook of social comparison: Theory and research*. New York: Plenum.

Hogg, M. A., & Abrams, D. 1988. *Social identifications: A social psychology of intergroup relations and group processes*. London: Routledge.

Hogg, M. A., & Abrams, D. 1990. Social motivation, self-esteem and social identity. In D. Abrams & M. A. Hogg (Eds.), *Social identity theory: Constructive and critical advances*: 28–47. London: Harvester Wheatsheaf.

Hogg, M. A., & Abrams. D. (Eds.). 1993a. *Group motivation: Social psychological perspectives*. London: Harvester Wheatsheaf.

Hogg, M. A., & Abrams, D. 1993b. Towards a single-process uncertainty-reduction model of social motivation in groups. In M. A. Hogg & D. Abrams (Eds.), *Group motivation: Social psychological perspectives*: 173–190. London: Harvester-Wheatsheaf.

Hogg, M. A., & Abrams, D. 1999. Social identity and social cognition: Historical background and current trends. In D. Abrams & M. A. Hogg (Eds.), *Social identity and social cognition*: 1–25. Oxford: Blackwell.

Hogg, M. A., & Abrams, D. In press. Social categorization, depersonalization and group behavior. In M. A. Hogg & R. S. Tindale (Eds.), *Blackwell handbook of social psychology: Group processes*. Oxford: Blackwell.

Hogg, M. A., Cooper-Shaw, L., & Holzworth, D. W. 1993. Group prototypicality and depersonalized attraction in small interactive groups. *Personality and Social Psychology Bulletin*, 19: 452–465.

Hogg, M. A., & Hains, S. C. 1996. Intergroup relations and group solidarity: Effects of group identification and social beliefs on depersonalized attraction. *Journal of Personality and Social Psychology*, 70: 295–309.

Hogg, M. A., & Hains, S. C. 1998. Friendship and group identification: A new look at the role of cohesiveness in groupthink. *European Journal of Social Psychology*, 28: 323–341.

Hogg, M. A.. Hains, S. C., & Mason, I. 1998. Identification and leadership in small groups: Salience, frame of reference, and leader stereotypicality effects on leader evaluations. *Journal of Personality and Social Psychology*, 75: 1248–1263.

Hogg, M. A., & Hardie, E. A. 1991. Social attraction, personal attraction and self-categorization: A field study. *Personality and Social Psychology Bulletin*, 17: 175–180.

Hogg, M. A., & Hardie, E. A. 1992. Prototypicality, conformity and depersonalized attraction: A self-categorization analysis of group cohesiveness. *British Journal of Social Psychology*, 31: 41–56.
Hogg, M. A., & Hardie, E. A. 1997. Self-prototypicality, group identification and depersonalized attraction: A polarization study. In K. Leung, U. Kim, S. Yamaguchi, & Y. Kashima (Eds.), *Progress in Asian social psychology*, vol. 1: 119–137. Singapore: Wiley.
Hogg, M. A., Hardie, E. A., & Reynolds, K. 1995. Prototypical similarity, self-categorization, and depersonalized attraction: A perspective on group cohesiveness. *European Journal of Social Psychology*, 25: 159–177.
Hogg, M. A., & McGarty, C. 1990. Self-categorization and social identity. In D. Abrams & M. A. Hogg (Eds.), *Social identify theory: Constructive and critical advances*: 10–27. London: Harvester Wheatsheaf.
Hogg, M. A., & Moreland, R. L. 1995. *European and American influences on small group research*. Paper presented at the small groups preconference of the Joint Meeting of the European Association of Experimental Social Psychology and the Society for Experimental Social Psychology, Washington, DC.
Hogg, M. A., & Mullin, B.-A. 1999. Joining groups to reduce uncertainty: Subjective uncertainty reduction and group identification. In D. Abrams & M. A. Hogg (Eds.), *Social identity and social cognition*: 249–279. Oxford: Blackwell.
Hogg, M. A., & Reid, S. In press. Social identity, leadership, and power. In J. Bargh & A. Lee-Chai (Eds.), *The use and abuse of power: Multiple perspectives on the causes of corruption*. Philadelphia: Psychology Press.
Hogg, M. A., & Terry, D. J. (Eds.). In press. *Social identity processes in organizational contexts*. Philadelphia: Psychology Press.
Hogg, M. A., Terry, D. J., & White, K. M. 1995. A tale of two theories: A critical comparison of identity theory with social identity theory. *Social Psychology Quarterly*, 58: 255–269.
Hornsey, M. J., & Hogg, M. A. 1999. Subgroup differentiation as a response to an overly-inclusive group: A test of optimal distinctiveness theory. *European Journal of Social Psychology*, 29: 543–550.
Hornsey, M. J., & Hogg, M. A. In press a. Subgroup relations: A comparison of mutual intergroup differentiation and common ingroup identity models of prejudice reduction. *Personality and Social Psychology Bulletin*.
Hornsey, M. J., & Hogg, M. A. In press b. Intergroup similarity and subgroup relations: Some implications for assimilation. *Personality and Social Psychology Bulletin*.
Ibarra, H. 1995. Race, opportunity, and diversity of social circles in managerial networks. *Academy of Management Journal*, 38: 673–703.
Janis, I. L. 1982. *Groupthink: Psychological studies of policy decisions and fiascoes* (2nd ed.). Boston: Houghton Mifflin.
Kandola, R. 1995. Managing diversity: New broom or old hat. *International Review of Industrial and Organizational Psychology*, 10: 131–168.
Kramer, R. M. 1991. Intergroup relations and organizational dilemmas: The role of categorization processes. In L. L. Cummings & B. M. Staw (Eds.), Research in organizational behavior, vol. 13: 191–228. Greenwich, CT: JAI Press.
Lalonde, R. N. 1992. The dynamics of group differentiation in the face of defeat. *Personality and Social Psychology Bulletin*, 18: 336–342.
Lester, R. E. 1987. Organizational culture, uncertainty reduction, and the socialization of new organizational members. In S. Thomas (Ed.), *Culture and communication: Methodology, behavior, artifacts, and institutions*: 105–113. Norwood, NJ: Ablex.
Levine, J. M., & Moreland, R. L. 1990. Progress in small group research. *Annual Review of Psychology*, 41: 585–634.
Levine, J. M., & Moreland, R. L. 1994. Group socialization: Theory and research. *European Review of Social Psychology*, 5: 305–336.
Long, K., & Spears, R. 1997. The self-esteem hypothesis revisited: Differentiation and the disaffected. In R. Spears, P. J. Oakes, N. Ellemers, & S. A. Haslam (Eds.), *The social psychology of stereotyping and group life*: 296–317. Oxford: Blackwell.

Lord, R. G., Foti, R. J., & DeVader, C. L. 1984. A test of leadership categorization theory: Internal structure, information processing, and leadership perceptions. *Organizational Behavior and Human Performance*, 34: 343–378.

Mael, F. A., & Ashforth, B. E. 1992. Alumni and their alma mater: A partial test of the reformulated model of organizational identification. *Journal of Organizational Behavior*, 13: 103–123.

Mael, F. A., & Ashforth, B. E. 1995. Loyal from day one: Biodata, organizational identification, and turnover among newcomers. *Personnel Psychology*, 48: 309–333.

Marcus-Newhall, A., Miller, N., Holtz, R., & Brewer, M. B. 1993. Crosscutting category membership with role assignment: A means of reducing intergroup bias. *British Journal of Social Psychology*, 32: 124–146.

Marques, J. M. 1990. The black-sheep effect: Out-group homogeneity in social comparison settings. In D. Abrams & M. A. Hogg (Eds.), *Social identity theory: Constructive and critical advances*: 131–151. London: Harvester Wheatsheaf.

Marques, J. M., & Paez, D. 1994. The "black sheep effect": Social categorization, rejection of ingroup deviates and perception of group variability. *European Review of Social Psychology*, 5: 37–68.

Marques, J. M., & Yzerbyt, V. Y. 1988. The black sheep effect: Judgmental extremity towards ingroup members in inter- and intro-group situations. *European Journal of Social Psychology*, 18: 287–292.

Marques, J. M., Yzerbyt, V. Y., & Leyens, J: P. 1988. The black sheep effect: Extremity of judgements towards in-group members as a function of group identification. *European Journal of Social Psychology*, 18: 1–16.

Marris, P. 1996. *The politics of uncertainty: Attachment in private and public life*. London: Routledge.

Moreland, R. L., & Levine, J. M. 1997. *Work group socialization: A social identity approach*. Paper presented at the Second Australian Industrial and Organizational Psychology Conference, Melbourne.

Moreland, R. L., & Levine, J. M. In press. Socialization in organizations and work groups. In M. Turner (Ed.), *Groups at work: Advances in theory and research*. Mahwah, NJ: Lawrence Erlbaum Associates.

Moreland, R. L., Hogg, M. A., & Hains, S. C. 1994. Back to the future: Social psychological research on groups. *Journal of Experimental Social Psychology*, 30: 527–555.

Morris, M. W., & Peng, K. 1994. Culture and cause: American and Chinese attributions for social and physical events. *Journal of Personality and Social Psychology*, 67: 949–971.

Mowday, R. T., & Sutton, R. I. 1993. Organizational behavior: Linking individuals and groups to organizational contexts. *Annual Review of Psychology*, 44: 195–229.

Mullen, B., & Copper, C. 1994. The relation between group cohesiveness and performance: An integration. *Psychological Bulletin*, 115: 210–227.

Nkomo, S. M., & Cox, T., Jr. 1996. Diverse identities in organizations. In S. R. Clegg, C. Hardy, & W. R. Nord (Eds.), *Handbook of organization studies*: 338–356. London: Sage.

Nye, J. L., & Forsyth, D. R. 1991. The effects of prototype-based biases on leadership appraisals: A test of leadership categorization theory. *Small Group Research*, 22: 360–379.

Nye, J. L., & Simonetta, L. G. 1996. Followers' perceptions of group leaders: The impact of recognition-based and inference-based processes. In J. L. Nye, & A. M. Bower (Eds.). *What's social about social cognition: Research on socially shared cognition in small groups*: 124–153. Thousand Oaks, CA: Sage.

Oakes, P. J., Haslam, S. A., & Turner, J. C. 1994. *Stereotyping and social reality*. Oxford: Blackwell.

Oakes, P. J., & Turner, J. C. 1990. Is limited information processing the cause of social stereotyping. *European Review of Social Psychology*, 1: 111–135.

Pettigrew, T. F. 1998. Intergroup contact theory. *Annual Review of Psychology*, 49: 65–85.

Pratt, M. G. In press. To be or not to be? Central questions in organizational identification. In D. Whetten & P. Godfrey (Eds.), *Identity in organizations: Developing theory through conversations*. Thousand Oaks, CA: Sage.

Pratto, F., Sidanius, J., Stallworth, L. M., & Malle, B. F. 1994. Social dominance orientation: A personality variable predicting social and political attitudes. *Journal of Personality and Social Psychology*, 67: 741–763.
Prentice, D. A., & Miller, D. T. (Eds.). 1999. *Cultural divides: Understanding and overcoming group conflict.* New York: Russell Sage Foundation.
Raven, B. H. 1965. Social influence and power. In I. D. Steiner & M. Fishbein (Eds.), *Current studies in social psychology*: 371–382. New York: Holt, Rinehart & Winston.
Reicher, S. D., Spears, R., & Postmes, T. 1995. A social identity model of deindividuation phenomena. *European Review of Social Psychology*, 6: 161–198.
Riordan, C. M., & Shore, L. M. 1997. Demographic diversity and employee attitudes: An empirical examination of relational demography within work units. *Journal of Applied Psychology*, 82: 342–358.
Robinson, W. P. (Ed.). 1996. *Social groups and identities: Developing the legacy of Henri Tajfel.* Oxford: Butterworth-Heinemann.
Ross, L. 1977. The intuitive psychologist and his shortcomings. *Advances in Experimental Social Psychology*, 10: 174–220.
Rubin, M., & Hewstone, M. 1998. Social identity theory's self-esteem hypothesis: A review and some suggestions for clarification. *Personality and Social Psychology Review*, 2: 40–62.
Rush, M. C., & Russell, J. E. A. 1988. Leader prototypes and prototype-contingent consensus in leader behavior descriptions. *Journal of Experimental Social Psychology*, 24: 88–104.
Saks, A. M., & Ashforth, B. E. 1997. Organizational socialization: Making sense of the past and present as a prologue for the future. *Journal of Vocational Behavior*, 51: 234–279.
Sherman, S. J., Hamilton, D. L., & Lewis, A. C. 1999. Perceived entitativity and the social identity value of group memberships. In D. Abrams & M. A. Hogg (Eds.), *Social identity and social cognition*: 80–110. Oxford: Blackwell.
Sigelman, L. 1986. Basking in reflected glory revisited: An attempt at replication. *Social Psychology Quarterly*, 49: 90–92.
Snyder, C. R., Lassegard, M.-A., & Ford, C. E. 1986. Distancing after group success and failure: Basking in reflected glory and cutting off reflected failure. *Journal of Personality and Social Psychology*, 51: 382–388.
Spears, R., Oakes, P. J., Ellemers, N., & Haslam, S. A. (Eds.). 1997. *The social psychology of stereotyping and group life*. Oxford: Blackwell.
Steiner, I. D. 1974. Whatever happened to the group in social psychology? *Journal of Experimental Social Psychology*, 10: 94–108.
Tajfel, H. 1972. Social categorization (English translation of "La categorisation sociale"). In S. Moscovici (Ed.), *Introduction a la psychologie sociale*, vol. 1: 272–302. Paris: Larousse.
Tajfel, H. 1974a. *Intergroup behavior, social comparison and social change. Unpublished Katz-Newcomb lectures*, University of Michigan, Ann Arbor.
Tajfel, H. 1974b. Social identity and intergroup behaviour. *Social Science Information*, 13: 65–93.
Tajfel, H. 1975. The exit of social mobility and the voice of social change. Social Science Information, 14: 101–118.
Tajfel, H., & Turner, J. C. 1979. An integrative theory of inter-group conflict. In W. G. Austin & S. Worchel (Eds.), *The social psychology of intergroup relations*: 33–47. Monterey, CA: Brooks-Cole.
Tajfel, H., & Turner, J. C. 1986. The social identity theory of intergroup behavior. In S. Worchel & W. G. Austin (Eds.), *The psychology of intergroup relations*: 7–24. Chicago: Nelson-Hall.
Taylor, D. M., & McKirnan, D. J. 1984. A five-stage model of intergroup relations. *British Journal of Social Psychology*, 23: 291–300.
Taylor, S. E., & Fiske, S. T. 1978. Salience, attention, and attribution: Top of the head phenomena. *Advances in Experimental Social Psychology*, 11: 249–288.
Terry, D. J., & Callan, V. J. 1998. Intergroup differentiation in response to an organizational merger. *Group Dynamics: Theory, Research, and Practice*, 2: 67–87.
Terry, D. J., Carey, C. J., & Callan, V. J. In press. Employee adjustment to an organizational merger: An intergroup perspective. *Personality and Social Psychology Bulletin*.

Terry, D. J., & Hogg, M. A. 1996. Group norms and the attitude-behavior relationship: A role for group identification. *Personality and Social Psychology Bulletin*, 22: 776–793.

Terry, D. J., & Hogg, M. A. (Eds.). 1999. *Attitudes, behavior, and social context: The role of norms and group membership*. Mahwah, NJ: Lawrence Erlbaum Associates.

Trope, Y., & Liberman, A. 1993. The use of trait conceptions to identify other people's behavior and to draw inferences about their personalities. *Personality and Social Psychology Bulletin*, 19: 553–562.

Tsui, A., Egan, T., & O'Reilly, C., III. 1992. Being different: Relational demography and organizational attachment. *Administrative Science Quarterly*, 37: 549–579.

Turner, J. C. 1975. Social comparison and social identity: Some prospects for intergroup behaviour. *European Journal of Social Psychology*, 5: 5–34.

Turner, J. C. 1982. Towards a cognitive redefinition of the social group. In H. Tajfel (Ed.), *Social identity and intergroup relations*: 15–40. Cambridge: Cambridge University Press.

Turner, J. C. 1985. Social categorization and the self-concept: A social cognitive theory of group behavior. In E. J. Lawler (Ed.), *Advances in group processes: Theory and research*, vol. 2: 77–122. Greenwich, CT: JAI Press.

Turner, J. C. 1991. *Social influence*. Milton Keynes, UK: Open University Press.

Turner, J. C. 1996. Henri Tajfel: An introduction. In W. P. Robinson (Ed.), *Social groups and identities: Developing the legacy of Henri Tajfel*: 1–23. Oxford: Butterworth-Heinemann.

Turner, J. C., Hogg, M. A., Oakes, P. J., Reicher, S. D., & Wetherell, M. S. 1987. *Rediscovering the social group: A self-categorization theory*. Oxford: Blackwell.

Turner, M. E., Pratkanis, A. R., Probasco, P., & Leve, C. 1992. Threat, cohesion, and group effectiveness: Testing a social identity maintenance perspective on groupthink. *Journal of Personality and Social Psychology*, 63: 781–796.

Tyler, T. R. 1990. *Why people obey the law*. New Haven, CT: Yale University Press.

van Knippenberg, A., & Ellemers, N. 1993. Strategies in intergroup relations. In M. A. Hogg & D. Abrams (Eds.), *Group motivation: Social psychological perspectives*: 17–32. London: Harvester Wheatsheaf.

van Knippenberg, D. 1997. *A social identity perspective on mergers and acquisitions*. Paper presented at the Second Australian Industrial and Organizational Psychology Conference, Melbourne.

Vaughan, G. M. 1978. Social change and intergroup preferences in New Zealand. *European Journal of Social Psychology*, 8: 297–314.

Wann, D. L, Hamlet, M. A., Wilson, T. M., & Hodges, J. A. 1995. Basking in reflected glory, cutting off reflected failure, and cutting off future failure: The importance of group identification. *Social Behavior and Personality*, 23: 377–388.

Wesolowski, M. A., & Mossholder, K. W. 1997. Relational demography in supervisor-subordinate dyads: Impact on subordinate job satisfaction, burnout, and perceived procedural justice. *Journal of Organizational Behavior*, 18: 351–362.

Williams, K. D., & Sommer, K. L. 1997. Social ostracism by coworkers: Does rejection lead to loafing or compensation? *Personality and Social Psychology Bulletin*, 23: 693–706.

Worchel, S., Morales, J. F., Páez, D., & Deschamps, J.-C. (Eds.). 1998. *Social identity: International perspectives*. London: Sage.

Yukl, G. A., & Falbe, C. M. 1991. Importance of different power sources in downward and lateral relations. *Journal of Applied Psychology*, 76: 416–423.

Zuckerman, M. 1979. Attribution of success and failure revisited, or: The motivational bias is alive and well in attributional theory. *Journal of Personality*, 47: 245–287.

47

Maintaining Masculinity: Men Who Do 'Women's Work'

Ben Lupton

Source: *British Journal of Management* 11 Special Issue (2000), S33–S48.

Introduction

The past decade has seen increasing interest in men's actions and experiences within the study of gender and organizations. As Collinson and Hearn (1994) have argued, earlier approaches, both mainstream management writing and feminist analyses, have tended to 'assume' male experience while simultaneously placing men at the centre of their arguments. Traditional management writing tended to treat organizations as non-gendered whilst at the same time conflating 'men' and 'management'. Feminist approaches on the other hand have demonstrated the gendered nature of organizations, showing how male values pervade the workplace and how men exercise power to the detriment of women. The main focus, however, has (legitimately) been on female experience within patriarchal societies and gendered organizations. The result is, as Collinson and Hearn argue, that 'categories of men and masculinity are frequently central to analyses yet . . . remain taken for granted, hidden and unexamined' (1994, p. 3). The developing literature around masculinities (for example Carrigan, Connell and Lee, 1985; Connell, 1987, 1992; Morgan, 1992; Kerfoot and Knights, 1993; Collinson and Hearn, 1994) has rendered men more 'visible' and provides a theoretical basis for the study of men in organizations as part of a contribution to the understanding of gendered work and organizations.

This literature offers a number of insights which are helpful in interpreting men's experience and behaviour in organizations. First, masculinity is usefully defined as a 'socially generated set of behaviours and practices surrounding the group named men' (Kerfoot and Knights, 1993, p. 661) rather than as a fixed attribute of someone of the male sex. Consequently it may be regarded as a role that is socially performed, enacted and reproduced through discourse (Collinson and Hearn, 1994). Adopting this view opens up a number of possibilities which are absent in a more essentialist position. Masculinity, as Kvande (1998) points out, can be performed by both men and women, is subject to change over time and, on account of its dynamic nature can be studied through observation of action and interpretation of discourse. Second, the use of the plural 'masculinities' as opposed to the singular 'masculinity' represents a recognition that masculinity is not homogenous (Cheng, 1996, p. xiii) but that 'multiple' masculinities exist, both historically and contemporaneously.

The third, and crucial, insight is contained within the concept of 'hegemonic masculinity' (Carrigan, Connell and Lee, 1985) which has been 'utilised extensively' (Kerfoot and Whitehead, 1998, p. 439) 'as a theoretical tool with which to understand men's behaviour and actions'. In developing this idea Carrigan, Connell and Lee depart from earlier work in arguing for a 'dynamic conception of masculinity as a structure of social relations' (1985, p. 587). Here, masculinities are not seen as fixed and neutral roles, but as practices through which a wider system of gendered power relations are maintained and reconstituted. Within this system, certain 'highly specific versions of what masculinity can be' are 'privileged' (Kerfoot and Knights, 1993, p. 60); or as Connell puts it 'certain constructions of masculinity are hegemonic, while others are subordinated or marginalised' (1992, p. 736). The term hegemonic masculinity has been used to refer to 'the currently dominant form of masculinity' in any society at any particular time (Cheng, 1996, p. xii). This may not be behaviour exhibited by any one man or group but an idealized 'masculine' way to behave (Fuller, 1996) against which male (and female) behaviour is judged as appropriate or otherwise, and through which 'unequal power relations between men and women are maintained' (Mills, 1998), and significantly, unequal power relations between dominant and subordinate men (Carrigan, Connell and Lee, 1985). Fuller (1996) identifies five core attributes of hegemonic masculinity in modern western societies; possession of work or money, heterosexuality, non-femininity, manliness and denial of vulnerability and emotion.

Once masculinity is located within a system of social relations it makes sense, as Carrigan *et al.* argue, to conceive of it as being 'embedded in the dynamics of institutions . . . quite as much as the personality of individuals' (1985, p. 591). Thus occupations and organizations are important arenas in which masculinities are defined and maintained and challenged. Notions of work are central to (hegemonic) masculine identities (Morgan, 1992), providing extrinsic and intrinsic rewards by which masculinity may be judged by self and others. Indeed, as Carrigan *et al.* argue, the reproduction of hegemonic masculinity underpins 'the social definition of some kinds of work as "men's work" or "women's work," and the definition of some kinds of work as more masculine than others' (1985, p. 594). It is no surprise, then, that as Collinson and Hearn observe, 'men in organizations often seem preoccupied with the creation and maintenance of various masculine identities and with the expression of gendered power and status in the workplace' (1994, p. 3).

Organizations may be seen, and studied, as places where masculinities are constructed, refined and put to the test. In doing so we can usefully draw on recent critical discussions of the concept of hegemonic masculinity (Kerfoot and Whitehead, 1998; Whitehead, 1999). These writers, whilst accepting its explanatory power, argue that it can lead to an underemphasis on 'individual action and subjectivity' (Kerfoot and Whitehead, 1998, p. 440). In particular they refer to the 'inevitability implicit in the concept' (p. 440) which may draw attention away from the possibility of resistance, reconstruction, rejection and even subversion of hegemonic masculinity. Kerfoot and Whitehead suggest that masculinity is most fruitfully regarded as an 'aspect of identity formation' (1998,

p. 440). As has been noted, this process is not played out in a vacuum, rather masculinities are produced and reproduced in relation to other masculinities or femininities, through struggles for power and resources within a wider system of gender relations (Connell, 1987; Kerfoot and Knights, 1993). Consequently masculinities are perpetually subject to reconstruction, renegotiation and challenge. Hegemonic masculinity can be regarded as 'especially vulnerable, given that it is generally articulated only in fundamental opposition to alternative, subordinated ways of being' (Kerfoot and Whitehead, 1998, p. 441).

Masculinities become highly visible when they are under challenge (Morgan, 1992), and for the researcher this offers an opportunity for their study. One of the circumstances where masculinity is 'on the line' is when men enter occupations that are traditionally populated by women. The long-standing segregation of labour markets by sex (Bradley, 1989; Reskin and Hartmann, 1986; Anker, 1997) has meant that most men and women work in occupations that are predominantly populated by members of the same sex. In Britain, two thirds of women and a slightly higher proportion of men work in occupations where they outnumber the opposite sex by a ratio of at least two to one, and many occupations are considerably more segregated than that (Hakim, 1992). There is an extensive literature in both psychology and sociology which examines the perpetuation of this pattern of segregation. Not only do occupations become sex-typed, seen as more appropriate for one or other sex to enter, but the segregation is maintained by active strategies of exclusion and demarcation (Witz, 1992). However, men and women do move into gender-atypical occupations. As Morgan (1992) and Bradley (1989) have shown that can take place in differing circumstances. The move may be part of a 'take-over' where one sex supplants the other, for example the feminization of clerical work and printing or the masculinization of spinning and midwifery. The process of infiltration involves men and women entering atypical occupations and remaining in a small minority. The entry of women into traditionally-male occupations in these circumstances has generated an extensive literature (e.g. Spencer and Podmore, 1987; Sheppard, 1989; Marshall, 1995), and this has drawn attention to the barriers to progress that women face in traditionally-male occupations. The study of male minorities is comparatively recent, but the developing literature has offered some insights, and has particular currency given realization that movement of men into non-traditional occupations is required for a reduction in segregation and also because of recruitment difficulties experienced by certain professions where women are in the majority, for example primary school teaching and nursing.

In particular this literature has highlighted the fact that male and female minorities in occupations have different experiences and face different challenges. Kanter (1977) had argued that minorities, where they represent less than 15% of the group, would experience particular treatment on account of their token status. This would result from their heightened visibility, contrast from the majority group and assimilation, which refers to their stereotyping as members of a particular group. Kanter showed how these 'perceptual tendencies', their effects and the responses they engendered were largely detrimental to the minority group. She argued that this would apply to all minority groups on account of their minority status, regardless of their identity. The implication of this being that male tokens could expect to experience the same effects as female tokens. Subsequent research, however, has shown that gender does play a role in the treatment and experience of minorities. Floge and Merrill (1986) showed that while the effects of heightened visibility were the same for female physicians and male nurses, contrast and assimilation worked to the advantage of male nurses and the disadvantage of female physicians. Male nurses were assumed to be more technically competent and have greater leadership skills, and thus had higher promotion chances than female nurses. They also benefited from a closer identification with male physicians. The fact that male tokens benefit from their token status has been repeatedly found in subsequent studies. Men tend to rise to the top of their occupations or professions more quickly (Williams, 1995; Pringle, 1993) riding a glass escalator (Kvande, 1998), congregate in particular specialities, usually higher status ones (Williams, 1995; Long, 1984), and find themselves channelled into administration or management roles (Williams, 1995), which tends to accelerate their progress.

However, there has been an increasing realization that while men benefit from taking their gender advantage into female occupations, they also suffer a challenge to their masculinity, both through working alongside women, and from performing a role which is regarded by society as one which women normally undertake. As Williams suggests, this may be a more fundamental challenge than women face when moving into traditionally male occupations, 'While many women may enjoy the "feminine" aspects of their work their femininity is not contingent on proving themselves competent in "gender-appropriate" work, which is often how masculinity is experienced by men' (1993, p. 15). Cockburn made a similar point, but rather more directly, suggesting that 'the handful of men who cross into traditional female areas of work at the female level will be written off as effeminate, tolerated as eccentrics or failures' (1988, p. 40). The suggestion of homosexuality or effemininity represents a challenge to the heterosexual and macho construction of hegemonic masculinity in western societies. As Morgan (1992) reports it is a common theme in response to male entry to female dominated occupations, both from within and outside the occupation. For example Heickes (1992) identified 'homosexual' as one of the 'role-entrapments' which male nurses have encountered. Allan (1993) has shown how the incongruity between elementary teaching and hegemonic masculinity has created a 'double-bind' for male teachers. By asserting a traditional masculinity, their competence and suitability to work in a caring and nurturing profession would be called into question, yet by adopting a more feminine approach they would invite challenge to their sexuality and masculinity.

The maintenance of masculinity under challenge may be seen as an example of the wider process of 'identity work' in organizations (Thompson and McHugh, 1990). Identity here is conceived as 'a tool which we use to present ourselves in, and possibly transform ourselves into, images appropriate to our social, cultural and work context' (1990, p. 287), a performance which we act out using or adapting available 'scripts'. The emphasis is on the process of identity-making rather than as a tangible identity located in a fixed self, reflecting the postmodernist critique of notions of the self. However as Casey (1995) argues there is some merit in retaining a notion of outcomes alongside processes in the discussion of identity and self. This allows us to conceive of conflicting identities as subjective experiences. These can be multiple and conflicting at the personal level in a Freudian or psychodynamic sense, but as Carrigan *et al.* (1985) suggest, their impact may spill over into the social and political domains. We might look for evidence of this where women and men enter occupations usually undertaken by the other sex. Here there is potentially a difference between what Thompson and McHugh (1990, p. 288) term 'personal identity' and 'social identity', in this context the individual's gender identity and that of their work role. We know something from existing work about how both women and men seek to resolve this. Studies of women managers (Marshall, 1995; Sheppard, 1989) have shown how women 'balance the conflicting statuses of "female" and "manager" ' (Sheppard, 1989, p. 145). As Marshall observes, in this context 'women find their own ways . . . of creating acceptable identities. For example, some stress their professional roles, some become "one of the boys", some emphasize caring and some use their sexuality overtly' (1995, p. 193). Where these 'deliberate behaviours' (Sheppard, 1989, p. 150) involve individuals compromising their sense of femininity, 'desexualizing' themselves, and making other concessions to role expectations, they can have 'harmful consequences for individual identity' (Marshall, 1995, p. 321). Marshall's case studies also leave the strong impression that these strategies involve hard, tiring and unrelenting work.

Studies of men in non-traditional occupations have also identified a range of strategies employed to manage the identity mismatch. What is interesting is that while women's strategies to a varying degree often involve compromising their individual gender identity to the demands of masculine work, male strategies give primacy to the preservation of masculine identity. One strategy identified in the existing literature is for men to be demonstrably careerist, emphasizing the career prospects rather than the job, with its gendered connotations. A second is to identify with other more powerful male groups (Floge and Merrill, 1986) seeking, as it were, hegemonic masculinity by association. A third strategy is to represent the work as more masculine, even to the extent of re-titling the job to avoid the non-masculine associations (Pringle, 1993). This may go further and involve doing the job differently, prioritizing the more masculine elements, for example the technical or managerial aspects. The challenge to sexuality may be dealt with by emphasizing one's heterosexuality (Morgan, 1992), or engaging in discourses which reinforce masculinity in relation to others, for example women and homosexuals (Barrett, 1996), or by 'impression management' (Thompson and McHugh, 1990) for example through dress (Collier, 1998).

Masculine identities, like other identities 'constantly have to be constructed, negotiated and reconstructed in routine social interaction' (Collinson and Hearn, 1994, p. 8). Given that such identities confer power and advantage within a patriarchy, this is important work for men to do, and becomes particularly important in environments where masculine identities are placed under challenge. The remainder of this paper examines the ways in which men in non-traditional occupations go about this work. The results of in-depth interviews with a sample of such men are used to illustrate the different types of challenge to their masculinity which they face and the varying strategies that they employ to meet these challenges. The paper concludes by considering the contrasting implications of these response strategies within the context of stasis and change in segregated labour markets.

Methodology

The study on which this paper is based is part of a wider research project examining the entry of men into traditionally female occupations. That project is concerned with identifying the 'kinds of men' who enter women's work and their reasons for doing so. One of the approaches adopted was to interview a number of men who had decided to enter a selection of non-traditional occupations. In asking men about their unusual career choices, discussion often turned to issues of masculinity and challenges to masculinity, and it is upon these discussions that the current paper is based.

The data presented in the paper are drawn from nine in-depth interviews with men who have entered, or are studying to enter, a traditionally female occupation. Four were junior level administrative and clerical staff in a large public-sector institution in the north-west of England. These served as pilot interviews for the main study. The other five represented the first interviews of the main study and these respondents were undertaking postgraduate courses at a major university in the same region. Three were studying to be human resource managers, one to be a primary school teacher and one to be a librarian. In each case these male students were outnumbered significantly by women on their course of study, the percentage of men being 10%, 13% and 30% respectively; and in the profession as a whole, the corresponding percentages being 22%, 15% and 16%.[1] Each of the student interviewees had undertaken a significant period of work placement in the relevant profession, and three had worked previously in that occupation. All nine interviewees had offered to assist with the research by responding to a circular asking for volunteers. Interviews were held at the respondent's place of work or study.

The interviews were largely unstructured and focused loosely around topic areas pertinent to the main study. In practice such structure that existed was based around the respondents' work and life history, and the researcher allowed interviewees a free rein to describe and interpret incidents and issues that were important to them. Interviewees were given guarantees of anonymity for themselves and their organizations. The interviews were tape-recorded and the respondent was given control of the tape. Sensitive topics, for example those around sexuality, were left to the end of the interviews when a rapport has been established. Interviewees appeared to speak freely and unguardedly.

There are two features of the interviewer/subject relationship that should be considered when reading the results. First, seven of the interviewees were known by the interviewer, though none closely. It is difficult to determine what effect this will have had on the responses, whether for example interviewees 'held back' thoughts that they might have shared with a stranger who they would not re-encounter, or whether, on the other hand they found it easier to confide in someone they were familiar with. I am not convinced that either effect was marked, all respondents spoke candidly and openly on a range of subjects which might be considered sensitive by many people. The particular frankness of the two respondents who were not known to me persuades me slightly towards the former suggestion. The second feature to consider is the fact that the interviewer is male. It is apparent from the data that the respondents said things to a male interviewer that would have been unlikely to have been revealed to a woman. In this sense the male interviewer may be regarded as privileged. On the other hand, in the context of a study on the performance of masculinities it is particularly important to bear in mind that the performance of masculinity is not suspended on entry to the interview room. Given the power imbalance inherent in any kind of interview there was potentially a risk that interviewees would 'play up' a version of masculinity that was thought to correspond to the interviewer's own outlook.

To guard against this I was extremely careful not to be seen to advance a particular view of masculinity, and to be neutral in response of whatever views were expressed by the respondents. Whilst it is impossible to achieve complete objectivity as a qualitative interviewer, or indeed to know whether one has achieved it, the diversity of views expressed on questions of masculinity is a tentative indication that respondents found it comfortable to relate their thoughts to me.

The data were analysed manually. This was manageable given the number of interviews undertaken. It also allowed the researcher to remain 'close' to the data. Issues that emerged in each interview were highlighted and cross-referenced until themes and patterns emerged. These themes and some of the data which illustrate them are presented below. In each case a pseudonym has been used to protect the anonymity of the respondent.

Masculinities at Work: Challenges and Negotiations

This section draws on the interview data to illustrate the ways in which men perceive a challenge to masculinity when entering female-dominated occupations, and also the different approaches that they use to reconcile their masculine identity with that of the gender typing of their occupation. It emerged from the interviews that entering a female-dominated occupation carried with it a perceived challenge to the respondents' identity as men. This threat was recognized in three different ways; first through the challenge to the workplace as an arena for exercising and regenerating masculinity (which I will call regendering the work place), second through a fear of being feminized through exposure to women (feminization), and third through the threat of being stigmatized as effeminate and/or homosexual through an association with women, or by doing a 'woman's job' (stigmatization). The interviews revealed that these challenges were met in two different ways. One approach was to reconstruct the job so as to minimize its non-masculine associations, and a second was to reconstruct one's masculinity so that it is congruent with, and appropriate to, a female-dominated work environment.

Challenges to Masculinity (1) — Regendering the Workplace

A number of writers have shown how the workplace is a site for defining and reproducing masculinities (Morgan, 1992; Collinson and Hearn, 1994). The strength of this conceptual link between work and 'being a man' is illustrated in the interview data. The respondents subscribed to a view of sex-roles which sees men as aspiring to the role of 'breadwinner' and women to that of mother and carer:

> to a certain extent some of it [the difference between male and female roles] might originate from very basic aspects of human behaviour, going back to the man going out to getting the food and the women staying at home, you know cavemen stuff, going and getting the dinosaur and bringing it back and as much as we progress that's always going to be within us. (John, Human Resource Management)

> blokes to be the provider, that's what your aim is, and women, it doesn't seem to be important to them, perhaps it is the mothers or the role models they have. (Mick, Human Resource Management)

It is interesting that these men subscribed to an essentialist view of masculinity which they saw as being grounded in 'biological fact', as a 'genetic thing', something which went 'back to the ark'. Men and women were seen to have different interests and abilities related to these roles and tended to make different occupational choices as a result.

The role of provider was clearly and directly linked to male 'esteem' by more than one interviewee; women on the other hand, it was claimed, would find esteem through nurturing activities or through establishing relationships with others. In the modern world, work (rather than catching dinosaurs) was seen as the route to masculine self-esteem, and the workplace as a place to maintain it. One man anticipated how he would work harder once he had a family to support. A second suggested that work was connected with acquiring a mate as well as providing for one:

> I was coming out of a long term relationship but I was thinking that it would be better to be in a job, you impress someone a bit more if you've got things going for you than if you haven't, so that was a motivational factor definitely. (Brian, Clerical Work)

When masculine identity is bound up with success at work, men working in female occupations may face a challenge arising from the fact that these occupations are generally of low status and attract lower pay. The opportunity to gain esteem and through playing the role of provider may be limited. As one interviewee commented:

Jim	my wife is the major breadwinner in our house.

Q	is that something that bothers you?

Jim	yeah it does actually yeah it does at times . . . I mean she doesn't ram it down my throat. (Jim, Clerical Work)

A second interviewee felt that his ability to attract a mate was compromised by the perceived low status of his job:

> if you are meeting a woman for instance it is not a good thing to show off to say 'I'm a teacher' . . . it's not a great pulling line, not that I use them anyway, it's [primary school teaching] not something that is particularly macho or high status (with irony, in falsetto voice) 'ooh he is a teacher, he is going to have a big car'. (Maurice, Primary School Teacher)

Other challenges result from a perception of work as a place for maintaining masculinities through engaging behaviour or discourse which reproduces or refines a sense of masculinity. Where men are not present in the workplace, or are in a small minority, those opportunities are restricted or denied. The men in the sample were clearly aware of this. One of the recurring themes was the absence of 'male' conversations through which masculinity could be displayed and developed; and instead the existence of 'female' conversations which did not afford this opportunity, and from which they were either excluded, or excluded themselves (Kanter, 1977; Floge and Merrill, 1986).

> the things they talk . . . about are more traditionally women's things, the classic shopping conversations and what not (laughs) and I don't ever seem to get included in them. (Malcolm, Clerical Work)

> sometimes I go in [to work] and I want to talk about whether or not that was offside and no one is going to be interested and so on. So yes I do miss that kind of thing sometimes, I think, definitely. (Maurice, Primary School Teacher)

> it's probably quite wearing at first I think, the conversation. My mates talk about football and going to the pub, and you wouldn't have those sorts of conversations; I mean the girls were talking about [sarcastic voice] 'buying a jumper' this morning and more women and girlie things . . . they talk about all sorts, shopping, haircuts and everything [laughs]. (Mick, Human Resource Management)

Such conversations were typically seen as 'banal' and the respondents reported that they either stayed out of them or lowered themselves to join in. Some men openly commented that they missed the 'macho environment', and regretted the restrictions of their normal behaviour that working with women entailed, even for those who did not consider themselves to hold traditional male values, 'I have to be even less macho than I'm not already, if you see what I mean'. Generally this restriction was seen to entail being 'less rude' in terms of language used and an inability to relax as one would in male company.

The restriction of masculinity, for one respondent, went beyond conversation. He reported a restriction on his ability to engage in sexual behaviour in the office environment in a way that he clearly expected to do in a mixed workplace:

> because of the situation I'm in I don't feel that I can really do the flirting, I can be flirted with but I can't do the flirting because it's a very female environment, I'd get a lot of shit for it. (Brian, Clerical Work)

His frustration did not seem to result from the absence of this kind of behaviour, but rather that he was the victim and not the perpetrator of the 'harassment':

> the sexual harassment thing is reversed, you know usually it's male harassing female, right?, though I haven't been harassed [laughs] though in a joking way there is always a threat of it and again I don't know whether that is actually joking you know, I mean if the situations were reversed, would the female consider it joking? (Brian, Clerical Work)

The respondent gave some examples, such as this one:

> and another time someone in the department put their arm around me from behind in a friendly way, and I was really quite shocked, I don't know whether anything was meant by it, sort of like being friendly or whatever, but if the situation had been reversed maybe it would have been a bit inappropriate, yeah. (Brian, Clerical Work)

The fact that this respondent worked with older women seemed to be a limiting factor. These women were seen as non-sexual and therefore not an outlet for flexing his masculinity through flirting. He informed me that he would be more comfortable working with a younger woman 'especially if she's female and attractive'.

In the same way that male sexualized behaviour directed at women is constrained in a predominantly female workplace, so too would be such behaviour between men. This is not a reference to such behaviour between homosexual men but between heterosexual men. One respondent, Peter, who is now embarked in a career as a librarian, recalled his time as a manual worker at a provincial airport:

Peter They [the other workers] used to, they used to . . . I think this is quite common really, talking to other people, they used to shag each other, they used to hold people down and shag them . . . for fun [simulating intercourse] . . . you know they would get hold of someone and oooaaaargh and I think that it is quite common if you start asking about . . . but it was sexual power definitely, holding them down [makes thrusting gesture] messing about like. Christ, I hope this is anonymous . . . they were always touching each other up and pinching each others' bums and things like that, quite bizarre.

Q Were these people who were openly heterosexual?

Peter Yes. . . . they were always going on about how many women they had had. (Peter, Library and Information)

Peter was the last respondent in this series of interviews and therefore I have been unable as yet to corroborate his assertion that such behaviour would be commonly found in some types of all-male workplace. It was Peter's interpretation that this type of sexualized behaviour was a form of exercising power over weaker, or more junior men, a means of reinforcing a dominant masculinity. This is plausible. Peter, unlike John (above), was not a participant in this behaviour and had no difficulty in moving out of the kind of environments in which it took place, though he was clear in his expectation that a number of his erstwhile colleagues would have.

It is apparent from these accounts that typically male work and predominantly male working environments allow for the performance of particular masculine roles. Traditionally female work and predominantly female workplaces do not.

Challenges to Masculinity (2) — Feminization

The second threat to masculinity may be described as a fear of feminization. This took three different forms, each embodying a greater challenge to masculine identity; becoming invisible as a man, being 'adopted' as a women and becoming feminine through working with women.

Becoming invisible as a man was mentioned by more than one respondent. For example one interviewee reported that when 'female conversations' went on in his presence 'you just kind of blend in and get on with it'. Being adopted as a woman was described by other respondents, who talked of themselves becoming 'one of the girls'.

John once you get into it [working with women] they start saying things to you as if you were one of the girls, which is bizarre, and you start talking back to them in the same way . . . you don't notice it happening to you and you do become one of them [the women].

Q And what kind of things are we talking about, can you give me one or two examples?

John well you get into the realm of women's problems, you can get into the realms of a light hearted conversation over a women's page in a coffee break. (John, Human Resource Management)

A different interviewee recognized the same phenomenon, but was distinctly less comfortable with it. In fact, this respondent was very uncomfortable in relating this incident, and stopped the interview for a minute of two while he prepared himself for disclosing it:

> I was walking down the corridor and I bumped into somebody who works [with me] and she didn't look happy and I said to her 'what's the matter' and she said 'it's just my period' . . . and that's like 'bloody hell' and I started thinking . . . And it was just er sort of unusual cos, yeah, women talk about their periods amongst themselves but not to men, yeah, usually (Brian, Clerical Work)

I spent some time trying to tease out the significance of this encounter to the respondent, who was evasive. Eventually I suggested a possible interpretation.

Q Well one interpretation you might put on it, and tell me if this is wrong, is that it was almost that you had become an honorary woman in the office.

Brian Yeah, I think working here, I feel slightly feminized. (Brian, Clerical Work)

This respondent later elaborated, saying that he had a concern in the back of his mind that working in a female environment might rub off on him and that he might have started picking up mannerisms and tones of voice which are typically female, and perhaps more significantly, that he worried that other people may have started to notice. His concern was expressed directly in this remark:

> in the environment I work in I'm worried about losing my male sexual identity you know.

Challenges to Masculinity (3) — Stigmatization

The third challenge to masculinity is the fear of stigmatization by other men for working in a female occupation. All respondents were of the opinion that for certain occupations, where men were in the minority, the men had a reputation for being effeminate or gay, or less masculine, or in other ways stigmatized for taking on a female role. Actors, hairdressers, airline stewards and nurses were mentioned by many, and librarians by one. Only two respondents were prepared to extend this to occupations that they had worked in or considered:

Brian I thought to myself once why not go for a secretarial job in a record company and I thought 'no' they wouldn't employ a male in that role, they would want a doll behind the desk

Q Are there other reasons why you wouldn't take a job like that?

Brian Yeah I'd feel a bit weird in that role, so I'd know that it would usually be filled by a female . . . just like a bloke doing a girl's job.

Q A stigma to it?

Brian Yeah, you know.

Q What kind of stigma is that?

Brian You might be called a 'big girl's blouse', mightn't you? (Brian, Clerical Work)

A second respondent reported his experience as a personnel officer in a retail company, a role almost exclusively occupied by women.

John I know that at Company X, personnel . . . was to be put down and ignored, and anyone who belonged to a personnel department was an inconsequential part of the business . . . so if you were then a male in that female handmaiden environment then you were pretty dismal, a pretty sad state of affairs.

Q Please expand on 'sad state of affairs', what kind of implications are there?'

John The function didn't contribute anything to the business in their [male line managers'] eyes and as a bloke in retail management you should be contributing something to the business, so if you . . . moved into personnel you were in some way deficient and not of any consequence.

Q So was it in a sense even worse than being a women in personnel?

John Yes I was, because you were letting the side down really. (John, Human Resource Management)

As this example illustrates, working in a traditionally female role exposes men to the charge that not only are they something less than a man, but that they are letting other men down.

Together with the degendering of work and feminization, stigmatization of this kind presents a challenge to masculinity. In the next section I will examine the ways in which men seek to meet these challenges.

Negotiating Masculinity Under Challenge

In dealing with these challenges, two very different strategies were used by the men that I talked to. The first was to redefine the job to fit more closely with an acceptable conception of masculinity. The second was to adapt one's masculinity, or one's perception of it, compromising it to work with women and accepting, and even promoting, that compromise.

Reconstructing the Occupation

There is some evidence in the sociological literature of how men 'reconstruct' their job to avoid perception of its negative associations with women. Pringle (1993) showed how male secretaries engaged in this kind of activity. First, they tended not to call themselves secretaries, but administrative assistants or another title without an overtly feminine association. Second, they tended to emphasize in discussion those tasks in the job which might be considered 'feminine', for example making little or no reference to typing while emphasizing the planning and organizing aspects of the job. Third, Pringle found evidence that male secretaries actually reconstructed their jobs so that they spent more time on the 'masculine' aspects and less on the 'feminine'. Evidence for all three of these approaches was found; an example of the first strategy is demonstrated in this exchange:

Malcolm I never sort of said I was a secretary. I sort of said I was a temp who did a lot of secretarial work.

Q Why did you not say that you were a secretary?

Malcolm Well for one thing there were some jobs that I did that involved more than being a secretary. Sort of admin work.

Q Were there women working alongside you doing the same thing?'

Malcolm Yeah

Q And how would they describe their job title?

Malcolm They would describe themselves as secretaries, yeah. And I suspect it goes back to being a woman's job, so I think it [being a secretary] did affect me really, but not much. (Malcolm, Clerical Work)

The second and third approaches were exemplified by the remarks of the following two clerical workers.

Jim described his job as a trouble shooter, 'whatever comes into the office I try and sort out' and then emphasized the marketing aspects of his job. Later the job was described as less glamorous, 'I suppose the day is sat behind a computer processing documents'. (It is interesting that the neutral word 'processing' is used here and not 'typing' which may be more clearly associated with women.) When asked about the effect of being employed in a predominantly female role, Jim said that he felt the disadvantage of suffering from '*the general perception by people is that it is a glorified secretarial position . . .*'.

Brian had a job that combined some marketing and some basic administrative and secretarial duties. He was keen to emphasize the marketing aspects and felt that the other areas were a waste of his skills and his time '*filing, putting letters in envelopes . . .*'

However, the strength of the need to reconcile masculine identity with role identity in the face of challenge, and range of rationalizations that some men employ to do so are particularly well-illustrated by the comments of the following interviewee, Mick. This man was engaged on a postgraduate course leading to entry to the personnel/human resources profession. Both the course and the profession are dominated, numerically, by women. This association with women stems from personnel's origins as a welfare function and has proved difficult to dislodge despite the more managerial and strategic orientation of the modern-day function.

This interviewee commenced by emphasizing the 'hard' edge of personnel work, and was anxious to play down its soft (feminine) image. Elaborating on the perception that personnel might be a soft option he stated:

Mick it's not soft at all, it is probably one of the hardest jobs to do in the organization telling people that they are going to be made redundant, I think that when people don't do the horrible jobs and don't take the hard decisions, then that is what undervalues the profession. (Mick, Human Resources Management)

By stressing the 'hard' masculine aspects of the job, this respondent then recognized the need to square this with the fact that so many women entered the profession. His rationalization was that women were not clear what the job really entailed:

Mick well I think a lot of them thought it was about the nicer jobs, one of the girls came up to me on the first day and said personnel is about 'the fluffy bits' and I was like 'oh my god' having come from a manufacturing background it's not really the nicer things, there is a lot of clamping down on people, sacking them, giving people a hard time that are off ill . . . (Mick, Human Resource Management)

His rationalization seemed to be that it was not that he was entering a job which is appropriate for women, but that women are mistakenly entering a career which is suitable for masculine men such as him:

Mick I was kind of thinking that people on the course didn't really know what it was that they were doing. I wasn't doubting my motives for being here, it was more the other people, do they know what they are doing here.

Q The women?

Mick Yeah the women, do they know much about the course, the profession, really?

Another explanation that he used was that women, unlike him, were seeking lower-level personnel jobs:

Mick I knew that there was a lot of women who were kind of personnel assistants and personnel advisors, but I never really saw myself as being in that role.

Here the rationalization was designed to play down gender incongruity by suggesting that men (particularly him) and women were doing different things in the profession. He later continued this line of argument when he distanced himself from the 'softer' training and development specialities in the profession. His third strategy to bolster his view of his masculinity was to question women's competence. Women were seen as 'not as tough' as interviewers and not determined enough to get to the root of problems:

Mick ... they [women] are not cut out to take the decisions ... It's a lot to do with business, and managing a resource, and perhaps men are just better at that, I don't know and maybe although all the nice things and doing things tactfully, may be the right ethical way to do it, maybe it is not the way to get business profits and maybe women don't run the business like that.

Furthermore the presence of women was seen as devaluing the profession. The personnel function was regarded as 'soft' and lacking in respect as a result:

Q Do you think the profession would be different if there were more men in it?

Mick Yes I think that it would probably command more respect within business.

The theme that women somehow 'drag down' the profession was a recurrent one in this discourse:

Mick a lot of personnel professionals aren't good at their jobs, and perhaps they tend to be women ... there are proportionately more bad personnel officers that are women.

This respondent's inability to find employment in a professional post was also construed as relating to the female control of the profession:

Mick if you go into a personnel department there is always four girls, a corps of administrators and the culture of women and they try and recruit more women and you have got a sort of closed shop ... there's no way of getting in.

He then described how he was contemplating an alternative route to the top in personnel which would avoid the barriers and stigmatization associated with women's domination of the profession, and their apparent unwillingness to employ him:

Mick I have thought of going into operations and using my qualifications as a way of getting across, the only way of getting in.

This man's strategy for meeting the challenge posed to his masculinity by his career choice, was to cling to his masculine identity and to reconstruct the identity of the career around it. Another radically different, approach also emerged and this involved renegotiating masculinity to match the challenge of the job.

Renegotiating Masculinity

In addition or instead of renegotiating the occupation, there was also evidence of renegotiating masculinity, tailoring it to working within a female environment. There were different, sometimes conflicting, perceptions amongst the respondents as to what this involved:

> I've just found that by taking the piss out of it [women's conversation or behaviour] or whatever and deliberately being overly macho, in a flirtatious way, that is a good way and that is the way I have sort of dealt with it. (Maurice, Primary School Teacher)

Other respondents felt that a more appropriate way to deal with working in a female environment was to play down overtly 'masculine' behaviour.

> Well I don't go round saying 'alright darling' and all that, you know, whistling from scaffolding . . . (Brian, Clerical Worker)

Others went further and described incorporating, at the level of presentation, aspects of 'female' behaviour into their social repertoire:

> and you get into daft things like they say 'we hate men' . . . and I say 'I'm glad I'm not one of them [a man] if I see one I'll warn them' . . . and you are accepted as one of them and they talk to you about their specific problems, whatever they may be, physical, mental or boyfriend. (John, Human Resource Management)

Interestingly, given his comment above, this was also an approach adopted by Maurice:

> I play up to the campness as well, like deliberately joining in the conversations like when they [women teachers] go 'Ugh, stockings' I go [camp voice] 'yes I can never get stockings to fit'. (Maurice, Primary School Teacher)

Maurice was employing two contrasting approaches to managing his identity in a female environment, one which reinforced his masculinity and one which allowed an acceptance into a female-dominated workplace. I put this to Maurice:

Q It is almost, as if it is, it is almost a conscious strategy?'

Maurice Oh I think so definitely yeah, and it is almost conscious but it is not conscious. (Maurice, Primary School Teacher)

Alongside this awareness amongst the respondents that there was some work to be done in managing their masculinity there was also a perception that an ability to work effectively in a female occupation could be regarded as a positive aspect of masculinity, and one which not all men were capable of embracing. There was a view expressed that some men can handle working with or for women and other men cannot.

Interviewees tended to contrast their version of masculinities with those, which in their eyes, were less developed:

> not myself so much, but certain people I know some friends of mine have very definite ideas about what are women's roles and whatnot and I think you'll find certainly the idea of working for a woman quite hard to take. (John, Human Resource Management)

The same respondent spent some time explaining the kind of behaviour that 'worked' in a female workplace, and he defined it in relation to other masculinities:

> and men resent, dislike . . . resent working with women, resent taking orders and directions from women, cos they can't get over this gender thing which surprised me to some extent in this day and age, but there is a lot of it out there . . . and another [problem] is that they have a problem giving direction to women . . . not appreciating that there are differences between men and women . . . not being aware of the nuances of the environment. (John, Human Resource Management)

Indeed, John along with a number of these respondents felt that his masculinity was better suited to a predominantly-female work environment:

Q you said that you were adept at working in a female environment, do you think that you would be less adept working in a male . . .

John Yes I have felt that it's difficult . . . once you get used to working in that [female] environment it can be difficult to adjust back the other way . . . you wouldn't be relaxed in a male environment, and you find you can talk straight to a man, you can say you don't like it do it right and why, but with a woman sometimes you have to be a little more circumspect and if you were that way in a male dominated environment, I don't think that they would respect that.

Q Do you think that you would have adapted if you had been on the shop floor in a very macho environment, would that be for you?

John I think that I would possibly struggle with that, I would maybe find that quite intimidating, I don't know.

It is clear that some men are able to develop a particular type of masculinity which enables them to move out of traditionally-male occupations, and still remain simultaneously comfortable with their maleness, their job and their working environment.

Discussion

The study shows that men who enter female occupations encounter three types of challenges to masculinity. First, working in a female occupation limits their ability to use work identity to confirm masculine identity, and working with women restricts opportunities to reinforce masculinity in the work environment. Second, they face a fear of feminization through working in a predominantly female environment, and third, they face the threat of being stigmatized as effeminate or homosexual by virtue of undertaking a role which is sex-typed by society as female. Might these not be regarded as directly equivalent in nature to the challenges that women face on entering traditionally male occupations? There are arguments to suggest that male experience may be qualitatively different. First, as Bradley has argued 'compromised femininity is still a possible female identity' (1993, p. 14), women who establish themselves in the male work environments are often admired or respected, and by adopting stereotypical male behaviours do not call their core identity as women into question. Second, as masculinity, particularly its hegemonic variety, confers such advantages both within and outside the workplace, men may face a particular imperative to preserve it.

The paper illustrates that there are two very different ways in which men deal with these challenges. The first of these is to reconstruct the occupation, either at the level of presentation or in reality, so that it is more congruent with their notions of masculinity; the second is to express and enact a different masculinity. These can be regarded as processes of realignment. When men enter 'female' occupations, it can be argued that their masculinity and their occupational identity are misaligned, and that work is required to bring them back into line.

It is important to consider the implications of the different strategies that men adopt to achieve this. The approach of reconstructing work has the effect of re-enforcing the stereotypical views of appropriate male and female work that cause it to be 'necessary' in the first place. Men who concentrate on the marketing or administrative aspects of their secretarial or clerical jobs perpetuate the notions of what work is appropriate for men and what is not. Moreover they leave the other less interesting and rewarding tasks for women to do. More importantly men's need to maintain masculinity can have important effects on the nature of jobs and the way they are done. The nature and definition of jobs, occupations and professions are not fixed in the order of things but are negotiated by political processes and through the exercise of power. We can see the implications of reconstructing the job around masculinity in the example of the prospective personnel manager, Mick. This individual seeks to create a personnel role which is congruent with his masculinity. As was noted by a different respondent, real men do not work in personnel, to do so was to be a 'sad state of affairs' and to 'let the [male] side down'. Faced with this challenge, personnel management has to be redefined so that it allows masculinity to be acted out in its practice. It becomes a tough and hard occupation, with an emphasis on trouble-shooting, disciplining and sacking people, identifying shirkers and malingerers and dealing with them. Other aspects of personnel management which do less to buttress this man's view of his masculinity, such as training, development, culture change, empowerment, involvement and ethical employment practices are played down or ignored. It is interesting, and perhaps significant, that this man's father is a personnel manager. In a revealing story the son describes his father taking on responsibility at work for a new part of the factory, regaining control and sacking five people. The father returned from work and described himself to his son as the 'new sheriff in town'. Personnel management offers the opportunity to be masculine, in the hegemonic sense, but only if construed in this kind of way with its connotations of control and power.

This logic, paradoxically given that they are in the majority, leaves no place for women in the profession. Women's presence needs to be rationalized away, either women have made the wrong career choice, are incompetent or are only appropriately employed at the lower levels. A second difficulty is for men to find an appropriate way to the top of the profession without being tainted by the female associations at the lower levels. Other routes, via male-dominated areas of management, are followed. This is one individual's construction of personnel management, and it is not possible to say that it represents a wider trend. However the reader may not be surprised to learn that men are (in fact) disproportionately represented at higher levels in the profession, women tend to be over-represented in training and under-represented in employee relations and strategic management, and that men often enter senior personnel roles from other management specialities in many cases without the professional grounding (Long, 1984; Legge, 1987). It is also consistent with this argument that writers on personnel management have argued that the 'hard' version of the new HRM has gained primacy over the 'soft' (Legge, 1995). The way in which occupations are constructed to achieve alignment with the accepted gender roles of those who undertake them has implications for the way work is organized and undertaken.

The second process of realignment identified in the study is to bring masculinity into line with the occupation. It is possible, and indeed likely, that many men who enter female-dominated occupations bring with them a less 'macho' masculinity into this work. Some of the men in the study perceived themselves as subscribing to a version of masculinity which was not predicated on notions of control, domination, and definition of female as 'other'. However there was also evidence that some men developed or worked at an expression of masculinity which would allow them to operate comfortably in a female domain. What is significant is that these men were able to do that without losing a sense of maleness. This masculinity was seen as positive by these men, and in fact viewed rather more positively than the 'hegemonic' type described above.

The fact that occupations remain highly segregated by gender has many implications for society and the economy. In the context of the widespread, long-standing and continuing disadvantages experienced by women in paid work, any discussion of men's experiences within this system of gender relations must retain a sense of perspective. However, a neglect of men's experience can lead to partial explanations. A health service spokesman recently argued on television that the recruitment crisis in nursing could be attributed to the fact that today's young women have a wider range of career opportunities than was the case in previous generations. The possibility that a contributory cause might be the fact that this occupation is effectively excluded to half of the population had clearly not occurred to him. The process of desegregation of labour markets not only requires that women enter male-dominated occupations, but that men enter traditionally-female work domains. We have seen in this paper that the association of work with traditional notions of masculinity can create strong disincentives and difficulties for those wishing to move across the traditional work/gender boundaries. Progress towards labour markets which are significantly less segregated by gender will require many social changes; the emergence and acceptance of masculinities which allow the coexistence of positive male identities with non-traditional work will need to be one of them.

Notes

1. *Source*: IPD membership figures, graduate level. Nov 1997; Labour Force Survey, March 1997 (rounded).

References

Allan, J. (1993). 'Male Elementary Teachers'. In: C. Williams (ed.), *Doing 'Women's Work': Men in Nontraditional Occupations.* Sage, London.

Anker, R. (1997). 'Theories of occupational segregation by sex: an overview', *International Labour Review*, 136(3), pp. 315–339.

Barrett, F. (1996). 'The Organizational Construction of Hegemonic Masculinity: The Case of the US Navy', *Gender, Work and Organization*, 3(3), pp. 129–141.

Bradley, H. (1989). *Men's Work, Women's Work.* Polity Press, Cambridge.

Bradley, H. (1993). 'Across the Great Divide'. In: C. Williams (ed.), *Doing 'Women's Work': Men in Nontraditional Occupations.* Sage, London.

Casey, C. (1995). *Work, Self and Society: After Industrialism.* Routledge, London.

Carrigan, T., R. Connell and J. Lee. (1985). 'Toward a New Sociology of Masculinity', *Theory and Society*, 14(5), pp. 551–604.

Cheng, C. (ed.) (1996). *Masculinities in Organisations.* Sage, Thousand Oaks CA.

Cockburn, C. (1988). 'The Gendering of Jobs'. In: S. Walby (ed.), *Gender Segregation at Work.* Open University Press, Milton Keynes.

Collier, R. (1998). ' "Nutty Professors", "Men in Suits" and "New Entrepreneurs": Corporeality, Subjectivity and Change in the Law School and Legal Practice', *Social and Legal Studies*, 7(1), pp. 27–53.

Collinson, D. and J. Hearn (1994). 'Naming Men as Men: Implications for Work, Organisation and Management', *Gender, Work and Organisation*, 1(1), pp. 2–20.

Connell, R. (1987). *Gender and Power.* Polity Press, Cambridge.
Connell, R. (1992). 'A very straight gay: masculinity, homosexual experience, and the dynamics of gender', *American Sociological Review*, 57, pp. 735–751.
Floge, L. and D. Merrill (1986). 'Tokenism Reconsidered: Male Nurses and Female Physicians in a Hospital Setting', *Social Forces*, 64(4), pp. 925–947.
Fuller, P. (1996). Masculinity, Emotion and Violence. In: L. Morris and E. Lyon (eds), *Gender Relations in Public and Private: New Research Perspectives.* Macmillan, Basingstoke.
Hakim, C. (1992). 'Explaining Trends in Occupational Segregation: The Measurement, Causes, and Consequences of the Sexual Division of Labour', *European Sociological Review*, 8(2), pp. 127–152.
Heickes, E. (1992). 'When Men are the Minority: The Case of Men in Nursing', *The Sociological Quarterly*, 32(3), pp. 389–401.
Kanter, R. (1997). *Men and Women of the Corporation.* Basic Books, New York.
Kerfoot, D. and D. Knights (1993). 'Management, Masculinity and Manipulation: From Paternalism to Corporate Strategy in Financial Services in Britain', *Journal of Management Studies*, 30(4), pp. 659–77.
Kerfoot, D. and S. Whitehead (1998). ' "Boys own" stuff: Masculinity and the management of further education', *Sociological Review*, 46(3), pp. 436–457.
Kvande, E. (1998). 'Doing Masculinities in Organizational Restructuring'. Paper presented to the Gender, Work and Organisation Conference, Manchester, January.
Legge, K. (1987). 'Women in Personnel Management: Uphill Climb of Downhill Slide?'. In: A. Spencer and D. Podmore (eds), *In a Man's World: Essays on Women in Male-Dominated Professions.* Tavistock, London.
Legge, K. (1995). *Human Resource Management: Rhetorics and Realities.* Macmillan, London.
Long, P. (1984). *The Personnel Specialists: A Comparative Study of Male and Female Careers.* IPM, London.
Marshall, J. (1995). *Women Managers Moving On.* Routledge, London.
Mills, A. (1998). 'Cockpits, Hangars, Boys and Galleys: Corporate Masculinities and the Development of British Airways', *Gender, Work and Organisation*, 5(3), pp. 173–188.
Morgan, D. (1992). *Discovering Men.* Routledge, London.
Pringle, R. (1993). Male Secretaries. In: C. Williams (ed.), *Doing 'Women's Work': Men in Nontraditional Occupations.* Sage, London.
Reskin, B. and H. Hartmann (1986). *Women's Work, Men's Work: Sex Segregation on the Job.* National Academy Press, Washington.
Sheppard, D. (1989). 'Organisation, Power and Sexuality: The Image and Self-Image of Women Managers'. In: J. Hearn, D. Sheppard, P. Tancred-Sherriff and G. Burrell (eds), *The Sexuality of the Organisation.* Sage, London.
Spencer, A. and D. Podmore (eds) (1987). *In a Man's World: Essays on Women in Male-Dominated Professions.* Tavistock, London.
Thompson, P. and D. McHugh (1990). *Work Organisations.* Macmillan, London.
Whitehead, S. (1999) 'Hegemonic Masculinity Revisited' (Review Article) *Gender, Work and Organization*, 6(1), pp. 58–62.
Williams, C. (ed.) (1993). *Doing 'Women's Work': Men in Nontraditional Occupations.* Sage, London.
Williams, C. (1995). *Still a Man's World: Men who do Women's Work.* UCP, Berkeley.
Witz, A. (1992). *Professions and Patriarchy.* Routledge, London.

48

The ASPIRe Model: Actualizing Social and Personal Identity Resources to Enhance Organizational Outcomes

S. Alexander Haslam, Rachael A. Eggins and Katherine J. Reynolds

Source: *Journal of Occupational and Organizational Psychology* 76 (1) (2003): 83–113.

In recent years, there has been a growing recognition of the fact that organizational success is not purely an economic issue — a question of financial capital. In particular, while an economic approach to organizational life (e.g., after Taylor, 1911) has suggested that management decisions needed to be guided by reverence for the financial bottom line, researchers are increasingly emphasizing the dangers of a failure to recognize and harness an organization's *social capital* (Burt, 1997; Leana & van Buren, 1999; Nahapiet & Ghoshal, 1998). Social capital can be defined as those resources inherent in the network of alliances and relationships within a workforce that contribute to, amongst other things, an organization's reputation, its members' *esprit de corps*, their loyalty and commitment.

An interest in these social aspects of organizational life and their contribution to organizational outcomes started around the middle of the last century with the emergence of the human relations school (after Mayo, 1933, 1949). However, it has recently been revitalized as (a) managers have become disillusioned with organizational practices that continue to rely on an economic analysis (e.g., downsizing, deskilling, delayering; e.g., see Micklethwait & Wooldridge, 1997; Mintzberg, 1996; Pfeffer, 1997; Thompson & Warhurst, 1998) and (b) researchers have developed theoretical and practical tools that allow them to elaborate upon the principles and processes that were alluded to — but not formalized or fleshed out empirically — in human relations work (e.g., see Haslam, 2001, for a review).

In this paper, we present a model for identifying and utilizing social capital that draws upon one such development. Based upon insights from social identity and self-categorization theories (e.g., Tajfel & Turner, 1979; Turner, Hogg, Oakes, Reicher, & Wetherell, 1987), the model is informed by work which suggests that an organization's social capital is partly determined by the *identity resources* of its employees. In particular, it focuses on two distinct resources: those associated with each employee's sense of their unique *personal identity* and their shared *social identity*. The former resource relates to employees' internalized definitions of themselves as individuals and the latter to their awareness that they are members of a common group (Tajfel, 1972; Turner, 1982). The model (and the research upon which it is predicated) suggests that appropriate identification and mobilization of these identity resources is a necessary component of organizational success. However, before outlining the model, we need to clarify some of its core theoretical underpinnings.

Social and Personal Identity

The distinction between personal and social identity developed from an early insight by Tajfel (1972, 1978) that, although they were ordered upon a continuum, there was a qualitative distinction between interpersonal and intergroup behaviour. Specifically, Tajfel argued that intergroup behaviour was associated with *social identity* — an individual's "knowledge that he [or she] belongs to certain groups together with some emotional and value significance to him [or her] of the group membership" (1972, p. 31). Together with empirical evidence from the so-called "minimal group studies" (Tajfel, Flament, Billig, & Bundy, 1971), in which individuals who had been randomly assigned to meaningless groups were found to favour members of their own ingroup to another outgroup, this assertion formed the basis for *social identity theory* (Tajfel & Turner, 1979). A core tenet of this was that, having defined themselves in terms of a particular social identity, individuals act to maintain or enhance the *positive distinctiveness* of the group with which that identity is associated. So, for example, once a person defines themselves as an employee of workgroup X, they should seek to maintain or increase the prestige, exclusivity and stature of that workgroup — particularly in relation to others with which it is compared (e.g., workgroups Y and Z).

In the 20 or so years since this prediction was first made, it has received an enormous amount of empirical support (e.g., see Brewer, 1979; Ellemers, Spears, & Doosje, 1999; Hogg & Abrams, 1988; Oakes, Haslam, & Turner, 1994; Turner *et al.*, 1987). When a given social identity is salient, individuals are thus found to act in ways that serve to advance group interests — striving to make their ingroup better than, and different from, salient outgroups — often at the expense of their own personal interests. These motivations have also been observed across a range of organizational contexts. So, where social identity is defined by membership of a given organization or work-team, the same desire to advance that unit's interests is apparent. Amongst other things, this leads to increases in (a) liking for the relevant organizational ingroup (Brown, 1978; Dutton, Dukerich, & Harquail, 1994; Terry & Callan, 1998), (b) organizational citizenship (Ouwerkerk, Ellemers, & de Gilder, 1999; van Knippenberg, 2000), (c) willingness to contribute to collective goals (Ellemers, de Gilder, & van den Heuvel, 1998; Tyler, 1999; Tyler & Blader, 2000), (d) collective action (Kelly & Kelly, 1994; Veenstra & Haslam, 2000), and (e) group productivity (James & Greenberg, 1989; Worchel, Rothgerber, Day, Hart, & Butemeyer, 1998).

Elaborating on these ideas, a more complete explanation of individuals' movement along Tajfel's interpersonal-intergroup continuum was provided by Turner (1982) in the process of developing *self-categorization theory* (Turner,

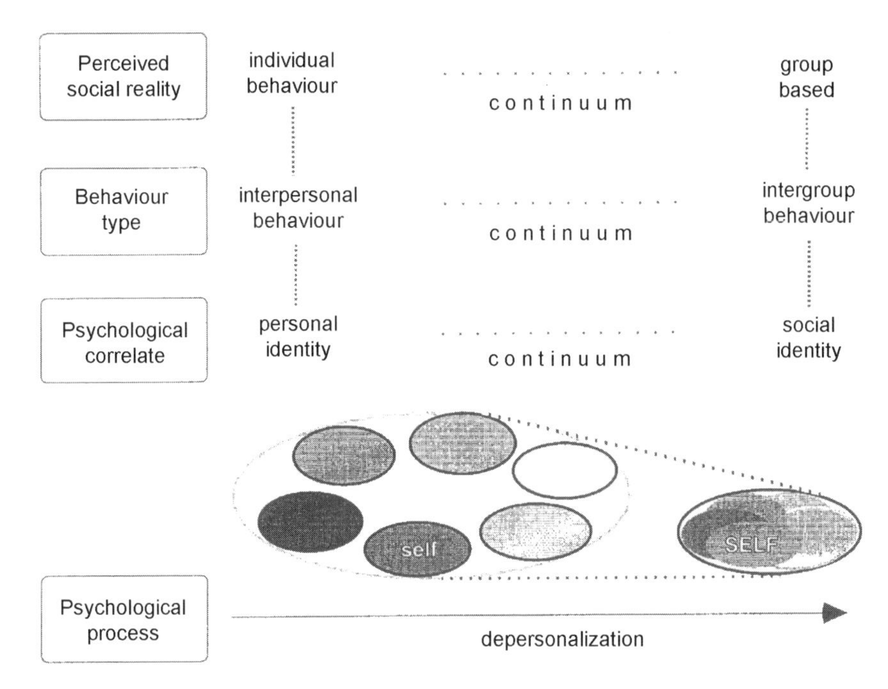

Figure 1: Variation in self-categorization as a function of depersonalization. self = self as unique individual with personal identity salient; SELF = self as interchangeable group member with social identity salient. The shift from self to SELF is produced by depersonalization (self-stereotyping).

1985; Turner *et al.*, 1987). This hypothesizes that an individual's self-concept can itself be defined along a continuum ranging from definition of the self in terms of personal identity to definition in terms of social identity. Moreover, it is proposed that the functioning of the self-concept is the cognitive mechanism that underpins the behavioural continuum described by Tajfel (1978). As Fig. 1 indicates, interpersonal behaviour is associated with a salient personal identity and intergroup behaviour with a salient social identity.

The psychological process that underpins social identity salience was referred to by Turner (1982) as *depersonalization*. This is a process of *self-stereotyping* through which the self comes to be perceived as categorically interchangeable with other ingroup members. So, when social identity is salient, it is predicted that individuals come to see other ingroup members as part of the self (redefining the self as 'we' rather than 'I'). Through this process, group members perceive similarities between their previously idiosyncratic perceptions, motivations, values, and goals. Moreover, they are more likely to see each other as valid sources of social influence who are qualified to provide valid information about identity-relevant features of social reality (Turner, 1987, 1991). For example, if membership of Company X becomes a salient social self-category for a particular employee, they should see other members of that company as important sources of information about work-related activity and turn to them for advice and guidance.

As well as this, when people who define themselves in terms of the same social identity disagree, they seek actively to reconcile discordant attitudes and actions. They can do this by redefining their identities, reinterpreting the social world, or engaging in *mutual social influence* (Turner, 1987; for examples, see Haslam, Turner, Oakes, Reynolds, & Doosje, 2003). In this way, social identity-based perceptual and interactive processes combine to improve communication and produce co-ordinated collective behaviour.

Over the last 15 years these predictions have also been subjected to rigorous empirical testing and they too have received broad support (e.g., for reviews see Ellemers *et al.*, 1999; Spears, Oakes, Ellemers, & Haslam, 1997). And again, much of this work has been conducted in the organizational domain. In these contexts, research has shown that factors which increase social identity salience (e.g., a history of collective interaction, outgroup threat, intergroup comparison or conflict) lead to (a) more homogeneous representations of relevant organizational ingroups (Oakes, Haslam, Morrison, & Grace, 1995), (b) greater trust (e.g., Kramer, Brewer, & Hanna, 1996), (c) better communication (Dovidio *et al.*, 1997; Postmes, 2003; Postmes, Tanis, & de Wit, 2001; Suzuki, 1998), and (d) improved co-operation (Kramer, 1993; Tyler & Blader, 2000; for reviews and discussion, see Ashforth & Mael, 1989; Haslam, 2001; Peteraf & Shanley, 1997).

In essence, then, social identifications can be thought of as particularly potent forms of social capital in the sense implied by social and organizational researchers. As Turner (2001, p. x) puts it, "the fact that individuals are able to act as both individuals and group members is a plus, adding immensely to the sophistication and possibilities of our social relationships". Yet, to date, little consideration has been given to the way in which this capital can be *developed*, *utilized* and *sustained* through organizational practice. How can the interests and perspectives of individuals be married with those of groups? How might conflict between groups whose members define themselves in terms of distinct identities be productive? How does an organization encourage social identification without making social identities stultifying and all consuming? In this paper, we attempt to address such questions within a proposed model for actualizing identity resources in organizations.

The ASPIRe Model

As suggested above, the theoretical and empirical grounding of social identity and self-categorization principles — in particular, those that rest on the distinction between personal and social identity — are well-established. As a result, these principles are now central to contemporary thinking in social psychology (e.g., see McGarty & Haslam, 1997; E. R. Smith & Mackie, 2000) and are having an increasing impact in the organizational domain (e.g., see Albert, Ashforth, & Dutton, 2000; Haslam, 2001; Haslam, van Knippenberg, Platow, & Ellemers, 2003; Hogg & Terry, 2001). In particular, this is because, in both disciplines, they have provided a productive framework for hypothesis generation and organizational analysis (e.g., see Ashforth & Mael, 1989; Dutton *et al., 1994;* Haslam, 2001; Hogg & Terry, 2000; Peteraf & Shanley, 1997; Turner & Haslam, 2001).

Yet, the failure systematically to integrate the insights of the theories within a programme of strategic organizational activity has meant that while the social identity approach points to the benefits (and shortcomings) of a range of organizational practices (e.g., participative decision making, distributed leadership, team-building), its practical implications remain to be both formalized and fully realized. The model proposed in this paper represents an initial attempt to redress this situation by integrating insights from research that has been conducted on a range of core organizational topics within a coherent model of organizational practice. Amongst other things, this points to ways in which social and personal identity resources can be realized and harnessed with a view to improving diversity management, employee well-being, and sustainable productivity.

As can be seen from the schematic representation in Fig. 2, the model has four temporal phases. Each of these takes place in the context of a superordinate identity that all the involved employees recognize and share (e.g., as members of a particular organization or organizational subunit). The phases move from an initial stage that ascertains which social identities employees use to define themselves (*AIRing*), through intermediate stages in which relevant subgroups and then the organization as a whole establish goals that are relevant to those identities (*Sub-Casing, Super-Casing*), to a final phase in which organizational planning and goal-setting is informed by the outcomes of the previous two phases (*ORGanizing*). Subsequent to this, employees' satisfaction with, and commitment to, the outcomes of the process are monitored as well as the organization's performance relative to emergent goals. The results of this then provide a basis for future iterations of the ASPIRe process.

The model is underpinned by an assumption that social and personal identities make distinct contributions to organizational life and that opportunities for organizational success and sustainability are maximized when structures allow for the expression and development of concerns and interests associated with each. Accordingly, the ASPIRe model incorporates distinct phases in which

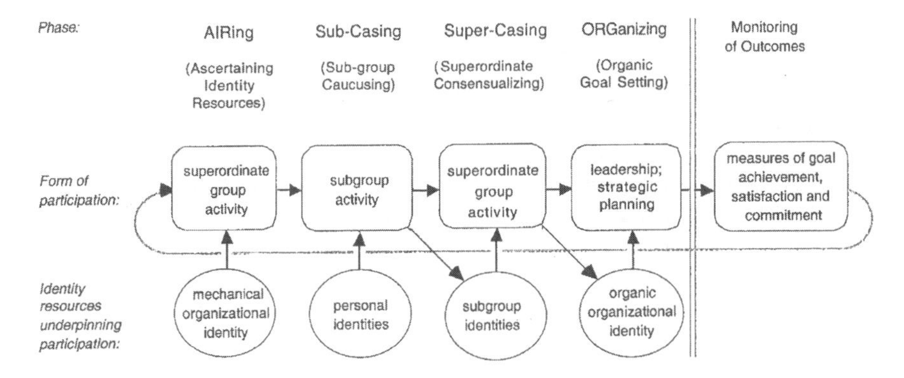

Figure 2: Actualizing Social and Personal Identity Resources: the ASPIRe model.

personal identities first contribute to the development of a subgroup social identity (in Sub-Casing), and then subgroup social identities contribute to the development of a superordinate organizational identity (in Super-Casing). The model is also based upon an assumption that because individual and social differences are an important feature of organizational life (e.g., see Jackson, 1992; Nkomo & Cox, 1996), a viable model also has to allow for the possibility of clashes of opinion at each of these levels. However, in order to harness the creative and innovative potential of such differences, successive phases of the ASPIRe process attempt to encourage the translation of lower-level heterogeneity into higher-order homogeneity. This strategy is consistent with the view that organizational success depends in part on an ability (a) to accommodate individual and group differences within a socially shared theory of the organization and then (b) to use that theory as a basis for action (along lines suggested by Lant & Baum, 1995; Moreland, Argote, & Krishnan, 1996; Weick & Roberts, 1993). In the following sections, we consider in more detail the content and rationale for each of the phases of the model.

Ascertaining Identity Resources (AIRing)

A central tenet of the social identity approach is that people's behaviour is guided by membership of particular personal and social categories only to the extent that they have internalized them as part of their self-concept (Turner, 1982). Psychologically, then, groups are defined not by the demographic features of their members (e.g., as women, managers, university graduates), but by the extent to which any such property forms a basis for shared self definition (a sense of 'we-ness'). This means that, psychologically, a group only exists if it provides its members with a sense of shared social identity (e.g., as 'us women', 'us doctors', 'us Americans'). By the same token, a sense of individuality (a sense of 'I-ness') derives psychologically from a person's sense that they have a personal identity that is uniquely theirs and that is *not* shared with other ingroup members (Turner & Onorato, 1999).

This being the case, it follows that if organizations are to capitalize upon people's social identities in order to enhance organizational outcomes (e.g., along lines suggested by Ellemers *et al.*, 1998; Haslam, 2001; Hogg & Terry, 2000; Tyler & Blader, 2000), the group memberships that are involved and mobilized in this process need to be seen by employees as *self-relevant* and *self-defining*. As Jackson (1992) notes, one of the problems associated with attempts to manage diversity in the workplace is that managers often *assume* that individual employees define themselves in terms of particular group memberships when in fact they do not (see also Brewer, 1995; Markoczy, 1997). Similarly, Peteraf and Shanley (1997, p. 182) observe that "a critical problem with the strategic groups literature to date has been a focus on clustering without attention to the actual relationships among group members".

Mismatched categorization of this form is particularly likely to occur when employees have a demographic profile that appears to qualify them for membership of a minority or disadvantaged group (e.g., as women, migrants, or aged). In part, this is because social identity theory predicts that when both (a) intergroup boundaries appear to be permeable and (b) individuals have a social mobility belief system, members of minority or low-status groups will seek actively to avoid being categorized in terms of these disadvantaged social identities and will instead attempt to 'pass' into the majority or high-status group (Tajfel & Turner, 1979, p. 35; see also Ellemers, van Knippenberg, & Wilke, 1990; Ellemers, Wilke, & van Knippenberg, 1993).

As a corollary of this point, it is also the case that individuals can self-categorize themselves in terms of group memberships that have no obvious or routinely accessed demographic basis. This is because there are as many possible social self categorizations as there are dimensions of human differentiation. In principle, then, a middle-aged Hispanic female production-line worker is as capable of defining herself in terms of her reading ability, her career aspirations, or her political views, as she is of defining herself in terms of any of her distinctive demographic attributes. Consistent with this view, Oakes and colleagues have shown that the critical determinant of self-category salience is not distinctiveness but the capacity for a given social category meaningfully to describe subjectively relevant features of social reality (Oakes & Turner, 1986; Oakes, Turner, & Haslam, 1991; Oakes *et al.*, 1994; see also McGarty, Haslam, Turner, & Oakes, 1993; Reynolds, Turner, & Haslam, 2000).

However, precisely because any individual worker has access to multiple social identities — many of which are not at all relevant to the organization's core activity — the AIRing process needs to be framed by sensitivity to the organization's broad agenda. We therefore suggest that this phase is best conducted at the level of the relevant superordinate organizational unit where it is likely to be informed by an existing (relatively undifferentiated or mechanical) organizational identity. This might be achieved, for example, by holding a general meeting of employees or administering an organization-wide survey whose rubric makes explicit reference to organizational identity (e.g., by alluding the organization's history, its core activities, its purpose and mission, its competitors; cf. Deal & Kennedy, 1982; Lant & Baum, 1995).

Whatever its precise form, the basic goal of AIRing is to identify those self-categorizations that are perceived by individual employees as most relevant to their ability to do their work and to distinguish these from identities that are *not* perceived to be self-relevant. Relevant identities, we suspect, would differ dramatically not just between organizations whose core business is quite different (e.g., a bank and a car rental company) but also between different organizations with the same core business (e.g., an American bank and a British bank) and even the same business at different times (e.g., a British bank the 1970s and the same bank in the 1990s).

Such variation is predictable and meaningful, since what defines a group will change as a function of the groups with which it compares itself and the broad social context in which it is located (Gioia, Schultz, & Corley, 2000; Haslam & Turner, 1992: Oakes *et al.*, 1994; Turner, 1985). For example, psychologists' definition of themselves as 'scientific' has been found to be more important in contexts where they compare themselves with actors than in those where they compare themselves with physicists (partly because being scientific contributes to a sense of positive distinctiveness in the former case but not the latter; Doosje, Haslam, Spears, Oakes, & Koomen, 1998). Similarly, the meaning of banking (and hence what it means to be a banker), should vary as a function, say, of whether banks are compared with schools or to shops. The core challenge, then, for organizations that want to work with identity resources is to be sensitive to such variation and to avoid prejudgement, unilateral imposition or reification of the identities in question (Reynolds, Eggins, & Haslam, 2001). Amongst other things, this is because (a) sensitivity of this form helps to communicate an organization's commitment to *procedural justice* (Tyler & Blader, 2000) and (b) the nature and meaning of employees' identities should change as the ASPIRe process progresses (Eggins, Reynolds, & Haslam, 2003, Haslam, 2001; see Fig. 1). Indeed, its success depends in part upon the organization's ability to allow this to happen.

The main outcome associated with the AIRing phase of the ASPIRe process is knowledge of the social identities that employees perceive to be relevant to their work-related activity (and of those that are irrelevant) and of the contours of those identities within the organization (cf. Lant & Baum, 1995; Lau & Murningham, 1998). At the end of this phase, employees and managers should have a sense of the key groupings that are psychologically meaningful within a particular workforce, how many such groupings there are, and which employees they contain. In an insurance company, for example, the emergent contours may differentiate between people in different areas (claims, investment, sales), between professionals and nonprofessionals, between senior and junior staff, or between office-based and mobile staff. Similarly, in a hospital, groups may be defined in terms professional affiliation (e.g., doctors, nurses, allied health professionals), years of service, or client base (oncology, rehabilitation, paediatrics).

Based on the outcomes of AIRing, a key task to be achieved by the end of this phase is a collective decision about which groups need to form the basis of the next phase of the ASPIRe process. In a manner consistent with self-categorization theory's principles of *comparative* and *normative fit* (Oakes, 1987; Oakes *et al.*, 1994; Turner, 1985), the objective here is to divide the superordinate organizational unit up into meaningfully distinct subcategories so as both (a) to maximize the perceived differences between the groupings and (b) to minimize the differences within them. For example, looking at the hypothetical company represented schematically in Fig. 3, it can be seen that it makes sense for this to be divided into subgroups on the basis of work domain (distinguishing between employees in marketing, service, and R&D areas) rather than on the basis of nature of employment (distinguishing between permanent, contract, and casual employees). This is because the former division is associated with a higher *metacontrast ratio* (MCR) of perceived intergroup differences to perceived intragroup differences (Turner, 1985; see also Campbell, 1958).

Illustrative of this process and its outcomes, we recently conducted a survey with employees of a large firm of electrical contractors in England (identified by the pseudonym ElectriCon), in order to identify which differences between

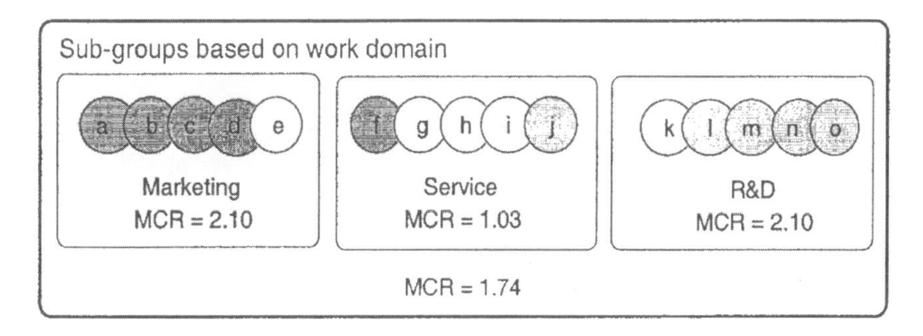

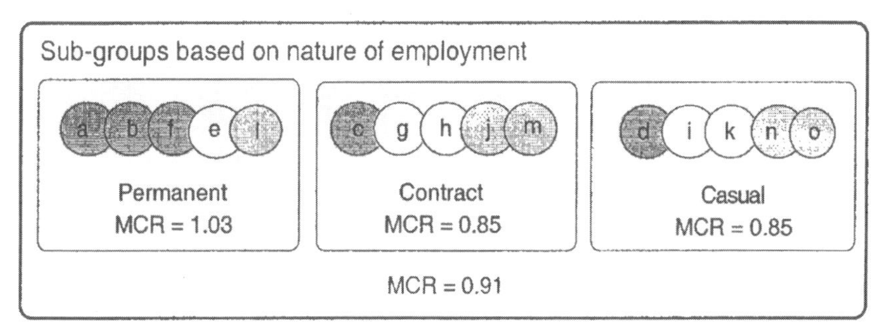

Figure 3: Comparative fit of two different bases for identifying subgroups in a hypothetical company. Different employees are identified by different letters. Shading reflects levels of subjectively perceived similarity between employees on dimensions most relevant to work-related activity in the organization (so that two people with the same form of shading perceive themselves to be very similar to each other and to be very different from people with other forms of shading). MCR = mean intergroup difference/mean intragroup difference (with similarly shaded others coded as 0 and differently shaded others coded as l). Following the metacontrast principle, it can be seen that, in this example, a division into subgroups based on work domain (Marketing, Service, R&D) is more fitting than one based on nature of employment (Permanent, Contract, Casual).

people in the organization were forming the basis of work-related social identities (Maple, Haslam, Reynolds, & Eggins, 2002). In related tasks, employees were first asked to identify whether people in the organization were different on each of 18 dimensions (e.g., age, years of employment, sex, salary, education level), and then to identify which of these dimensions of difference had the greatest impact on their working lives.

Responses to the first task were factor-analysed, and this procedure identified six factors that we labelled *status* (explaining 20.2% of variance), *ethnicity* (13.5%), *age* (8.6%), *education* (7.4%), *cronyism* (6.8%), and *work area* (5.6%). Subsequent analysis of the frequency with which the key items contributing to these factors were selected as having an impact on work activities indicated that status was by far the most important factor (an item contributing to this factor was seen as the most important basis of difference by 55% of employees). Moreover, this was the only factor that was correlated with organizational identification in the sample as a whole ($r = .22$, $p < .05$). On this basis, it seems reasonable to conclude that of all the social and demographic variables that might be used to identify salient subgroupings within ElectriCon, those related to status (e.g., seniority, salary, managerial responsibility) are psychologically most important and hence most worthwhile focusing on in the ASPIRe process.[1]

To summarize some of the above ideas, key arguments relating to the potential benefits of AIRing can be presented in terms of the following hypotheses:

> H1.1: Social identities that are identified on the basis of an AIRing process are more likely to be psychologically meaningful for employees than those identified on the basis of demographics.
>
> H1.2: The social identities revealed by AIRing are more likely to determine employees' actual behaviour than those identified on a demographic basis.
>
> H1.3: Organizational planning and development that take account of the social identities revealed by AIRing are more likely to be (and be perceived to be) worthwhile and fruitful than those based on demographic analysis.

Subgroup Caucusing (Sub-Casing)

Having made a decision about the basis on which the focal organizational unit should be divided, the next phase of the ASPIRe process involves providing a separate forum for independent subgroup *caucusing*. Here, each of the subgroups identified at the end of the AIRing phase engages in internal discussion and debate. This has a threefold purpose. First, it should allow subgroup members to identify and agree upon shared goals which will allow them to perform their work better. Second, it should help identify structural and other barriers which obstruct the achievement of these goals but which might be surmounted. Third, it should contribute to the development of a shared identity which is relevant to these goals and which subgroup members will internalize and carry over into (and beyond) subsequent stages of the ASPIRe process.

These ideas about the importance of procedures in which organizations develop collective strategies that build upon explicit recognition of subgroup difference have recently been developed and tested more rigorously by Eggins and colleagues in experimental studies that examine responses to negotiated outcomes under conditions where groups are or are not given an opportunity to engage in subgroup caucusing before coming together as a superordinate group (Eggins, Haslam, & Reynolds, 2002; Eggins, Haslam, & Ryan, 2000).

Consistent with the ASPIRe model, far more positive outcomes were observed in these studies where the opportunity to caucus was provided. For example, in Eggins, Haslam, & Reynolds (2002) first experiment, groups containing two men and two women were asked to negotiate a policy on the treatment of men in higher education. Prior to this, they were given the opportunity to caucus in two-person groups containing either two members of the same sex or one man and one woman. In the former condition — which provided an opportunity for caucusing in terms of a social identity relevant to the gender issue at hand —

group members (a) had a stronger sense of gender-based subgroup identity (*M*s [on 7-point scales, from 1 to 7] = 5.60, 5.13, respectively; $p < .05$), (b) had a greater expectation of subgroup consensus (*M*s = 5.33, 4.87, respectively; $p < .05$), (c) felt that group members worked better together (*M*s = 6.35, 5.96, respectively; $p < .05$), and (d) found it easier to agree (*M*s = 6.33, 5.89, respectively; $p < .05$). These and other findings (e.g., Eggins, Haslam, & Reynolds, 2002, experiment 2) confirm the point that subgroup caucusing enables subgroups to develop and express their identity in an environment in which they feel comfortable and where they find negotiation and decision-making easy- rather than heavy-going.

The general process that is envisioned in this phase of activity is one that we have referred to elsewhere as *group consensualization* (Haslam, Turner, Oakes, McGarty, & Reynolds, 1998; see also Turner, 1991). Specifically, because the broad context for Sub-Casing is intergroup (with each group aware that they are one of many engaging in this activity), it follows from social identity and self-categorization theories (e.g., Turner, 1985, 1991) (a) that individual group members should be motivated to develop a shared identity that differentiates their subgroup from others in the organization (Hogg, Turner, & Davidson, 1990), (b) that this process should tend to accentuate distinctive ingroup features leading to polarized self-representation and goals (Turner, Wetherell, & Hogg, 1989), and (c) that group members should be motivated to engage in mutual social influence so that they strive to agree upon this polarized identity and these goals (Haslam, Oakes, Reynolds, & Turner, 1999).

In important ways, this phase of the process is akin to the broad process of *team-building* through which a collection of individuals identify distinctive values, interests, and goals that they share but also work out ways of developing a shared theory of their ingroup that accommodates their idiosyncratic personal identities (Moreland *et al.*, 1996). In effect, then, the process contributes to the development of a low-level organic social identity on which basis group members come to trust each other and communicate more effectively (Haslam, 2001). At the same time, though, the fact that the evolution of this identity is not directly controlled by management allows greater freedom to explore the creative outcomes of the consensualization process and to identify (and attempt to correct) limitations of current organizational functioning without fear of personal reprisal. In this regard, a critical feature of Sub-Casing is that by being group-based, it allows individuals to give voice to those of their values and concerns that are shared with others and to do so in a supportive rather than an intimidatory environment (Eggins, Haslam, & Reynolds, 2002). And because the forum serves to identify goals which are associated with salient dimensions of employees' self-concepts, they should (a) be more involved in the process (Lewin, 1956; Shadur, Kienzle, & Rodwell, 1999; van Knippenberg & van Schie, 2000), (b) feel more respected by the organization (Smith & Tyler, 1997; Tyler, 1999; Tyler & Blader, 2000), and (c) be more committed to the goals that are agreed upon (Ellemers *et al.*, 1998; Wegge, 2000).

From management's perspective, of course, one potential concern with this phase is that the polarized and consensualized identities and goals that emerge as a result of the Sub-Casing process may be construed as dangerously conflictual and extreme. In our ElectriCon case, for example, if relations between the groups are insecure (i.e., perceived to be illegitimate and unstable), it may be the case that to conduct Sub-Casing would lead their members to display forms of ingroup favouritism and outgroup derogation (cf. Tajfel & Turner, 1979). Here, those in the high-status group might perceive members of the low-status group to be indolent and underskilled, while its members in turn would be seen by those in the low-status group as overpaid and arrogant. To the extent that these perceptions are promoted by Sub-Casing, they could clearly be quite damaging.

Similar concerns are reflected in Janis's (1982) groupthink model, which focuses on the capacity for highly cohesive groups to develop and pursue radical (and seemingly irrational) goals when 'the we-feeling of solidarity' runs too high (p. vii). For this reason, Stein (1982) counsels that:

> Managers need . . . to be aware that their subordinates might well profit from the facilitating effects of group membership. At the same time, however, they need to be aware of steamroller tactics, in which the group may become overstimulated and oversell itself. (p. 146)

Indeed, sensitivity to this point accounts for the fact that many organizations strive to develop participatory practices that are tightly controlled or nominal and where the rhetoric of empowerment fails to match the psychological reality (Harley, 1999; Kelly & Kelly, 1991; Mintzberg, 1996).

We believe that these various concerns would be more serious, however, if Sub-Casing were to occur at the *end* of the ASPIRe process and not to be framed by an overarching superordinate identity. Indeed, where no superordinate identity is available to inform Sub-Casing (and subsequent phases of the ASPIRe process), the dangers associated with proceeding beyond the AIRing phase may justify the postponement of Sub-Casing until such time as a superordinate identity has emerged or can be developed. However, where this superordinate identity is available, and can be used as a basis for subsequent phases of the ASPIRe process, this should mean that during Sub-Casing, organizations are able effectively to harness the creative potential of groups (of the form which recent analysis confirms to be present in groupthink; e.g., Turner & Pratkanis, 1998) without unmediated translation of its outcomes into organizational policy. Indeed, we would argue that *without* processes of this form (which allow for the emergence of conflicting social identities and goals), there is a far greater risk that the organization *as a whole* will display adverse symptoms of groupthink (in the form, for example, of generalized conformity, mindguarding, and excessive concurrence-seeking; e.g., see Locke *et al.*, 2001). Moreover, evidence from a range of studies (e.g., Eggins, Haslam, & Reynolds, 2002; Kelly & Kelly, 1994; Morley & Stephenson, 1979; Haslam & Reicher, 2002; Tyler & Blader, 2000) suggests that unless they allow for genuine collective participation and representation of the form envisaged here, employees are likely to feel alienated by strategic planning and organizational change programmes and are likely to react to them with cynicism and disengagement rather than enthusiasm and commitment (Jetten, O'Brien, & Trindall, 2002; Jetten, Duck, Terry, & O'Brien, 2002; King & Anderson, 1995; Mintzberg, 1996; Terry, Carey, & Callan, 2001; Wanous, Reichers, & Austin, 2000).

Again, then, the potential advantages of Sub-Casing can be summarized in the following hypotheses:

> H2.1: Members of the subgroups within an organization that participate in Sub-Casing should show a better awareness of (a) shared subgroup goals, (b) barriers to subgroup goal achievement, and (c) their organic subgroup identity than those that do not.
>
> H2.2: Members of subgroups that participate in Sub-Casing should perceive themselves to be more respected and more empowered, and the organization to be more just, than those that do not.
>
> H2.3: Within subgroups that participate in Sub-Casing, there should be evidence of (a) more trust, (b) superior communication, (c) more enthusiasm, and (d) more creativity than within subgroups that do not go through this process.

Superordinate Consensualizing (Super-Casing)

All in all, then, the consensualization of subgroup goals has the threefold attraction of optimizing an organization's potential for *social engagement*, *social creativity*, and *social correction*. However, at the end of Sub-Casing, this potential will be reflected in goals that are not only diverse but also quite polarized. The next phase of the ASPIRe process involves providing a common organizational forum that brings together the different subgroups (or multiple representatives of each of them) to engage in further discussion and debate. This forum again has three main purposes that are similar to those associated with Sub-Casing — although, here, contributions should be informed more by the outcomes of the previous phase and the subgroup identities it consolidated and less by personal identities. First, it should allow employees in general to identify and agree upon shared goals which will allow them to improve their work. Second, it should help them to identify structural and other barriers which obstruct the achievement of these goals but which might be surmounted. And third, these activities should contribute to the development of a shared organizational identity which is relevant to these goals and which subgroup members internalize and carry over beyond the ASPIRe process. Importantly though, this social identity should *differ* from that which informed the initial AIRing phase in so far as it *builds upon and explicitly recognizes the subgroup identities that emerged during Sub-Casing*.

In this way, while the organizational identity that underpinned the AIRing phase may have been relatively *mechanical* (in the sense that its content incorporated an undifferentiated view of group membership; Durkheim, 1933;

Haslam, 2001), the organizational identity that emerges in Super-Casing needs to be *organic*. That is, *its content needs to define the superordinate group in a way that allows for, and incorporates, subgroup difference*. Indeed, this process of *organizational identity change* is one of the main hallmarks of the ASPIRe process: for such change is simultaneously both brought about by the process and the basis of its efficacy.

Examples of organic identities of this form are fairly easy to find. Most obviously, they are apparent in multi-ethnic societies that work to develop an ideology of multiculturalism (or biculturalism), which celebrates and provides a place for people of different ethnic and cultural backgrounds. Such a philosophy can be contrasted with ideologies of assimilation (where only a single mechanical identity is promoted) and separatism (in which two distinct identities are promoted; Berry, 1984, 1991; Hornsey & Hogg, 2000a,b). Significantly, too, research by Huo, Smith, Tyler, and Lind (1994) and Jetten, O'Brien, & Trindall (2002) suggests that of these three ideologies, multiculturalism is most likely to be associated with an ability to deal satisfactorily with conflict and change in the workplace — largely because it steers a delicate course between conflict escalation (produced by separatism) and conflict avoidance (produced by assimilation).

The nature of the identity changes that occur between Sub-Casing and Super-Casing is represented schematically in Fig. 4. Here, it can be seen that the main goal of Super-Casing is to move towards a situation in which employees define themselves in terms of a relatively complex superordinate identity (as members of the focal organizational unit), but are simultaneously aware of the subgroup memberships from which that identity has been forged. In effect, this involves the resolution of what Huczynski and Buchanan (2001, p. 9) identify as the fundamental organizational dilemma and represents a recasting of insights first contained in Blake and Mouton's (1964) *dual concern model* of negotiation, in which the key to long-term success was found to lie in parties' capacity to marry a high level of concern for their own position and outcomes with high concern for the position and outcomes of the other party (see also Haslam, 2001, pp. 190–194; Pruitt & Rubin, 1986). In research (e.g., see Pruitt & Carnevale, 1993), each of the other three combinations of high and low concern has been found to have clear limitations as an end-point in the negotiation process: concern simply for the other party (akin to assimilation) leads to collusive outcomes, which ultimately contribute to resentment and a feeling of having been 'sold out'; concern only for one's own party (akin to separatism) serves primarily to exacerbate conflict; and low concern for both parties leads to inaction. Dual concern, however, is typically associated with a process of *integrative problem-solving*, in which parties are sensitive to their own needs and interests but strive to develop creative ('win-win') solutions that are also compatible with the values and interests of the other party and motivated by the need to reach a higher-order agreement.

Essentially, the same process as this is envisaged in the ASPIRe process. However, the key practical development here is that the model specifies *temporal sequencing* procedures that help to create the appropriate climate for an organic

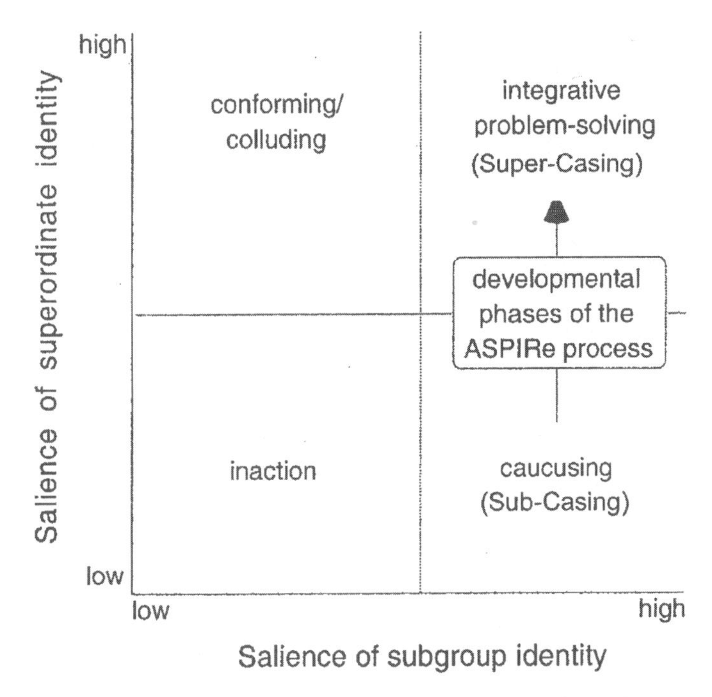

Figure 4: Salient social identities associated with Sub-Casing and Super-Casing phases of the ASPIRe process. In these two phases of the ASPIRe process, employees progress from a phase of *caucusing* associated with distinct subgroup identities to a secondary phase of *integrative problem-solving* that makes sense of those distinct identities in terms of a superordinate social identity.

organizational identity to develop. Specifically, the model builds upon pioneering work by Douglas (1957) and work in the social identity tradition by Morley and Stephenson (1979) and Stephenson (1981, 1984), which suggests that superordinate identities of this form are most likely to develop and to be internalized by all subgroups in organizations where those subgroups have *first* had the opportunity to explore their differences through effective caucusing (as provided in Sub-Casing; see also Olekalns, Smith, & Walsh, 1996; Pruitt, Peirce, Zubek, McGillicuddy, & Welton, 1993).

Again, these ideas have received support in empirical research reported by Eggins and colleagues in which groups are or are not given an opportunity to engage in subgroup caucusing before coming together as a superordinate group (Eggins, Haslam, & Reynolds, 2000, 2002). Following on from the data reported in the previous section, Eggins, Haslam, & Reynolds (2002, experiment 1) found that the positive outcomes associated with subgroup caucusing carried over to a subsequent phase in which members of the subgroups came together to negotiate a collective strategy. Compared with groups that had not participated in subgroup caucusing, members of groups that had participated in such activity retained (a) a stronger sense of gender-based sub-group identity (*M*s [on 7-point scales, from 1 to 7] = 5.54, 4.95, respectively; $p < .05$) and (b) a greater expectation of subgroup consensus (*M*s = 5.33, 4.87, respectively; $p < .05$). However, they also (c) felt that the people involved in this phase of negotiation worked better together (*M*s = 6.35, 5.96, respectively; $p < .05$) and (d) were less inclined to see the other subgroup as biased (*M*s = 2.19, 2.80, respectively; $p < .05$).

In a follow-up study, Eggins, Haslam, & Reynolds (2002, experiment 2) also found that when groups had participated in Sub-Casing prior to Super-Casing, they (e) enjoyed negotiation more (*M*s = 3.52, 3.20, respectively; $p < .05$) and (f) felt that their input was more valued (*M*s = 3.31, 3.02, respectively; $p < .05$). These and other findings suggest that, where Sub-Casing precedes Super-Casing, group members are likely to feel more comfortable with negotiation and decision-making processes, less hostile to the other parties engaged in them, and more valued. Moreover, in both these studies, mediational analysis provided support for the hypothesis that the relationship between experimental condition (Sub-Casing, no Sub-Casing) and group functioning was mediated by increased identification at the subgroup level. In other words, following Sub-Casing, groups worked better at the Super-Casing phase *because* Sub-Casing had allowed them to develop relevant subgroup identities.

From the standpoint of self-categorization theory (e.g., Turner, 1987), the main intellectual dividends associated with this strategy can be seen to flow (a) from the capacity for the two Casing phases to provide an opportunity for full expression of the personal and collective self and (b) from the *creative tension* that the two phases combine to create. Specifically, because employees define themselves at multiple levels (as individuals, as subgroup members and as members of a superordinate organizational unit), and self-definition at each of these levels has distinct consequences (e.g., van Knippenberg & van Schie, 2000), the Casing process works by providing a forum in which the implications and demands of *all* these levels of self-categorization can be explored. At the same time, though, in both phases, the operative social identity (subgroup in Sub-Casing and superordinate in Super-Casing) *provides employees with the psychological motivation to reconcile inconsistencies between adjacent levels of self-categorization* (Eggins, Haslam, & Reynolds, 2000; Turner, Oakes, Haslam, & McGarty, 1994). That is, in Sub-Casing self-categorization as a subgroup member motivates employees to develop a shared working understanding of that subgroup that accommodates individual differences (e.g., in competency and expertise), and in Super-Casing, self-categorization as an organizational member motivates employees to develop a shared working understanding of the organization that accommodates sub-group differences. In each case, then, the emergence of a more inclusive identity stimulates creativity and helps overcome cognitive inertia (Hodgkinson, 1997) by (a) encouraging a complex and elaborated cognitive understanding of the organization (Hodgkinson, 2001a,b), and (b) creating intellectual and material challenges that would not otherwise exist (Turner, 1991). Importantly too, this identity also brings with it new collective goals and suggests new individual and group activities. The objective of the final phase of the ASPIRe process is to establish which of these are viable and worth pursuing.

As with previous phases, the theorized benefits of Super-Casing can be summarized in the following hypotheses:

> H3.1: Members of an organization that participate (or are represented) in Super-Casing (following Sub-Casing) should show a better awareness of (a) shared organizational goals, (b) barriers to organizational goal achievement, and (c) the organic nature of the organization's identity than those that do not.
>
> H3.2: Members of an organization that participate (or are represented) in Super-Casing (following Sub-Casing) should perceive themselves to be more respected and more empowered, and the organization (and its decisions) to be more just than those that do not.
>
> H3.3: Within organizations whose members participate (or are represented) in Super-Casing (following Sub-Casing), there should be evidence of (a) more trust, (b) superior communication, (c) more enthusiasm, and (d) more creativity than within organizations that do not go through this process.

Organic Goal-Setting (ORGanizing)

The three phases outlined above incorporate procedures that seek to empower employees by allowing them to act in terms of self-determined personal and social identities (Reynolds & Platow, 2003). In order for this to happen, of course, the process needs to be set in train by personnel who are familiar with the logic and details of the model (information to which all participants should have access). Nevertheless, although managers will be part of subgroups (or may form a subgroup of their own), direct management control of the *specific content* of these phases should be minimal. Again, this is because the benefits of the model derive from its capacity to allow employees collectively to question and develop alternatives to existing organizational arrangements (arrangements that management may be quite attached to (for a range of reasons). Direct management intervention is counselled against, then, because in contrast to many popular management models (e.g., see Thompson & Warhurst, 1998), the goal of the ASPIRe model is *not* simply to reinforce managers' prejudices by telling them what they already know — but instead to promote *bottom-up* organizational creativity.

Having said that, at the end of the Super-Casing phase, there are clearly strategic decisions that need to be made relating to the direction in which the organization as a whole is to develop. It is at this stage that the issue of *leadership* becomes more important as decision-makers must now play a key role in determining how the organization as a whole can capitalize on the outcomes of AIRing and Casing phases through a process that we refer to as *organic goal setting* (ORGanizing). This process focuses on evaluating the appropriateness for the organization of the superordinate goals that emerged from Super-Casing. This decision needs to be informed by awareness of the identity and other resources available to the organization and of the obstacles that stand (and in the Casing phases were perceived to stand) in the way of goal achievement.

Following the basic tenets of goal-setting theory (e.g., Locke & Latham, 1990), it is important that the goals decided upon here are both *specific* and reasonably *hard to achieve*. Although goal-setting theory has traditionally defined the individual as the main unit of organizational analysis (and intervention), the relevance of these principles to group-level activities has been confirmed in recent research by Wegge and colleagues (Wegge, 2000; Wegge & Haslam, 2003; Wegge & Kleinbeck, 1996). In this research, it is found that groups generally perform significantly better on brainstorming tasks where they are given specific goals (e.g., to improve upon previous performance by a certain amount) rather than non-specific goals (to 'do your best'; see also O'Leary-Kelly, Martocchio, & Frink, 1994). For example, Wegge and Haslam (2003) and Haslam, Eggins, and Reynolds (2002) found that, compared with baseline testing (in which groups on average identified 19.6 uses for familiar objects), groups that were given goals to improve by 20–40% in a second phase of testing generated 4.0 more solutions, while those without such goals generated only 0.5 more solutions ($p < .01$). In a third phase of testing (where task fatigue might be expected to impact on performance) goals were doubled (with groups required to improve by 40–80% on their baseline performance). Here, groups with goals generated 2.7 more suggestions than baseline, while those without goals generated 2.6 *fewer* suggestions, $F(1, 70) = 10.30$, $p < .01$.

Moreover, consistent with the basic tenets of the social identity approach (and the rationale for previous phases of the ASPIRe process), this research also suggests that, like leadership itself, group goal-setting is most likely to be effective where members of the relevant organizational unit are encouraged to *participate* in the process and feel *collectively represented* by it (Turner & Haslam, 2001). Again, we expect this to be the case because it enhances feelings of procedural justice (and an associated sense of respect) and increases self-involvement and commitment (e.g., Eggins, Haslam, & Reynolds, 2002; Ellemers *et al.*, 1998; Tyler & Blader, 2000).

Consistent with this idea, in the study discussed above, Wegge and Haslam (2003; Haslam *et al.*, 2002) found that as goals became progressively more difficult over the second and third phases of testing, the performance of groups who were simply *given* goals by the experimenter started to decline, while that of groups who had developed their own goals continued to improve. In Phase 2 (where the goals were to improve by 20–40%), groups with experimenter-set goals generated 3.6 more suggestions, and those with self-set goals generated 4.3 more suggestions ($p > .5$), but in Phase 3 (where goals were to improve by 40–80%), groups with experimenter-set goals generated only 0.1 more suggestions, while those with self-set goals generated 5.4 more suggestions ($p < .02$). On the basis of these findings, it appears that when the going gets tough, the goals required to get groups going need to be ones in which they have some sense of ownership.

For this reason, it is important that even though ORGanizing can be defined as a executive activity, employees should continue to be (and feel) involved in this phase. As with the Super-Casing process, in order to preserve (and signal) ongoing commitment to the identities that were actualized in previous phases, it is thus recommended that, at the very least, it involve multiple representatives from various parties, and preferably those individuals who emerged from previous phases as subgroup representatives or leaders (i.e., prototypical subgroup members; Haslam & Platow, 2001a,b; Turner, 1991).

However, even if this does not occur, it is expected that through their own involvement in AIRing and Casing, leaders will have come to embrace a different understanding of the organization from that which they held previously (for related arguments, see Scott & Lane, 2000). In particular, this change should have been brought about through self-stereotyping in terms of a superordinate identity whose normative content is more complex and has a greater respect for the contribution of subgroups to the organization as whole (cf. Hodgkinson, 2001b). Indeed, in itself, this may constitute one of the most important contributions of the whole ASPIRe process, for, by creating identity-based bonds between leaders and followers (where previously there may only have been social categorical division), it should help to create *a new identity resource* from which the whole organization stands to prosper (Haslam & Platow, 2001a; Turner & Haslam, 2001).

Of course, whether or not the ASPIRe process does prove to be worthwhile is an empirical question, and (again in contrast to the greater body of contemporary management theory; Micklethwait & Wooldridge, 1997: Mintzberg, 1996), we are certainly not inviting theorists or practitioners to commit themselves to the model as an article of blind faith. Accordingly, the capacity of the organization to meet those goals that emerge from the ORGanizing process needs to be monitored closely, together with the contribution that the ASPIRe process makes to the organization's overall culture and performance. Such monitoring should also be as broad-based as possible: focusing on implications for both economic *and* social capital. In particular, attempts should be made to assess employees' satisfaction with, and commitment to, the identities and goals associated with ORGanizing. As well as helping to assess the viability and utility of the model, activity of this form is needed in order to monitor changes in the underlying identity resources of the organization (changes that inevitably will occur in response to changing social circumstances; Turner & Haslam, 2001) so that, if and when necessary, the ASPIRe process can be conducted anew.[2]

The potential benefits of ORGanizing — which, in many ways, summarize the proposed merits of the ASPIRe model as a whole — can be stated in the following hypotheses:

H4.1: Members of organizations that participate (or are represented) in ORGanizing (following Sub-Casing and Super-Casing) are more likely (a) to have a sense of ownership of the organization's decisions, goals, and plans, and (b) to perceive them to be fair and (c) appropriate, than those that do not.

H4.2: Members of organizations that participate (or are represented) in ORGanizing (following Sub-Casing and Super-Casing) are more likely (a) to be committed to the organization's decisions, goals, and plans, and (b) to use them as a guide for their own action, than those that do not.

H4.3: Organizations that engage in ORGanizing (following Sub-Casing and Super-Casing) are more likely to be (a) harmonious, (b) creative, and (c) productive than those that do not.

Implications of the Model and Points of Contact with Other Approaches

There are a number of novel and distinctive features of the ASPIRe model that derive from the fact that it attempts to translate social identity and self-categorization principles into a logical sequence of organizational activities. However, in presenting any new theoretical or practical model, it is usually a mistake to suggest that its contribution and insights are entirely original and its vision thoroughly iconoclastic. Accordingly, in the present case, it is important to note that there are significant points of contact between our model and other influential approaches to organizational practice. Indeed, as we see it, much of the model's appeal derives from its ability (with some recasting) to incorporate and lend coherence to insights from research and practice across a range of areas.

At a general level, then, the model is consistent with growing recognition of the need for organization and leadership practices that *empower* workers and allow them a level of autonomy and self-determination in the workplace (e.g., Kanter, 1979; Pfeffer, 1981, 1992), thereby demonstrating a commitment to principles of procedural justice that help instil a sense of pride and respect (Smith, Tyler, Huo, Ortiz, & Lind, 1998; Tyler & Blader, 2000). This call is reflected in practices such as *participative decision-making* (PDM; Yukl, 1989; see also Lewin, 1956; Morris, Hulbert, & Abrams, 2000), *participative group goal-setting (PGGS;* Wegge, 2000; Wegge & Kleinbeck, 1996) and *total quality management* (TQM; Lawler, Mohrman, & Ledford, 1992). The ASPIRe model is clearly consistent with these trends (associated, for example, with the increased use of delphi groups and quality circles in the workplace; Drafke & Kossen, 1998), in so far as it places an emphasis on employees' involvement in organizational decision-making and goal-setting, and places their own personal and social identities at the centre of this process.

At the same time, though, it is also the case that — successful as self-management processes have been in increasing organizational productivity and profit — critics of such developments point to the increasing human cost of such practices (e.g., Harley, 1999; Parker, 1993; Rees & Rodley, 1995; Sewell & Wilkinson, 1992). In particular, criticism centres on the fact that, in line with Mayo's (1949) human relations model, empowerment often involves conformity to the norms of an all-consuming monolithic team or organizational identity that limits employees' opportunity to criticize the superordinate unit or collectively pursue alternative courses of action. This can have a range of negative consequences, including the stifling of creativity, a tendency to compromise rather than challenge, and the emergence of an acquiescent organizational culture (cf. Janis, 1982). Such criticism lies at the heart of Locke's objection to the broad concept of teamwork and social capital (Locke, 1999; Locke *et al.*, 2001) and was the essence of early parodies of human relations ideals which equated its prescriptions for organizational life with a loss of (personal) identity (notably Whyte, 1960). Consistent with such concerns, evidence from some of our own recent research suggests that social identification of this form has a clear propensity to induce disenchantment and burnout as employees strive to live up to demanding group norms (e.g., to do more work with fewer resources) and, in the process exclude other potential avenues for (collective) self-actualization (e.g., Haslam, Powell, & Turner, 2000).

In this regard, a significant feature of the ASPIRe model is that it attempts to counteract the negative consequences of superordinate identities by ensuring that they are explicitly premised upon lower-level group memberships (which themselves provide an opportunity for the personal identities of employees to be expressed and affirmed). As we have already observed, a strategy of this form was first countenanced in the pioneering intergroup work of Douglas (1957) and formed the basis of Morley and Stephenson's (1979) and Stephenson's (1981, 1984) social identity model of negotiation. This approach is also consistent with the insights of the dual concern model (Pruitt & Carnevale, 1993; Pruitt & Rubin, 1986) in which the success of any negotiation process is seen to depend on an ability to take the perspective of, and show concern for, both self and other (and which suggests a framework conceptually similar to that presented in Figure 4).

In a similar vein, the ASPIRe model is consistent with the model of conflict resolution recently advocated by proponents of *dual identity theory* (e.g., Gaertner, Dovidio, Anastasio, Bachman, & Rust, 1993; Gaertner, Rust, Dovidio, Bachman, & Anastasio, 1996; Gonzalez & Brown, 1999; Hogg & Terry, 2000; Hornsey & Hogg, 1999, 2000a,b). Contrary to early assertions that conflict was best managed by encouraging participants to act purely in terms of (a) personal identities (Brewer & Miller, 1984), (b) distinct social identities (Hewstone & Brown, 1986), or (c) a shared superordinate identity (Gaertner *et al.*, 1993), this suggests that the most successful forms of conflict management involve *simultaneous recognition* of both the distinct social identities that give rise to conflict (e.g., between management and unions) and an overarching superordinate identity (e.g., as employees of the same company).

In relation to this work, the distinct contribution of the ASPIRe model (and of the empirical work that supports it, e.g., Eggins, Haslam, & Reynolds, 2003; Eggins, Haslam, & Ryan, 2000; Haslam & Reicher, 2002; Reynolds, Haslam, Turner, Eggins, & Ryan, 2000) is that it translates these insights into a practical organizational programme whose relevance is not restricted to conflict management. Part of this broader relevance derives from the fact that we see the utility of dual-identity strategies as a specific illustration of the benefits of a broad organizational (and social) philosophy that we have termed *organic pluralism* (Eggins *et al.*, 2003; Haslam, 2001; Haslam *et al.*, 2000). This suggests that a range of positive organizational outcomes flow from a superordinate organizational identity that recognizes, accommodates, and encourages subgroup identities that reflect the shared self-determined interests and aspirations of employees. These include the productivity dividends associated with the common ingroup identity model (Gaertner *et al.*, 1993) but also the encouragement of creativity, error rectification, and collective resistance to injustice (Veenstra & Haslam, 2000). Creativity is increased because shared superordinate social identification motivates employees to find novel ways of translating different subgroup goals into common organizational activity (as in Super-Casing; van Knippenberg & Haslam, 2003); error and injustice are more likely to be addressed because, where these are detected, individuals are not forced to combat them alone (a prospect that often leads individuals to capitulate; Asch, 1951) but can turn to others for intellectual, emotional, and practical support. In this way, organic pluralism is intended to engender the sense (and reality) of authentic social and organizational *choice* (cf. Kekes, 1993).

By capitalizing on inherent diversity (and resisting a monocultural model of the organization), the ASPIRe model therefore helps to ensure that productivity is sustainable and enduring rather than short-lived. In this, the model is envisaged as a tool that has the capacity to enhance diversity management (Reynolds *et al.*, 2000) but which can also contribute to the management of change, industrial relations, and employee well-being. And in these various roles, it should prove to be just as useful where there is a marked *absence* of conflict (and where there may be a high level of organizational torpor or uncritical compliance) as it does when conflict and discontent abound.

This suggestion is itself consistent with increasing recognition of the fact that, in principle, industrial harmony is not necessarily to be preferred to industrial conflict (De Dreu, Harinck, & Van Vianen, 1999; Eggins, Haslam, & Reynolds 2002; Kelly & Kelly, 1992; Stephenson, 1981, 1984; van Knippenberg & Haslam, 2003; Worchel, Coutant-Sassic, & Wong, 1993; for earlier statements to this effect, see Cooley, 1918; Coser, 1956; Katz, 1964; Kerr, 1954; Tannenbaum, 1965). This is partly because conflict and co-operation are effectively two sides of the same coin but also because a complete lack of conflict is typically a sign of apathy, disengagement, and entropy. A key feature of the ASPIRe model is thus that it provides a mechanism for organizations to move away from these states without precipitating anarchy or destructive conflict of the form that can flow from simple (nonorganic) pluralism (e.g., see Berry, 1984).

Finally, it is worth noting that the ASPIRe model is also consistent with a number of contemporary models of leadership. In particular, it accords with Deal and Kennedy's (1982) analysis of the role that *symbolic management* plays in creating cultures that recognize and celebrate the contribution that distinct subgroups and cultures make to an organization (Reynolds *et al.*, 2001). These authors argue that:

> Rather than being afraid of subcultures pulling apart, a symbolic manager will seek to strengthen each subculture as an effective cabal within the overall

> culture [and] will generally endorse the subculture's existence and meaning within the larger culture. (Deal & Kennedy, 1982, p. 153).

And they add:

> The symbolic manager will go out of his or her way to point out how each subculture brings unique strengths and values to the overall culture and how the subcultures all add value. Thus the whole experience of subculture clash will be used as a forum for enriching the culture and bonding the groups together.

Such arguments are clearly compatible with the general model we have outlined and also with general recommendations that effective organizational leadership needs to centre around the affirmation rather than negation of the social identities in terms of which employees define themselves (Haslam & Platow, 2001a,b; see also Adair, 1983).

They are broadly consistent, too, with House's (1971) *path-goal theory*, which suggests that the key to effective leadership lies in an ability to identify employees' personal goals and to provide paths for them to achieve these goals that are consistent with, and themselves contribute to, the superordinate goals of the organization. Significantly, though, the ASPIRe model suggests not only that work goals are personal (in the sense of being unique to the individual), but that many are also shared with other subgroup members. Accordingly, we see the ASPIRe model as an extension of House's theory (along lines suggested by House himself; see House, 1996), which suggests that, beyond the recognition of personal goals, it is important (and often much more practical; Tyler, 1999) for leaders to display commitment to a process that acknowledges and respects subgroup goals (together with the distinct identities that underpin them) and reconciles these with those of the organization.

Conclusion

The model of organizational practice proposed in this paper represents an initial attempt to demonstrate how social and personal identity resources might be assessed and utilized by organizations in an attempt to measure and make the most of their social capital. Key features of the model are that it assumes (a) that identities must be accepted as self-relevant and self-defining by employees rather than imposed upon them (e.g., on the basis of demography), (b) that, where they make an important contribution to employees' self concept, organizations must allow subgroup identities to be voiced and to have an impact on higher-level decisions, and (c) that a range of positive organizational outcomes are contingent upon an ability to develop shared goals and a shared identity that are premised upon lower-level goals relevant to the personal and subgroup identities of employees.

In so far as this model is premised upon the capacity, and need, for organizational outcomes to be collectively self-determined, it clearly conflicts with Taylorist wisdom, which asserts that organizational efficacy is contingent upon managers' 'right to manage' (e.g., Taylor, 1911). Along related lines, it also conflicts with models of top-down strategic planning that proliferate in the management literature and set management activities apart both from general organizational operations and from the general workforce (Mintzberg, 1990; Mintzberg & Lampel, 1999). However, it challenges the human relations view that organizations are most effective when employees simply define themselves in terms of a single organizational or team identity (e.g., Mayo, 1933, 1949; see also Gaertner *et al.*, 1993, 1996).

As we see it, all of the above models are limited by the fact that they engender either compliance (under [neo-]Taylorism) or conformity (under the common ingroup model) that limits the possibility for creative alternatives to the status quo (Turner & Oakes, 1997), rectification of collective errors (Locke *et al.*, 2001) or genuine empowerment in terms of self-relevant subgroup identities (Harley, 1999). Consistent with the tenets of organic pluralism, the major attraction of the present model is therefore that it overcomes the limitations of organizational monocultures in which there is a dangerous singularity of perspective and instead points to ways in which diversity, change, and conflict at both individual and group levels can be managed in a manner that contributes to an organization's vitality and long-term sustainability.

There are, of course, at least three obvious objections to the model. The first is that the model provides a vehicle for group-level dissent and conflict where otherwise an organization might encourage obedience or acquiescence. The second is that the model will be seen as problematic by members of those groups who feel that their interests are already well served either by an ability to exercise managerial prerogative or by the norms and values of all-embracing organizational identities. More generally, then, it can be argued that the model denies the various political realities that necessarily constrain (and in many case drive) organizational planning and decision-making (Pettigrew, 1973; Pfeffer & Salancik, 1974).

Both of these concerns are well founded, reflecting as they do the fact that organizations and organizational life *are* inherently political and, moreover, that this is often reflected in a conservative orientation that prefers *status quo*-reinforcing harmony and individualism to more radical alternatives (Pfeffer, 1997; Reicher, 1982; Tajfel, 1978). However, we would argue that far from being blind to the political dynamics of organizations, the ASPIRe model is an overt attempt to recognize and *change* those dynamics — in particular, by challenging individualistic policies or those which simply reinforce the interests of powerful groups.

A third objection to the model is that while we are aware of a broad body of social psychological research that points to the validity of the assumptions upon which it is based, there is, as yet, little evidence for the validity or utility of the ASPIRe process in its entirety (although it has been supported by evidence from preliminary experimental tests; Eggins, Haslam, & Reynolds, 2002). Nonetheless, largely because the model represents a reasonably elegant synthesis of theoretical and practical primitives, we think that it provides a sensible blueprint for practice and for further empirical work. In this regard, we consider the following to be the most important hypotheses to test in future research:

> H5.1: Positive organizational outcomes (on relevant individual and group, attitudinal, and behavioural measures) will be enhanced where organizational planning and development involve social identities that have emerged from an AIRing process rather than from purely demographic analysis.
>
> H5.2: Positive organizational outcomes will be enhanced where subgroup caucusing involves employee-relevant rather than employee-irrelevant social identities.
>
> H5.3: Positive organizational outcomes will be enhanced where subgroup caucusing occurs before, rather than after, superordinate consensualization.
>
> H5.4: Positive organizational outcomes will be enhanced where subgroup caucusing and superordinate consensualization involve, rather than avoid, goal setting.
>
> H5.5: Positive organizational outcomes observed among those who participate in ASPIRe processes will generalize to other organizational observers to the extent that those observers feel that self-relevant social identities have been represented in the process.

Research to test these hypotheses is currently underway (e.g., Eggins, Reynolds & Haslam, 2002; Eggins, Reynolds, Haslam & Veenstra, 2002). The aim here is to combine laboratory-based experimental studies with organization-based intervention studies to gain insights into both the process-based mechanics of ASPIRe and its long-term organizational impact. In particular, several large-scale organizational studies are currently underway in order to investigate the model in its entirety. Working in large public-sector organizations, these aim to take key sectors of an organization through the ASPIRe process, measuring participant attitudes and behaviour as well as overall organizational functioning at key stages during the study. The intention is to use noninvolved sectors (e.g., at different sites) or other similar organizations as controls. Field-based investigations of the model's efficacy will thus compare groups that have or have not been exposed to particular ASPIRe-related interventions (e.g., as specified under H5.1 to H5.5) to assess the relative importance and utility of each component.

We certainly do not rule out the possibility that the model will need to be modified on the basis of the findings that research and organizational activity generates. Indeed, given that a core implication of the social identity approach is that social and organizational processes are context-dependent and hence that prescriptions for practice need to be context specific rather than 'off the peg' (Turner & Haslam, 2001), we think that the model will always need to be employed with considerable sensitivity to the local organizational environment (e.g., along lines envisaged in the AIRing phase).

However, these caveats notwithstanding, there is clearly a pressing need to formalize the implications of the social identity approach and to translate its social and cognitive insights into useful organizational tools. For this reason, we hope that, at the very least, the present model serves to focus researchers on that goal and to stimulate further advances in this direction.

Notes

1. As we see it, ASPIRe is most likely to be successful where it is implemented and monitored by (or at least in collaboration with) human resources personnel with on-the-ground knowledge of and commitment to, the organization in question. This is because, as well as commitment to workplace democracy, it requires sensitivity to local organizational features and involves long-term monitoring of outcomes.

2. A further purpose of the AIRing process is to establish that employees' identification with the superordinate organization is sufficiently high to warrant continuing with the ASPIRe proecess. As noted below, this is because in the absence of superordinate identification, subsequent phases of the process could prove counter-productive. In the ElectriCon case, despite the fact that employees were sensitive to differences in status, it appeared that identification with the organization as a whole was sufficiently high to justify proceeding to the Sub-Casing phase (3.39 on a scale from 1 to 5).

References

Adair, J. (1983). *Effective leadership*. London: Pan.

Albert, S., Ashforth, B. E., & Dutton, J. E. (2000). Organizational identity and identification: Charting new waters and building new bridges. *Academy of Management Review*, 25, 13–17.

Asch, S. E. (1951). Effects of group pressure upon the modification and distortion of judgements. In H. Guetzkow (Ed.), *Groups, leadership and men* (pp. 177–190). Pittsburgh, PA: Carnegie Press.

Ashforth, B. E., & Mael, F. (1989). Social identity theory and the organization. *Academy of Management Review*, 14, 20–39.

Berry, J. W. (1984). Cultural relations in plural societies: Alternatives to segregation and their sociopsychological implications. In N. Miller & M. B. Brewer (Eds.), *Groups in contact: The psychology of desegregation* (pp. 11–27). Orlando, FL: Academic Press.

Berry, J. W. (1991). Understanding and managing multiculturalism: Some possible implications of research in Canada. *Psychology and Developing Societies*, 3, 17–49.

Blake, R. R., & Mouton, J. S. (1964). *The managerial grid*. Houston, TX: Gulf.

Brewer, M. B. (1979). Ingroup bias in the minimal intergroup situation: A cognitive-motivational analysis. *Psychological Bulletin*, 86, 307–324.

Brewer, M. B. (1995). Managing diversity: The role of social identities. In S. E. Jackson & M. N. Ruderman (Eds.), *Diversity in work teams: Research paradigms for a changing workplace* (pp. 47–68). Washington, DC: American Psychological Association.

Brewer, M. B., & Miller, N. (1984). Beyond the contact hypothesis: Theoretical perspectives on desegregation. In N. Miller & M. B. Brewer (Eds.), *Groups in contact: The psychology of desegregation* (pp. 281–302). Orlando, FL: Academic Press.

Brown, R. J. (1978). Divided we fall: Analysis of relations between different sections of a factory workforce. In H. Tajfel (Ed.), *Differentiation between social groups: Studies in the social psychology of intergroup relations* (pp. 395–429). London: Academic Press.

Burt, R. S. (1997). The contingent value of social capital. *Administrative Science Quarterly*, 42, 339–365.

Campbell, D. T. (1958) Common fate, similarity, and other indices of the status of aggregates of persons as social entities. *Behavioural Science*, 3, 14–25.
Cooley, C. H. (1918). *Social process*. New York: Scribners and Sons.
Coser, L. (1956). *The functions of conflict*. Glencoe, IL: Free Press.
De Dreu, C. K. W., Harinck, F., & Van Vianen, A. E. M. (1999). Conflict and performance in groups and organizations. In C. L. Cooper & I. Robertson (Eds.), *International review of industrial and organizational psychology* (Vol. 14, pp. 369–414). Chichester, UK: Wiley.
Deal, T. E., & Kennedy, A. A. (1982). *Corporate cultures: The rites and rituals of corporate life*. Harmondsworth, UK: Penguin.
Dovidio, J. F., Gaertner, S. L., Validzic, A., Matoka, K., Johnson, B., & Frazier, S. (1997). Extending the benefits of recategorization: Evaluations, self-disclosure, and helping. *Journal of Experimental Social Psychology*, 33, 401–420.
Doosje, B., Haslam, S. A., Spears, R., Oakes, P. J., & Koomen, W. (1998). The effect of comparative context on central tendency and variability judgements and the evaluation of group characteristics. *European Journal of Social Psychology*, 28, 173–184.
Douglas, A. (1957). The peaceful settlement of industrial intergroup disputes. *Journal of Conflict Resolution*, 32, 489–510.
Drafke, M. W., & Kossen, S. (1998). *The human side of organizations* (7th ed.) Reading, MA: Addison-Wesley.
Durkheim, E. (1933). *The division of labour in society*. London: Macmillan.
Dutton, J. E., Dukerich, J. M., & Harquail, C. V. (1994). Organizational images and member identification. *Administrative Science Quarterly*, 39, 239–263.
Eggins, R. A., Haslam, S. A., & Reynolds, K. J. (2002). Social identity and negotiation: Subgroup representation and superordinate consensus. *Personality and Social Psychology Bulletin*, 28, 887–899.
Eggins, R. A., Haslam, S. A., & Ryan, M. K. (2000). *Negotiation between groups in conflict: The importance of group representation and superordinate framing*. Unpublished manuscript, The Australian National University.
Eggins, R. A., Reynolds, K. J., & Haslam, S. A. (2002). *The temporal effects of subgroup representation on superordinate identification, group functioning and intergroup relations*. Unpublished manuscript, The Australian National University.
Eggins, R. A., Reynolds, K. J., & Haslam, S. A. (2003). Working with identities: The ASPIRe model of organizational planning, negotiation and development. In S. A. Haslam, D. van Knippenberg, M. J. Platow, & N. Ellemers (Eds.), *Social identity at work: Developing theory for organizational practice* (pp. 241–257). New York: Taylor & Francis.
Eggins, R. A., Reynolds, K. J., Haslam, S. A., & Veenstra, K. (2002). *Evidence that subgroup consensualization plays a positive role in organizational planning and development: An experimental test of the ASPIRe model*. Unpublished manuscript. The Australian National University.
Ellemers, N., de Gilder, D., & van den Heuvel, H. (1998). Career-oriented versus team-oriented commitment and behaviour at work. *Journal of Applied Psychology*, 83, 717–730.
Ellemers, N., Spears, R., & Doosje, B. (1999). *Social identity: Context, content and commitment*. Oxford: Blackwell.
Ellemers, N., van Knippenberg, A., & Wilke, H. A. (1990). The influence of permeability of group boundaries and stability of group status on strategies of individual mobility and social change. *British Journal of Social Psychology*, 29, 233–246.
Ellemers, N., Wilke, H., & van Knippenberg, A. (1993). Effects of the legitimacy of low group or individual status on individual and collective status-enhancement strategies. *Journal of Personality and Social Psychology*, 64, 766–778.
Gaertner, S. L., Dovidio, J. F., Anastasio, P. A., Bachman, B. A., & Rust, M. C. (1993). The common ingroup identity model: Recategorization and the reduction of intergroup bias. *European Review of Social Psychology*, 4, 1–25.
Gaertner, S. L., Rust, M. C., Dovidio, J. F., Bachman, B. A., & Anastasio, P. A. (1996). The contact hypothesis: The role of a common ingroup identity on reducing intergroup bias among minority and majority group members. In J. Nye & A. Brower (Eds.), *What's social about social cognition? Research on socially shared cognition in small groups* (pp. 230–260). Newbury Park, CA: Sage.

Gioia, D. A., Schultz, M., & Corley, K. G. (2000). Organizational identity, image and adaptive instability. *Academy of Management Review*, 25, 63–81.

Gonzalez, R., & Brown, R. J. (1999). *Maintaining the salience of subgroup and super-ordinate group identities during intergroup contact.* Paper presented at the Small Groups Preconference to the Annual Meeting of the Society of Experimental Social Psychology, St. Louis, MI, October 14–16.

Harley, B. (1999). The myth of empowerment: Work organization, hierarchy and employee autonomy in contemporary Australian workplaces. *Work, Employment and Society*, 13, 41–66.

Haslam, S. A. (2001). *Psychology in organizations: The social identity approach.* London: Sage.

Haslam, S. A., Eggins, R. A., & Reynolds, K. J. (2002). *ASPIRing to be better: An integrated approach to issues of diversity management, organisational development and creative productivity.* Paper presented at the British Psychology Society Occupational Psychology Conference, Blackpool, January 3–5.

Haslam, S. A., McGarty, C., Brown, P. M., Eggins, R. A., Morrison, B. E., & Reynolds, K. J. (1998). Inspecting the emperor's clothes: Evidence that randomly-selected leaders can enhance group performance. *Group Dynamics: Theory, Research and Practice*, 2, 168–184.

Haslam, S. A., Oakes, P. J., Reynolds, K. J., & Turner, J. C. (1999). Social identity salience and the emergence of stereotype consensus. *Personality and Social Psychology Bulletin*, 25, 809–818.

Haslam, S. A., & Platow, M. J. (2001a). Your wish is my command: How a leader's vision becomes a follower's task. In M. A. Hogg & D. J. Terry (Eds.), *Social identity processes in organizational contexts* (pp. 213–228). New York: Taylor & Francis.

Haslam, S. A., & Platow, M. J. (2001b). The link between leadership and followership: How affirming a social identity translates vision into action. *Personality and Social Psychology Bulletin*, 27, 1469–1479.

Haslam, S. A., Powell, C., & Turner, J. C. (2000). Social identity, self-categorization and work motivation: Rethinking the contribution of the group to positive and sustainable organizational outcomes. *Applied Psychology: An International Review*, 49, 319–339.

Haslam, S. A., & Reicher, S. D. (2002). *A user's guide to The Experiment-exploring the psychology of groups and power* (manual to accompany BBC video). London: BBC Worldwide.

Haslam, S. A., & Turner, J. C. (1992). Context-dependent variation in social stereotyping 2: The relationship between frame of reference, self-categorization and accentuation. *European Journal of Social Psychology*, 22, 251–278.

Haslam, S. A., Turner, J. C., Oakes, P. J., McGarty, C., & Reynolds, K. J. (1998). The group as a basis for emergent stereotype consensus. *European Review of Social Psychology*, 9, 203–239.

Haslam, S. A., Turner, J. C., Oakes, P. J., Reynolds, K. J., & Doosje, B. (in press). From personal pictures in the head to collective tools in the world: How shared stereotypes allow groups to represent and change social reality. In C. McGarty, V. Y. Yzerbyt, & R. Spears (Eds.), *Stereotypes as explanations: The formation of meaningful beliefs about social groups.* Cambridge: Cambridge University Press.

Haslam, S. A., van Knippenberg, D., Platow, M. J., & Ellemers, N. (Eds.) (in press). *Social identity at work: Developing theory for organizational practice.* New York: Taylor & Francis.

Hewstone, M., & Brown, R. (1986). Contact is not enough: An intergroup perspective on the 'Contact hypothesis'. In M. Hewstone & R. Brown (Eds.), *Contact and conflict in intergroup encounters.* Oxford: Basil Blackwell.

Hodgkinson, G. P. (1997). Cognitive inertia in a turbulent market: The case of UK residential estate agents. *Journal of Management Studies*, 34, 921–945.

Hodgkinson, G. P. (2001a). The psychology of strategic management: Diversity and cognition revisited. In C. L. Cooper & I. T. Robertson (Eds.), *International review of industrial and organizational psychology* (Vol. 16, pp. 65–119). Chichester, UK: Wiley.

Hodgkinson, G. P. (2001b). Cognitive processes in strategic management: Some emerging trends and future directions. In N. Anderson, D. S. Ones, H. K. Sinangil, & C. Viswesvaranan (Eds.), *Handbook of industrial, work and organizational psychology: Volume 2 — Organizational psychology* (pp. 416–440). London: Sage.

Hogg, M. A., & Abrams, D. (1988). *Social identifications: A social psychology of intergroup relations and group processes.* London: Routledge.
Hogg M. A., & Terry, D. J. (2000). Social identity and self-categorization processes in organizational contexts. *Academy of Management Review*, 25, 121–140.
Hogg, M. A., & Terry, D. J. (Eds.) (2001). *Social identity processes in organizational contexts.* New York: Taylor & Francis.
Hogg, M. A., Turner, J. C., & Davidson, B. (1990). Polarized norms and social frames of reference: A test of the self-categorization theory of group polarization. *Basic and Applied Social Psychology*, 11, 77–100.
Hornsey, M. J., & Hogg, M. A. (1999). Subgroup differentiation as a response to an overly-inclusive group: A test of optimal distinctiveness theory. *European Journal of Social Psychology*, 29, 543–550.
Hornsey, M. J., & Hogg, M. A. (2000a) Assimilation and diversity: An integrative model of subgroup relations. *Personality and Social Psychology Review*, 4, 143–156.
Hornsey, M. J., & Hogg, M. A. (2000b). Subgroup relations: Two experiments comparing subgroup differentiation and common ingroup identity models of prejudice reduction. *Personality and Social Psychology Bulletin*, 26, 242–256.
House, R. J. (1971). A path-goal theory of leader effectiveness. *Administrative Science Quarterly*, 16, 321–339.
House, R. J. (1996). Path-goal theory of leadership: Lessons, legacy and a reformulated theory. *Leadership Quarterly*, 7, 323–352.
Huczynski, A., & Buchanan, D. (2001). *Organizational behaviour* (4th ed.). London: Pearson.
Huo, Y. J., Smith, H. J., Tyler, T. R., & Lind, E. A. (1996). Superordinate identification, subgroup identification, and justice concerns: Is separatism the problem? Is assimilation the answer? *Psychological Science*, 7, 40–45.
Jackson, S. E. (1992). Team composition in organizational settings: Issues in managing an increasingly diverse workforce. In S. Worchel, W. Wood, & J. A. Simpson (Eds.), *Group processes and productivity* (pp. 136–180). Newbury Park, CA: Sage.
James, K., & Greenberg, J. (1989). Ingroup salience, intergroup comparison and individual performance and self-esteem. *Personality and Social Psychology Bulletin*, 15, 604–616.
Janis, I. L. (1982). *Groupthink: Psychological studies of policy decisions and fiascoes* (2nd ed.). Boston, MA: Houghton Mifflin.
Jetten, J., Duck, J., Terry, D., & O'Brien, A. (2002). Being attuned to intergroup differences in mergers: The role of aligned leaders for low status groups. *Personality and Social Psychology Bulletin*, 28, 1194–1201.
Jetten, J., O'Brien, A., & Trindall, N. (2002). Changing identity: Predicting adjustment to organizational restructure as a function of subgroup and superordinate identification. *British Journal of Social Psychology*, 41, 281–297.
Kanter, R. (1979). Power failure in management circuits. *Harvard Business Review*, July–August, 65–75.
Katz, D. (1964). Approaches to managing conflict. In R. L. Kahn & E. Boulding (Eds.), *Power and conflict in organizations.* London: Tavistock.
Kekes, J. (1993). *The morality of pluralism.* Princeton, NJ: Princeton University Press.
Kelly, C., & Kelly, J. (1991). 'Them and us': Social psychology and 'the new industrial relations'. *British Journal of Industrial Relations*, 29, 25–48.
Kelly, C., & Kelly, J. (1994). Who gets involved in collective action? Social psychological determinants of individual participation in trade unions. *Human Relations*, 47, 63–88.
Kelly, J., & Kelly, C. (1992). Industrial action. In J. F. Hartley & G. M. Stephenson (Eds.), *Employment relations: The psychology of influence and control at work* (pp. 246–268). Oxford: Blackwell.
Kerr, C. (1954). Industrial conflict and its mediation. *American Journal of Sociology*, 60, 230–245.
King, N., & Anderson, N. (1995). Innovation and change in organizations. London: Routledge.
Kramer, R. M. (1993). Cooperation and organizational identification. In J. K. Murnigham (Ed.), *Social psychology in organizations: Advances in theory and research* (pp. 244–268). Engelwood Cliffs, NJ: Prentice Hall.

Kramer, R. M., Brewer, M. B., & Hanna, B. A. (1996). Collective trust and collective action: The decision to trust as a social decision. In R. M. Kramer & T. R. Tyler (Eds.), *Trust in organizations: Frontiers of theory and research* (pp. 357–389). Thousand Oaks, CA: Sage.

Lant, T. K., & Baum, J. A. C. (1995). Cognitive sources of socially constructed competitive groups: Examples from the Manhattan hotel industry. In W. R. Scott & S. Christensen (Eds.), *The institutional construction of organizations: International and longitudinal studies* (pp. 15–38). Thousand Oaks, CA: Sage.

Lau, J., & Murningham, J. K. (1998). Demographic diversity and faultlines: The compositional dynamics of organizational groups. *Academy of Management Review*, 23, 325–340.

Lawler, E. E., Mohrman, S. A., & Ledford, G. E. (1992). *Employee involvement and total quality management*. San Francisco: Jossey-Bass.

Leana, C. R., & van Buren, H. J. (1999). Organizational social capital and employment practices. *Academy of Management Review*, 24, 538–555.

Lewin, K. (1956). Studies in group decision. In D. Cartwright & A. Zander (Eds.), *Group dynamics: Research and theory* (2nd ed., pp. 287–301). Evanston, IL: Row Peterson.

Locke, E. A. (1999). Some reservations about social capital. *Academy of Management Review*, 24, 8–11.

Locke, E. A., & Latham, G. P. (1990). *A theory of goal setting and task performance*. Englewood Cliffs, NJ: Prentice-Hall.

Locke, E. A., Tirnauer, D., Roberson, Q., Goldman, B., Latham, M. E., & Weldon, E. (2001). The importance of the individual in an age of groupism. In M. E. Turner (Ed.), *Groups at work: Advances in theory and research* (pp. 501–528). Hillsdale, NJ: Erlbaum.

Maple, N., Haslam, S. A., Reynolds, K. J., & Eggins, R. A. (2002). *The practicalities of AIRing: A preliminary study*. Unpublished manuscript, University of Exeter.

Markoczy, L. (1997). Measuring beliefs: Accept no substitutes. *Academy of Management Journal*, 40, 1228–1242.

Mayo, E. (1933). *The human problems of an industrial civilisation*. Cambridge, MA: Macmillan.

Mayo, E. (1949). *The social problems of an industrial civilisation*. London: Routledge and K. Paul.

McGarty, C., & Haslam, S. A. (Eds.) (1997). *The message of social psychology: Perspectives on mind in society*. Oxford: Blackwell.

McGarty, C., Haslam, S. A., Turner, J. C., & Oakes, P. J. (1993). Illusory correlation as accentuation of actual intercategory difference: Evidence for the effect with minimal stimulus information. *European Journal of Social Psychology*, 23, 391–410.

Micklethwait, J., & Wooldridge, A. (1997). *The witch doctors: What the management gurus are saying, why it matters and how to make sense of it*. London: Random House.

Mintzberg, H. (1990). The design school: Reconsidering the basic premises of strategic management. *Strategic Management Journal*, 11, 171–195.

Mintzberg, H. (1996). Ten ideas designed to rile everyone who cares about management. *Harvard Business Review*, 74, 61–67.

Mintzberg, H., & Lampel, G. (1999). Reflections on the strategy process. *Sloan Management Review, Spring*, 21–30.

Moreland, R. L., Argote, L., & Krishnan, R. (1996). Socially shared cognition at work: Transactive memory and group performance. In J. Nye & A. Brower (Eds.), *What's social about social cognition? Research on socially shared cognition in small groups* (pp. 57–84). Newbury Park, CA: Sage.

Morley, I. E., & Stephenson, G. (1979). *The social psychology of bargaining*. London: Allen & Unwin.

Morris, L., Hulbert, L., & Abrams, D. (2000). An experimental investigation of group members' perceived influence over leaders' decisions. *Group Dynamics: Theory, Research and Practice*, 4, 157–167.

Nahapiet, J., & Ghoshal, S. (1998). Social capital, intellectual capital and the intellectual advantage. *Academy of Management Review*, 23, 242–266.

Nkomo, S. M., & Cox, T., Jr. (1996). Diverse identities in organizations. In S. R. Clegg, C. Hardy, & W. R. Nord (Eds.), *Handbook of organization studies* (pp. 338–356). London: Sage.

O'Leary-Kelly, A. M., Martocchio, J. J., & Frink, D. D. (1994). A review of the influence of group goals on group performance. *Academy of Management Journal*, 37, 1285–1301.

Oakes, P. J. (1987). The salience of social categories. In J. C. Turner, M. A. Hogg, P. J. Oakes, S. D. Reicher, & M. S. Wetherell (Eds.), *Rediscovering the social group: A self-categorization theory* (pp. 117–141). Oxford: Blackwell.

Oakes, P. J., Haslam, S. A., Morrison, B., & Grace, D. (1995). Becoming an ingroup: Re-examining the impact of familiarity on perceptions of group homogeneity. *Social Psychology Quarterly*, 58, 52–61.

Oakes, P. J., Haslam, S. A., & Turner, J. C. (1994). *Stereotyping and social reality.* Oxford: Blackwell.

Oakes, P. J., & Turner, J. C. (1986). Distinctiveness and the salience of social category memberships: Is there a perceptual bias towards novelty? *European Journal of Social Psychology*, 16, 325–344.

Oakes, P. J., Turner, J. C., & Haslam, S. A. (1991). Perceiving people as group members: The role of fit in the salience of social categorizations. *British Journal of Social Psychology*, 30, 125–144.

Olekalns, M., Smith, P. L., & Walsh, T. (1996). The process of negotiating: Strategy and timing as predictors of outcomes. *Organizational Behavior and Human Decision Processes*, 68, 68–77.

Ouwerkerk, J. W., Ellemers, N., & de Gilder, D. (1999). Social identification, affective commitment and individual effort on behalf of the group. In N. Ellemers, R. Spears, & B. J. Doosje (Eds.), *Social identity: Context, commitment, content* (pp. 184–204). Oxford: Blackwell.

Parker, M. (1993). Industrial relations myth and shop floor reality: The team concept in the auto industry. In N. Lichtenstein & J. H. Howell (Eds.), *Industrial democracy in America* (pp. 249–274). Cambridge: Cambridge University Press.

Peteraf, M., & Shanley, M. (1997). Getting to know you: A theory of strategic group identity. *Strategic Management Journal*, 18, 165–186.

Pettigrew, A. M. (1973). *The politics of organizational decision-making.* London: Tavistock.

Pfeffer, J. (1981). *Power in organizations.* Boston, MA: Pitman.

Pfeffer, J. (1992). *Managing with power.* Boston, MA: Harvard Business School Press.

Pfeffer, J. (1997). *New directions for organization theory: Problems and prospects.* New York: Oxford University Press.

Pfeffer, J., & Salancik, G. R. (1974). Organizational decision making as a political process: The case of a university budget. *Administrative Science Quarterly*, 19, 135–151.

Postmes, T. (2003). A social identity approach to communication in organizations. In S. A. Haslam, D. van Knippenberg, M. J. Platow, & N. Ellemers (Eds.), *Social identity at work: Developing theory for organizational practice* (pp. 81–97). Philadelphia, PA: Taylor & Francis.

Postmes, T., Tanis, M., & de Wit, B. (2001). Communication and commitment in organizations: a social identity approach. *Group Processes and Intergroup Relations*, 4, 227–246.

Pruitt, D. G., & Carnevale, P. J. (1993). *Negotiation in social conflict.* Milton Keynes, UK: Open University Press.

Pruitt, D. G., Peirce, R. S., Zubek, J. M., McGillicuddy, N. B., & Welton, G. L. (1993). Determinants of short-term and long-term success in mediation. In S. Worchel & J. A. Simpson (Eds.), *Conflict between people and groups: Causes, processes, and resolutions. Nelson Hall series in psychology.* Chicago, IL: Nelson-Hall.

Pruitt, D. G., & Rubin, J. Z. (1986). *Social conflict: Escalation, stalemate and settlement.* New York: McGraw-Hill.

Rees, S., & Rodley, G. (Eds.) (1995). *The human costs of managerialism: Advocating the recovery of humanity.* Leichhardt, Australia: Pluto.

Reicher, S. D. (1982). The determination of collective action. In H. Tajfel (Ed.), *Social identity and intergroup relations* (pp. 41–84). Cambridge: Cambridge University Press.

Reynolds, K. J., Eggins, R. A., & Haslam, S. A. (2001). *Uncovering diverse identities in organizations: AIRing versus auditing.* Manuscript submitted for publication.

Reynolds, K. J., Haslam, S. A., Turner, J. C., Eggins, R. A., & Ryan, M. K. (2000). A social psychological analysis of diversity: The importance of working with group differences. *Paper presented at Current Research on Diversity in Australia's workshop. October 12, Canberra.*

Reynolds, K. J., & Platow, M. J. (2003). Why power in organizations really should be shared: Understanding power through the perils of powerlessness. In S. A. Haslam, D. van Knippenberg, M. J. Platow, & N. Ellemers (Eds.), *Social identity at work: Developing theory for organizational practice* (pp. 173–188). Philadelphia, PA: Psychology Press.

Reynolds, K. J., Turner, J. C., & Haslam, S. A. (2000). When are we better than them and they worse than us? A closer look at social discrimination in positive and negative domains. *Journal of Personality and Social Psychology*, 78, 64–80.

Scott, S. G., & Lane, V. R. (2000). A stakeholder approach to organizational identity. *Academy of Management Review*, 25, 43–62.

Sewell, G., & Wilkinson, B. (1992). 'Someone to watch over me': Surveillance, discipline and the just-in-time labour process. *Sociology*, 26, 271–289.

Shadur, M. A., Kienzle, R., & Rodwell, J. J. (1999). The relationship between organizational climate and employee perceptions of involvement. *Group and Organization Management*, 24, 479–503.

Smith, E. R., & Mackie, D. (2000). *Social psychology* (2nd ed.). New York: Taylor & Francis.

Smith, H. J. & Tyler, T. R. (1997). Choosing the right pond: The influence of the status of one's group and one's status in that group on self-esteem and group-oriented behaviours. *Journal of Experimental Social Psychology*, 33, 146–170.

Smith, H. J., Tyler, T. R., Huo, Y. J., Ortiz, D., & Lind, E. (1998). The self-relevant implications of the group-value model: Group membership, self-worth and treatment quality. *Journal of Experimental Social Psychology*, 34, 47–493.

Spears, R., Oakes, P. J., Ellemers, N., & Haslam S. A. (Eds.) (1997). *The social psychology of stereotyping and group life.* Oxford: Blackwell.

Stein, M. I. (1982). Creativity, groups, and management. In R. A. Guzzo (Ed.), *Improving group decision making in organizations: Approaches from theory and research* (pp. 127–155). New York: Academic Press.

Stephenson, G. M. (1981). Intergroup bargaining and negotiation. In J. C. Turner & H. Giles (Eds.), *Intergroup behaviour.* Oxford: Basil Blackwell.

Stephenson, G. M. (1984). Interpersonal and intergroup dimensions of bargaining and negotiation. In H. Tajfel (Ed.), *The social dimension: European developments in social psychology* (pp. 646–667). Cambridge: Cambridge University Press.

Suzuki, S. (1998). In-group and out-group communication patterns in international organizations: Implications for social identity theory. *Communication Research*, 25, 154–182.

Tajfel, H. (1972). La categorisation sociale (Social categorization). In S. Moscovici (Ed.), *Introduction a la psychologie sociale* (pp. 272–302). Paris: Larouse.

Tajfel. H. (1978). Interindividual behaviour and intergroup behaviour. In H. Tajfel (Ed.), *Differentiation between social groups: Studies in the social psychology of intergroup relations.* London: Academic Press.

Tajfel, H., Flament, C., Billig, M. G., & Bundy, R. F. (1971). Social categorization and intergroup behaviour. *European Journal of Social Psychology*, 1, 149–177.

Tajfel, H., & Turner, J. C. (1979). An integrative theory of intergroup conflict. In W. G. Austin & S. Worchel (Eds.), *The social psychology of intergroup relations* (pp. 33–47). Monterey, CA: Brooks/Cole.

Tannenbaum, A. S. (1965). Unions. In J. G. March (Ed.), *Handbook of organizations* (pp. 710–763). Chicago: Rand McNally.

Taylor, F. W. (1911). *Principles of scientific management.* New York: Harper.

Terry, D. J., & Callan, V. J. (1998). Ingroup bias in response to an organizational merger. *Group Dynamics: Theory, Research and Practice*, 2, 67–81.

Terry, D. J., Carey, C. J., & Callan, V. J. (2001). Employee adjustment to an organizational merger: An intergroup perspective. *Personality and Social Psychology Bulletin*, 27, 267–290.

Thompson, P., & Warhurst, C. (Eds.) (1998). *Workplaces of the future.* Basingstoke, UK: Macmillan.

Turner, J. C. (1982). Towards a cognitive redefinition of the social group. In H. Tajfel (Ed.), *Social identity and intergroup relations* (pp. 15–40). Cambridge: Cambridge University Press.

Turner, J. C. (1985). Social categorization and the self-concept: A social cognitive theory of group behaviour. In E. J. Lawler (Ed.), *Advances in group processes* (Vol. 2, pp. 77–122). Greenwich, CT: JAI Press.

Turner, J. C. (1987). The analysis of social influence. In J. C. Turner, M. A. Hogg, P. J. Oakes, S. D. Reicher, & M. S. Wetherell (Eds.), *Rediscovering the social group: A self-categorization theory*. Oxford: Blackwell.

Turner, J. C. (1991). *Social influence*. Buckingham, UK: Open University Press.

Turner, J. C. (2001). Foreword to S. A. Haslam. *Psychology in organizations: The social identity approach* (pp. x–xiii). London: Sage.

Turner, J. C., & Haslam, S. A. (2001). Social identity, organizations and leadership. In M. E. Turner (Ed.), *Groups at work: Advances in theory and research* (pp. 25–65). Hillsdale, NJ: Erlbaum.

Turner, J. C., Hogg, M. A., Oakes, P. J., Reicher, S. D., & Wetherell, M. S. (1987). *Rediscovering the social group: A self-categorization theory*. Oxford, UK: Blackwell.

Turner, J. C., & Oakes, P. J. (1997). The socially structured mind. In C. McGarty & S. A. Haslam (Eds.), *The message of social psychology: Perspectives on mind in society* (pp. 355–373). Oxford: Blackwell.

Turner, J. C., Oakes, P. J., Haslam, S. A., & McGarty, C. A. (1994). Self and collective: Cognition and social context. *Personality and Social Psychology Bulletin*, 20, 454–463.

Turner, J. C., & Onorato, R. (1999). Social identity, personality and the self-concept: A self-categorization perspective. In T. R. Tyler, R. Kramer, & O. John (Eds.), *The psychology of the social self.* Hillsdale, NJ: Erlbaum.

Turner, J. C., Wetherell, M. S., & Hogg, M. A. (1989). Referent informational influence and group polarization. *British Journal of Social Psychology*, 28, 135–147.

Turner, M. E., & Pratkanis, A. R. (1998). Twenty-five years of groupthink theory and research: Lessons from an evaluation of the theory. *Organizational Behaviour and Human Decision Processes*, 2/3, 210–235.

Tyler, T. R. (1999). Why people co-operate with organizations: An identity-based perspective. In B. M. Staw & R. Sutton (Eds.), *Research in organizational behaviour* (Vol. 21, pp. 201–246). Greenwich, CT: JAI Press.

Tyler, T. R., & Blader, S. (2000). *Co-operation in groups: Procedural justice, social identity and behavioral engagement.* Philadelphia, PA: Psychology Press.

van Knippenberg, D. (2000). Work motivation and performance: A social identity perspective. *Applied Psychology: An International Review*, 49, 357–371.

van Knippenberg, D., & van Schie, E. C. M. (2000). Foci and correlates of organizational identification. *Journal of Occupational and Organizational Psychology*, 73, 137–147.

van Knippenberg, D., & Haslam, S. A. (2003). Realizing the diversity dividend: Exploring the subtle interplay between identity, ideology and reality. In S. A. Haslam, D. van Knippenberg, M. J. Platow, & N. Ellemers (Eds.), *Social identity at work: Developing theory for organizational practice* (pp. 61–77). Philadelphia, PA: Taylor & Francis.

Veenstra, K., & Haslam, S. A. (2000). Willingness to participate in industrial protest: Exploring social identification in context. *British Journal of Social Psychology*, 39, 153–172.

Wanous, J. P., Reichers, A. E., & Austin, J. T. (2000). Cynicism about organizational change. *Group and Organization Management*, 25, 132–153.

Wegge, J. (2000). Participation in group goal setting: Some novel findings and a comprehensive model as a new ending to an old story. *Applied Psychology: An International Review*, 49, 497–515.

Wegge, J., & Haslam, S. A. (2003). Group goal-setting, social identity and self-categorization: Engaging the collective self to enhance group performance and organizational outcomes. In S. A. Haslam, D. van Knippenberg, M. J. Platow, & N. Ellemers (Eds.), *Social identity at work: Developing theory for organizational practice* (pp. 43–59). Philadelphia, PA: Taylor & Francis.

Wegge, J., & Kleinbeck, U. (1996). Goal-setting and group performance: Impact of achievement and affiliation motives, participation in goal-setting, and task interdependence of group members. In T. Gjesme & R. Nygard (Eds.), *Advances in motivation* (pp. 145–177). Oslo: Scandinavian University Press.

Weick, K. E., & Roberts, K. H. (1993). Collective mind in organizations: Heedful interrelating on flight decks. *Administrative Science Quarterly*, 38, 357–381.

Whyte, W. H. (1960). *The organization man.* Harmondsworth, UK: Penguin.

Worchel, S., Coutant-Sassic, D., & Wong, F. (1993). Toward a more balanced view of conflict: There is a positive side. In S. Worchel & J. Simpson (Eds.), *Conflict between people and groups* (pp. 73–89). Chicago: Nelson-Hall.

Worchel, S., Rothgerber, H., Day, A., Hart, D., & Butemeyer, J. (1998). Social identity and individual productivity within groups. *British Journal of Social Psychology*, 37, 389–413.

Yukl, G. A. (1989). *Leadership in organizations* (2nd ed.). Englewood Cliffs, NJ: Prentice Hall.

Part Nine

Teamwork

49

Time and Transition in Work Teams: Toward a New Model of Group Development

Connie J. G. Gersick

Source: *Academy of Management Journal* 31 (1) (1988): 9–41.

Groups are essential management tools. Organizations use teams to put novel combinations of people to work on novel problems and use committees to deal with especially critical decisions; indeed, organizations largely consist of permanent and temporary groups (Huse & Cummings, 1985). Given the importance of group management, there is a curious gap in researchers' use of existing knowledge. For years, researchers studying group development — the path a group takes over its life-span toward the accomplishment of its main tasks — have reported that groups change predictably over time. This information suggests that, to understand what makes groups work effectively, both theorists and managers ought to take change over time into account. However, little group-effectiveness research has done so (McGrath, 1986).

One reason for the gap may lie in what is unknown about group development. Traditional models shed little light on the triggers or mechanisms of change or on the role of a group's environment in its development. Both areas are of key importance to group effectiveness (Gladstein, 1984; Goodstein & Dovico, 1979; McGrath, 1986). This hypothesis-generating study, stimulated by an unexpected set of empirical findings, proposed a new way to conceptualize group development. It is based on a different paradigm of change than that which underlies traditional models, and it addresses the timing and mechanisms of change and groups' dynamic relations with their environments.

Traditional Models of Group Development

There have been two main streams of research and theory about group development. The first stream deals with group dynamics, the other with phases in group problem solving. Group dynamics research on development began in the late 1940s, with a focus on the psychosocial and emotional aspects of group life. Working primarily with therapy groups, T-groups, and self-study groups, researchers originally saw a group's task in terms of the achievement of personal and interpersonal goals like insight, learning, or honest communication (Mills, 1979). They explored development as the progress, over a group's life-span, of members' ability to handle issues seen as critical to their ability to work, such as dependency, control, and intimacy (Bennis & Shepard, 1956; Bion, 1961; Mann, Gibbard, & Hartman, 1967; Slater, 1966).

In 1965, Tuckman synthesized this literature in a model of group development as a unitary sequence that is frequently cited today. The sequence, theoretically the same for every group, consists of forming, storming, norming, and performing. Tuckman and Jensen's 1977 update of the literature on groups left this model in place, except for the addition of a final stage, adjourning. Models offered subsequently have also kept the same pattern. Proposed sequences include: define the situation, develop new skills, develop appropriate roles, carry out the work (Hare, 1976); orientation, dissatisfaction, resolution, production, termination (LaCoursiere, 1980); and generate plans, ideas, and goals; choose/agree on alternatives, goals, and policies; resolve conflicts and develop norms; perform action tasks and maintain cohesion (McGrath, 1984).

The second stream of research on group development concerns phases in group problem solving, or decision development. Researchers have typically worked with groups with short life-spans, usually minutes or hours, and studied them in a laboratory as they performed a limited task of solving a specific problem. Studies have focused on discovering the sequences of activities through which groups empirically reach solutions — or should reach solutions — and have used various systems of categories to analyze results. By abstracting the rhetorical form of group members' talk from its content and recording percentages of statements made in categories like "agree" and "gives orientation," researchers have portrayed the structure of group discussion. The classic study in this tradition is Bales and Strodtbeck's (1951) unitary sequence model of three phases in groups' movement toward goals: orientation, evaluation, and control.

Though they differ somewhat in the particulars, models from both streams of research have important similarities. Indeed, Poole asserted that "for thirty years, researchers on group development have been conducting the same study with minor alterations" (1983b: 341). The resultant models are deeply grounded in the paradigm of group development as an inevitable progression: a group cannot get to stage four without first going through stages one, two, and three. For this reason, researchers construe development as movement in a forward direction and expect every group to follow the same historical path. In this paradigm, an environment may constrain systems' ability to develop, but it cannot alter the developmental stages or their sequence.

Some theorists have criticized the validity of such models. Research by Fisher (1970) and by Scheidel and Crowell (1964) suggested that group discussion proceeds in iterative cycles, not in linear order. Bell (1982) and Seeger (1983) questioned Bales and Strodtbeck's methodology. Poole (1981, 1983a, 1983b) raised the most serious challenge to the problem-solving models by demonstrating that there are many possible sequences through which decisions can develop in groups, not just one.[1] Despite these critiques, however, the classic research continues to be widely cited, and the traditional models continue to be widely presented in management texts as the facts of group development (Hellriegel,

Slocum, & Woodman, 1986; Szilagy & Wallace, 1987; Tosi, Rizzo, & Carroll, 1986).

Apart from the question of validity, there are gaps in all the extant models, including those of the critics, that seriously limit their contribution to broader research and theory about groups and group effectiveness. First, as Tuckman pointed out in 1965 and others have noted up to the present (Hare, 1976; McGrath, 1986; Poole, 1983b), they offer snapshots of groups at different points in their life-spans but say little about the mechanisms of change, what triggers it, or how long a group will remain in any one stage. Second, existing models have treated groups as closed systems (Goodstein & Dovico, 1979). Without guidance on the interplay between groups' development and environmental contingencies, the models are particularly limited in their utility for task groups in organizations. Not only do organizational task groups' assignments, resources, and requirements for success usually emanate from outside the groups (Gladstein, 1984; Hackman, 1985), such groups' communications with their environments are often pivotal to their effectiveness (Katz, 1982; Katz & Tushman, 1979).

The Approach of this Study

The ideas presented here originated during a field study of how task forces — naturally-occurring teams brought together specifically to do projects in a limited time period — actually get work done. The question that drove the research was, what does a group in an organization do, from the moment it convenes to the end of its life-span, to create the specific product that exists at the conclusion of its last meeting? I was therefore interested not just in interpersonal issues or problem-solving activities, the foci of past research, but in groups' attention to outside resources and requirements, their temporal pacing, and in short, in whatever groups did to make their products come out specifically the way they did, when they did. Since the traditional models do not attend to these issues, I chose an inductive, qualitative approach to increase the chances of discovering the unanticipated and to permit analysis of change and development in the specific content of each team's work.

This study was designed to generate new theory, not to test existing theory, and the paper is organized to present a new model, not to refute an old one. For clarity, however, differences between the proposed and traditional models of group development are noted after each segment of the Results section.

Methods

Because this study was somewhat unconventional, it may help to start with an overview. I observed four groups (A, B, C, and D in Table 1) between winter 1980 and spring 1981, attending every meeting of every group and generating complete transcripts for each. This observation was done as part of a larger study of group effectiveness (Gersick, 1982; Hackman, forthcoming). I also prepared a detailed group-project history for presentation to each team.

After I had completed studies of the four groups it was evident that their lives had not gone the way the traditional models predicted. Not only did no single developmental model fit all the teams, the paradigm of group development as a universal string of stages did not fit the four teams taken together. The sequences of activities that teams went through differed radically across groups. Moreover, activities and issues that most theories described as sequential progressions were in some cases fully simultaneous or reversed.

Those findings prompted me to reexamine the groups' transcripts. I began formulating a tentative new model of group development through the method of grounded theory (Glaser & Strauss, 1967), identifying similarities and differences across the histories and checking emerging hypotheses against original raw data. The results were rewarding, but since three of the four groups were from the same setting, it seemed important to continue to expand the data base. I sought groups that fit into the research domain but that varied as much as possible in project content and organizational setting. As Harris and Sutton pointed out, "Similarities observed across a diverse sample offer firmer grounding for . . . propositions [about the constant elements of a model] than constant elements observed in a homogeneous sample" (1986: 8). Four additional groups (E, F, G, and H in Table 1) were studied in 1982–83. In line with Glaser and Strauss's suggestion, I stopped after observing the second set of groups because all the results were highly consistent.

The Research Domain

Several features distinguish the groups included in the domain of this research. They were real groups — members had interdependent relations with one another and developed differentiated roles over time, and the groups were perceived as such both by members and nonmembers (Alderfer, 1977). Each group was convened specifically to develop a concrete piece of work; the groups' lives began and ended with the initiation and completion of special projects. Members had collective responsibility for the work. They were not merely working side by side or carrying out preset orders; they had to make interdependent decisions about what to create and how to proceed. The groups all worked within ongoing organizations, had external managers or supervisors, and produced their products for outsiders' use or evaluation. Finally, every group had to complete its work by a deadline.

Data Sources

The eight groups in the study (see Table 1) came from six different organizations in the Northeast; the three student groups came from the same university. Their life-spans varied in duration from seven days to six months. I did not select groups randomly but did choose them carefully to ensure that they fit within the research domain and that all meetings could be observed from the

Table 1: The Groups Observed

Teams[a]	Task	Time-span	Number of meetings
A. Graduate management students: 3 men	Analyze a live management case	11 days	8
B. Graduate management students: 2 men, 3 women	Analyze a live management case	15 days	7
C. Graduate management students: 3 men, 1 woman	Analyze a live management case	7 days	7
D. Community fundraising agency committee: 4 men, 2 women	Design a procedure to evaluate recipient agencies	3 months	4
E. Bank task force: 4 men	Design a new bank account	34 days	4
F. Hospital administrators: 3 men, 2 women	Plan a one-day management retreat	12 weeks	10
G. Psychiatrists and social workers: 8 men, 4 women[b]	Reorganize two units of a treatment facility	9 weeks[c]	7
H. University faculty members and administrators: 6 men	Design a new academic institute for computer sciences	6 months[c]	25

[a]The three student groups were from one large, private university. Team H was from a small university.
[b]Two other members attended only once; one other member attended two meetings.
[c]The actual time-span (shown) differed from the initially expected span (see Table 2).

start to the finish of their projects. The management students were recruited from graduate courses that required group projects. After describing the study to each class, I asked the groups to volunteer. I gained entry to the other five groups through referrals to individual members. Team members were provided with information about the study and with opportunities to ask questions; no team was included without all its members' permission. All teams except team D permitted audio taping.

Data Collection

Every meeting of every team was observed, and handwritten transcripts were made during each meeting to back up the audio tapes. In addition to records of members' verbal communication, the handwritten notes included group-level indicators of the energy members applied to their work (attendance, scheduling, and duration of meetings), the use of physical devices to structure work (writing on blackboards and taking notes), and routines (meeting times, locations, and seating patterns). For the second four groups, I also interviewed members after their projects were over to address aspects of each project's development that I did not directly observe: the project's history, events that happened outside meetings, and members' expectations, perceptions, and evaluations of the project.

Data Analysis

This study follows the tradition of group dynamics research in its qualitative analytical approach. I developed a case history for each of the first four groups after its product was completed, the unit of analysis being the group meeting. I did not reduce teams' activities to a priori categories for three reasons. (1) Existing category systems have measured the frequency of groups' activities without necessarily indicating their meaning; a large percentage of problem-orientation statements, for example, could mean either that a group did a careful job or that it had great difficulty defining its task. (2) A priori categories would have been unable to capture qualitative, substantive revisions in groups' product designs. (3) Category systems may be used for specific hypothesis testing but are inappropriate for inductive discourse analysis in theory development (Labov & Fanshel, 1977: 57).

Instead of using a priori categories, I read transcripts repeatedly and used marginal notes to produce literal descriptions of what was said and done at each meeting that were much like detailed minutes. These descriptions encompassed modes of talk, like production work, arguing, and joking; topics covered; teams' performance strategies, that is, implicit or explicit methods of attacking the work; any immediate or long-term planning they did; patterns of relations among members, such as roles, coalitions, and conflicts; and teams' discussions about or with outside stakeholders and authorities.

The entire course of meetings was searched to pinpoint milestones in the design of the products. This process was similar to that usually followed implicitly when a scholar develops a history of the body of work of an artist, writer, or scientist. I identified ideas and decisions that gave the product its basic shape or that would be the fundamental choices in a decision tree if the finished product were to be diagrammed. I also identified points at which milestone ideas were first proposed, whether or not they were accepted at that time. The expression of agreement to adopt a proposal and evidence that the proposal had been adopted were the characteristics of milestone decisions. When a proposal was adopted, either subsequent discussion was premised on it or concrete action followed from it. The milestones added precision to the qualitative historical portrait of each team's product. I searched the complete string of each team's meetings to identify substantive themes of discussion and patterns of group behavior that persisted across meetings and to see when those themes and patterns ceased or changed.

After the first four histories were complete, I searched them for general patterns by isolating the main points from each team's case, forming hypotheses based on the similarities and differences across groups, and then returning to the data to assess and revise the hypotheses. Analysis of data from the first four groups suggested a new model of group development, which I explored and refined in the second stage of the study.

Analysis of the second set of groups again began with the construction of a detailed project history for each team, but construction of the second set of histories was more systematic. To help preserve the literal completeness of project histories and to forestall premature closure on the developmental model, I condensed each team's transcripts in three successive steps. Every turn members took to speak was numbered and the content condensed to retain the literal meaning in a streamlined form; for example, "628: Rick role-plays president's reaction to the idea of tiering the account." I then condensed these documents by abstracting members' exchanges, a few statements at a time, into a detailed topic-by-topic record of the meeting; for example, "646–656: strategizing how to get soundings from outsiders on whether or not to tier the account." The third condensation produced a concise list of the events — the discussions, decisions, arguments, and questions — of each meeting. The following is a sample item: "Team estimates outsiders' reactions to tiering account. Decides to test the waters before launching full design effort; plans how to probe without losing control over product design." The condensation process reduced transcripts of 50 or more pages to 1-page lists, concise enough to allow an overall view of teams' progress across all meetings, yet documented minutely enough to trace general observations back to the numbered transcripts for concrete substantiation or refutation.

After the second four teams' histories were complete, I used them for another iteration of theory-building work. Transcripts of meetings and interviews were searched to see whether or not features common to the first four groups appeared. Again, similarities and differences among all eight groups were used to extend and refine the model.

Presentation of Results

Qualitative research permits wide exploration but forgoes the great economy and precision with which quantified results can be summarized and tested. This study employed description and excerpts from meetings and interviews to document, in members' words as often as possible, what happened in the teams and how they progressed over time.

Results

An Overview of the Model

The data revealed that teams used widely diverse behaviors to do their work; however, the *timing* of when groups formed, maintained, and changed the way they worked was highly congruent. If the groups had fit the traditional models, not only would they have gone through the same sequence of activities, they would also have begun with an open-ended exploration period. Instead, every group exhibited a distinctive approach to its task as soon as it commenced and stayed with that approach through a period of inertia[2] that lasted for half its allotted time. Every group then underwent a major transition. In a concentrated burst of changes, groups dropped old patterns, reengaged with outside supervisors, adopted new perspectives on their work, and made dramatic progress. The events that occurred during those transitions, especially groups' interactions with their environments, shaped a new approach to its task for each group. Those approaches carried groups through a second major phase of inertial activity, in which they executed plans created at their transitions. An especially interesting discovery was that each group experienced its transition at the same point in its calendar — precisely halfway between its first meeting and its official deadline — despite wide variation in the amounts of time the eight teams were allotted for their projects.

This pattern of findings did not simply suggest a different stage theory, with new names for the stages. The term "stage" connotes hierarchical progress from one step to another (Levinson, 1986), and the search for stages is an effort to "validly distinguish . . . types of behavior" (Poole, 1981: 6–7), each of which is indicative of a different stage. "Stage X" includes the same behavior in every group. This study's findings identified temporal periods, which I termed phases, that emerged as bounded eras within each group, without being composed of identical activities across groups and without necessarily progressing hierarchically. It was like seeing the game of football as progressing through a structure of quarters (phases) with a major half-time break versus seeing the game as progressing in a characteristic sequence of distinguishable styles of play (stages). A different paradigm of development appeared to be needed.

The paradigm through which I came to interpret the findings resembles a relatively new concept from the field of natural history that has not heretofore been applied to groups: *punctuated equilibrium* (Eldredge & Gould, 1972). In this paradigm, systems progress through an alternation of stasis and sudden appearance — long periods of inertia, punctuated by concentrated, revolutionary periods of quantum change. Systems' histories are expected to vary because situational contingencies are expected to influence significantly the path a system takes at its inception and during periods of revolutionary change, when systems' directions are formed and reformed.

In sum, the proposed model described groups' development as a punctuated equilibrium. *Phase 1*, the first half of groups' calendar time, is an initial period of inertial movement whose direction is set by the end of the group's first meeting. At the midpoint of their allotted calendar time, groups undergo a *transition*, which sets a revised direction for *phase 2*, a second period of inertial movement. Within this phase 1-transition-phase 2 pattern, two additional points are of special interest: the first meeting, because it displays the patterns of phase 1; and the last meeting, or completion, because it is a period when groups markedly accelerate and finish off work generated during phase 2.

Special Aspects of the Model

The importance of the first meeting was its power to display the behaviors (process) and themes (content) that dominated the first half of each group's life. Each group appears to have formed almost immediately a framework of givens about its situation and how it would behave. This framework in effect constituted a stable platform from which the group operated throughout phase 1.

Members occasionally clearly indicated their approach to something, stating their premises and how they planned to behave ("The key issue here is X; let's work on it by doing Y"); however, teams seldom formulated their frameworks through explicit deliberation. Instead, frameworks were established implicitly, by what was said and done repeatedly in the group. That phenomenon was observable on several fronts. The themes, topics, and premises of discussions provided evidence; for example, a group might take as given that its organization's staff is not talented and discuss every project idea in terms of how hard it would be to explain to the sales force. Members' interaction patterns — the roles, alliances, and battles members took on — also revealed implicit frameworks. Performance strategies, or methods of attacking the work, were another indicator. A group's behavior toward its external contexts — for example, acting dependent or acting assertive about outside stakeholders — provided evidence as well. Finally a group's overall standing on its task — whether it was confident of a plan and working on it, deadlocked in disagreement over goals, or explicitly opposed to the assignment and unwilling to begin work[3] — helped to establish its implicit framework.

Central approaches and behavior patterns that appeared during first meetings and persisted during phase 1 disappeared at the halfway point as groups explicitly dropped old approaches and searched for new ones. They revised their frameworks. The clearest sign of transition was the major jump in progress that each group made on its project at the temporal midpoint of its calendar. Further comparisons, across meetings within groups and across groups, revealed five empirical earmarks of the transition, a set of events uniquely characteristic of midpoint meetings. The frameworks that groups formed at transition carried them through a second period of momentum, phase 2, to a final burst of completion activities at their last meetings.

Illustration of the Model

Three groups will serve as examples to illustrate each part of the model. Each is representative of the overall model, yet each shows some aspects especially concisely, and the differences between the groups show the diversity within the pattern.

First Meeting and Phase 1

Almost immediately, in every team studied, members displayed the framework through which they approached their projects for the first half of their calendar time. Excerpts show the scope, variety, and nature of those frameworks.

Excerpt 1 (E1)

A team of three graduate management students start their first, five-minute encounter to plan work on a group case assignment, defined by the professor as an organizational design problem.[4]

1. Jack: We should try to read the [assigned] material.

2. Rajeev: But this isn't an organizational design problem, it's a strategic planning problem.

3. (Jack and Bert agree.)

4. Rajeev: I think what we have to do is prepare a way of growth [for the client].

5. (Nods, "yes" from Jack and Bert.)

Excerpt 1, representing less than one minute from the very start of a team's life, gives a clear view of the opening framework. The team's approach toward its organizational context (the professor and his requirements) is plain. The members are not going to read the material; they disagree with the professor's definition of the task and will define their project to suit themselves.

Their pattern of internal interaction is equally visible. When Rajeev made three consequential proposals — about the definition of the task, the team's lack of obligations to the professor, and the goal they should aim for — everyone concurred. There was no initial "storming" (Tuckman, 1965; Tuckman & Jensen, 1977) in this group. The clip also shows this team's starting approach toward its task: confidence about what the problem is, what the goal ought to be, and how to get to work on it. The team's stated performance strategy was to use strategic planning techniques to "prepare a way of growth."

Excerpt 2 (E2)

The following excerpt of the team's next work session, two days later, shows how well the minute of dialogue from the first meeting indicated lasting patterns.

1. Jack: I have not looked at any of the readings — did you look at all?

2. (Bert and Rajeev laugh.)

3. Jack: . . . I was thinking . . . we could do alternatives — different ways to grow . . . like a prospectus for a consulting study.

4. Bert: That's exactly the way I'd go. (Restates Jack's position.)

5. Rajeev: Well . . . we are thinking mostly in the same manner. My idea was . . . . (He states the same plan.)

(After five minutes of discussion about the client and his situation, Rajeev suggests they start work.)

6. Jack: We've got some more time . . . I think it would be premature to describe alternative goals yet. . . .

7. Rajeev: If we can generate some of the assumptions now and talk about the alternatives later — it's a two-step thing.

8. Jack: OK, that's fine. Let's start that.

9. Rajeev: (at blackboard) What are the things on which the business depends?

The dialogue shows that the team is still disregarding the professor (E2, 1 & 2),[5] still working in easy agreement (E2, 4, 5, & 8), and still taking the same approach to the task (E2, 3). It also shows the group acting on its expressed intentions, employing a logical, orderly technique to construct its product (E2, 6–9). The team worked within this framework for two full meetings.

Rajeev led the group through a structured set of strategic planning questions. At that point, the team had a complete draft outline of a growth plan for its client.

Excerpt 3 (E3)

A group of four bank executives open their first meeting to design a new type of account.

1. Don: What do you think we ought to do to start this, Rick? Just go through each of these? (Referring to a written list of topics.)

2. Rick: Well, I want to explain to Gil and Porter — we had a little rump session the other day just to say "What the hell *is* this thing? What does it *say*, and what are the things that we have to decide?" And what we did was run through a group of 'em. . . . These are not necessarily in order of importance — they're in order of the way we thought of 'em, really. . . .

This excerpt of the first 25 seconds in the life of another task force, showing a quite different beginning, also illustrates the team's approach toward its task, and its performance strategy. This team did not choose a product through the whole first half of its life. Given a new set of federal rules, the team's reaction was to ask the questions "What the hell *is* this thing? What does it say?" The team was uncertain, and as the project began they approached the task as a job of mapping out "the things we have to decide."

The excerpt is also an elegant summary of the group's performance strategies. It shows that the leader prepared for the meeting with one other member, that the preparation consisted of generating a list of topics to be covered, and that this list was arranged only "in order of the way we thought of 'em." This general strategy was followed for every one of the group's meetings. Before each meeting, a pair of members prepared skeletal documents for the group to work from. Items were checked off the documents as they were covered, but discussions were more like pinball games than orderly progressions. Each question ricocheted the conversational ball onto several new questions, and occasionally bells and lights went off as the team made a decision about a specific point.

The link between the team's pinball-style performance strategies and its approach to its task as "mapping" was strong. As one member, trying to keep track of the discussion, said to another, "It's all intertwined."

For the first two of its four meetings, the dominant activity of this team's members was to generate the questions that needed to be settled in a loosely structured format and to go as far as they could in answering each. Their own definition of where they were, from inception through the end of this period, was that they did not yet know "what we're planning to offer. We're still thinking."

Excerpt 4 (E4)

Five hospital department heads are a few minutes into their first meeting to plan the fourth in a series of management retreats for their peers and division chief. They have just chosen a date and place.

1. Nancy: So, in order of preference, the [dates we want are] the tenth, third, and ninth.

2. Sandra: Sounds great. . . . (to Bernard): I think you probably should talk to the division chief about — did he give you any thoughts about what we should do next?

3. Bernard: I'd say — that's on us. . . .

4. Sandra: Um hum. The only thing I feel strongly about is — it's not time to have an outside [facilitator].

5. Bill: Well, I'm not for or against [that] but — what are we trying to achieve? Trust among — people? . . . the highest value [on the participants' critique of the previous retreat was] developing trust among the managers themselves . . . and not only trust among ourselves. . . . I think there has to be trust — upward.

6. Sandra: And *that's* the issue we talk about, and walk around the edges of. . . . We say, "Yeah, Tom [division chief], we trust you," but we don't trust you very *well*, 'cause we don't dare say we don't trust *you*, Tom.

7. Bill: Yep. The sacred cow, like you said earlier.

8. Bernard: There's three levels, aren't there? The people we supervise, peers that we work with, as well as. . . .

The hospital administrators team began at an impasse. After they had swiftly decided where and when to hold the retreat, the pace plummeted with the question of what to do with the event. The team's opening framework shows the problem. The members' position toward their organizational context was complicated because the final product had to please the task delegator, the division chief, but he had given no indication of what they "should do next," and the team leader was unwilling to ask (E4, 2–4). The team's approach to its task was closely related to that ambivalence. Members' opening premises were that the retreat ought to deal with trust, especially with regard to the division chief, and that they should run it themselves without bringing in an outside facilitator (E4, 4–6). Those premises put the team in a self-imposed bind as evidence by the statement " 'cause we don't dare say we can't trust *you*, Tom." The team's key phase 1 question was "What are we trying to achieve?" (E4, 5).

The concern with intradivisional relationships and the feelings of directionlessness in the group continued for the first six weeks of the team's 12-week life-span. In a later interview, a member said, "From [the beginning] to the [end of October] all I can remember is talking. With absolutely *no* idea of what was going to happen. None." This was so even though members were concerned and *wanted* meetings to be different. Another member said, "It was very frustrating from September 20 until maybe November 1 for me [the first through the sixth weeks]. That's a long time to be frustrated." A third member noted, "It was very difficult to get the work going. We had no direction, only to put together a retreat. . . . Nothing was happening! I was very frustrated." The group made no decisions about what to do at the retreat during its first phase.

Table 2 summarizes the findings about first meetings and phase 1. Column 1 presents each team's starting approach toward its task, and column 2 summarizes the central task activity of phase 1, including the first meeting.

Each group immediately established an integrated framework of performance strategies, interaction patterns, and approaches toward its task and outside context. The most concise illustration of this finding comes from the student group, whose (1) easy agreement on (2) a specific plan for its work represented (3) a decision to ignore the outside requirements for its task — all within the same minute of group discussion. Such frameworks embodied the central themes that dominated all through the first half of groups' calendar time, even for teams that were frustrated with the paths they were following. This finding contradicts traditional models, which pose teams' beginnings as a discrete stage of indeterminate duration during which teams orient themselves to their situation, explicitly debating and choosing what do do.

Though each team began with the formation of a framework, each framework was unique as illustrated by the contrast between the students' instant confidence and the hospital administrators' directionlessness. Some teams began with harmonious internal interaction patterns; others, with internal storms. Teams took very different approaches to authority figures from their outside contexts, as evidenced by the hospital administrator's preoccupation with the division chief versus the students' cheerful disregard for the professor. These findings contradict the typical stage theory paradigm in which it is assumed that all teams essentially begin with the same approach toward their task (e.g., orientation), their team (e.g., forming then storming), and toward authority (e.g., dependency).[6]

The Midpoint Transition

As each group approached the midpoint between the time it started work and its deadline, it underwent great change. The following excerpts from transitional meetings illustrate the nature and depth of this change. Particular points to

Table 2: An Overview of the Groups' Life Cycles

Teams	First meeting	Phase 1	Transition	Phase 2	Completion
A. Student team A	Agreement on a plan.	Details of plan worked out: client's ''growth options.''	First draft revised; second draft planned.	Details of second plan worked out: organization design.	Homework compiled into paper, finished, and edited.
B. Student team B	Disagreement on task definition.	Argument over how to define task: challenge vs. follow client's problem statement.	Task defined; case analysis rough-outlined.	Details of outline worked out: affirmative action plan, following client's request.	Paper (drafted by one member) finished; edited.
C. Student team C	One member proposes concrete plan; others oppose it.	Argument over details of competing plans (''structured'' vs. ''minimal'') but no discussion of goals.	Goals chosen; case analysis outlined.	Details of outline worked out: ''minimalist'' U.S. trade policy.	Homework compiled into paper, finished, and edited.
D. Community fundraising agency committee	Agreement on a plan.	Details of plan worked out: ''nonthreatening'' self-evaluation for member agencies.	First draft revised; second draft planned.	Details of second plan worked out: explicitly allocations-related evaluation plan.	Report (drafted by two members) edited.
E. Bank task force	Uncertainty about new product; federal regulations unclear.	Team ''answers questions''; maps possible account features.	Account completely outlined.	Members work throughout bank on systems, supplies for account.	Account finalized for advertising; bank-wide training planned.
F. Hospital administrators	Team fixes on ''trust'' theme; uncertain what to do with it for program.	Unstructured trial and rejection of program possibilities; disagreement about goals.	Complete program outlined.	Consultant hired to plan program; team arranges housekeeping details.	Responsibility for final preparations delegated.
G. Psychiatrists and social workers	Leader presents ''the givens''; team opposes project.	Subgroup reports presented; members object to all plans; leader rebuts objections.	Disagreement persists; leader picks one plan; redelegates task; dissolves team.		
H. University faculty members and administrators	Team divided on whether to accept project; leader proposes diagnosis as first step.	Structured exploration; diagnosis of situation.	Team redefines task; commits to project.	Computer institute designed (original task) plus system for university computer facilities planning.	Report (written by leader from members' drafts) edited and approved.

notice are members' comments about the time and their behavior toward external supervisors.

Excerpt 5 (E5)

The students begin their meeting on the sixth day of an 11-day span.

1. Rajeev: I think, what he said today in class — I have, already, lots of criticism on our outline. What we've done now is OK, but we need a lot more emphasis on organization design than what we — I've been doing up to now.

2. Jack: I think you're right. We've already been talking about [X]. We should be talking more about [Y].

3. Rajeev: We've done it — and it's super — but we need to do other things, too.

4. (Bert agrees.)

5. Jack: After hearing today's discussion — we need to say [X] more directly. And we want to say more explicitly that. . . .

6. Rajeev: . . . should we be . . . organized and look at the outline? . . . We should know where we're going.

(The group goes quickly through the outline members had prepared for the meeting, noting changes and additions they want to make.)

7. Rajeev: The problem is, we're very short on time.

The students came to this meeting having just finished the outline of the strategic plan they had set out to do at their opening encounter (see E1). At their midpoint, they stopped barreling along on their first task. They marked the completion of that work, evaluated it, and generated a fresh, significantly revised agenda. The team's change in outlook on its task coincided with a change in stance toward the professor. Revisions were made that were based on "what he said today in class" and "hearing today's discussion." Having reaffirmed the value of their first approach to the case, members reversed their original conviction that it was "not an organizational design problem." This was the first time members allowed their work to be influenced by the professor, and at this point, they accepted his influence enthusiastically.

It is significant that Rajeev's remark, "we're very short on time," was only the second comment about the adequacy of the time the group had for the project, and it marked a switch from Jack's early sentiment that "we've got some more time" (E2, 6). A new sense of urgency marked this meeting.

The students knew what they wanted to create at their first meeting; the bank team members started much closer to scratch and they were not nearly as far along as the students at the midpoint. Their transitional meeting was different from the students' in character but similar in scope and magnitude.

Since the bankers scheduled each meeting ad hoc, it is noteworthy that the third one fell on the 17th day of a 34-day span. As he convened the group, Don worried that if they continued their present course, they might not finish on time: "We can explore all the ramifications [of the regulations], but I just hope we don't get *stuck*, toward the end, without. . . ." In the first minutes of this meeting, members confirmed their intentions to move to the next step: "Basically, we're gonna lay out the characteristics of the account." The next two hours were spent problem solving with two staff experts who had been invited to the meeting to make sure the account design would fit the bank's computer systems. By the end of that time, the basic design was finished.

The leap forward on the task coincided with a change in the team's relationship with its organizational context. At first, the group decided its meetings would be closed to staff people "[until] . . . we know how we want to handle this. . . ." The third meeting marked that shift. Moreover, one of the members had a key meeting with the bank chairman that afternoon, to argue for the extra resources he now felt were needed to market the team's product successfully.

If the bank team started out less far along in its work than the students, the hospital team was even a step behind the bankers. Though everyone said that intradivision relationships were the key topic to address, the team could not agree on a goal for the retreat and spent the first half of its life describing and rejecting a series of ideas. Statements 1–3 of the following excerpt show how little concrete progress the team had made halfway through its calendar; the remaining lines show how much they then accomplished at the meeting.

Excerpt 6 (E6)

The hospital administrators hold their fifth meeting, in the sixth week of a 12-week span.

1. Bernard (to Bill, just before the meeting): I'm gonna bring Tom [the division chief] to the next meeting, Bill. . . . Last time we were struggling like we are here — Tom [really helped] to sort things out. . . .

2. Bernard (convening meeting): . . . I think we need to . . . brainstorm about [the program] — see what we might come up with, and bounce it off Tom next time. (He recaps an idea he brought to the previous meeting.)

3. Sandra: We'd each be responsible for an hour of the program? As facilitators, or role playing — whatever we decided to do?

(Later in the meeting, there was a dramatic shift in the discussion when Nancy described a management simulation program on the problems of middle managers, run by a consultant who worked nearby.)

4. Sandra: If awareness is all that comes out of the day . . . I think that's a good — a reasonable goal.

5. Nancy: Understanding, too, some of the forces that operate on us as middle managers — that's where we are, in our relationship with the top manager . . . .

6. Bernard: Yeah . . . that's the thing that we all share together, with the exception of Tom — is that we're in the middle, and it's a difficult spot to be in. And this would show that. . . .

7. Sandra: (adds up the time that the simulation and debriefing would take) So — there's the rest of the day! . . . I think that's reasonable to run by Tom.

(The team endorsed the program and decided to invite the consultant Nancy mentioned to run it. The following are from the close of the meeting.)

8. Bill: We are making progress! I was afraid we weren't moving fast enough!

9. Sandra: I had the same problem! . . . I felt . . . in the beginning, there was a lot of talk. . . . That's necessary in some degree — then, I think, you gotta move on it.

10. Bernard: We've made *progress*, folks. . . . [Next week] Tom'll be here, we'll throw those ideas out to him — Monday, we're going to look at the [conference center] — so we've made progress.

This team's midpoint anxiety about finishing on time showed in the meeting and in interviews: "I was uncomfortable that time was going to run out and we were not going to have it done." "I called Nancy and said 'Look — this needs to start going, or we're going to get to [the program date] wondering what we're going to do!' " Yet in a single session, the team managed to solve all the major problems it had struggled with for six weeks. The theme of the new program design, "being in the middle," was actually not new to the group. It had come up in the very first meeting (see E4, 7) and had been discussed with some enthusiasm at the fourth meeting. But members had been preoccupied trying to make the "trust" idea work. Because "being in the middle" did not fit into the team's original framework, it did not lead to a program design during phase 1.

Two more major changes show in excerpt 6. One was the reversal of the first-meeting approach that members had to run the program themselves (E4, 4) with the decision to get an outside facilitator. Members said in interviews that this change made a tremendous difference. One person captured the whole transition: "The [mood in the team] went *down* . . . and then all of a sudden, it took kind of a swoop . . . 'Ah! It's going to happen!' We decided what we were going to do. . . . The decision to bring in a facilitator was a great relief! Then we got the division chief — he said 'OK, go ahead,' and the rest was just mechanics."

The second change occurred in the team's approach toward its task delegator, when Bernard reversed his early decision not to ask the division chief for help (E4, 3; E6, 1). Indeed, the anticipation of talking *to* Tom appeared to spur the team's work at the same time that it marked the end of the talk *about* him.

The structure of the transition period was similar for all the teams, even though the specific details differed widely. Table 3 shows the timing of each team's transition meeting, describes the changes that occurred in the work at that point, and documents those changes in members' words. Five major indicators, or earmarks, of the transition are reviewed below.[7]

First, teams entered transition meetings at different stages in their work, but for each, progress began with the completion or abandonment of phase 1 agendas. For example, groups A and D entered transition meetings with complete drafts of plans that had been hatched when they started, and team H finished a system diagnosis just before its midpoint (see Table 1). The hospital administrators dropped key premises that the program would be about trust and run by themselves. Team G's leader unexpectedly pronounced the group's task complete at its midpoint (G, 2),[8] but interviews indicated that members, too, felt it was time to move dramatically: "At that point . . . there was a need to go up. But instead of going up, we stopped."

Second, team members expressed urgency about finishing on time. At this time — and no other — members expressed explicit concern about the pace and timeliness of their work: "We ought to be conscious of deadlines" (Team H, transition meeting; see also Table 3: A, 2; B, 1 & 2; D, 4; E, 1; and F, 2). Group G, dissolved with no prior warning (or protest) at its midpoint, was the only team that did not fit this pattern.

Third, teams' transitions all occurred at the midpoints of their official calendars, regardless of the number or length of meetings teams had before or after that.

Fourth, new contact between teams and their organizational contexts played important roles in their transitions. Most often, this contact was between the team and its task delegator. Sometimes it was initiated by the team (E and F), sometimes by both at once (A, D, and H),[9] and sometimes by the task delegators (B and C).

These contacts both fostered decision making and influenced decision outcomes. Five groups showed explicit new interest in the match between their product and outside resources and requirements. Excerpts A and D and the bank's work with computer experts show how groups shaped their products specifically to contextual resources and requirements. The bank group also illustrates the other side of the coin — a team member took his new assessment of the project out to the organization to request more resources. The importance of this contact is highlighted by the exception, team G, whose lack of information about outside requirements exacerbated its inability to choose. A member stated during its pretransition meeting: "*If* we are expected [to do X] then there is no [way to support plan A over plan B, but] . . . that may not be the demand. Obviously, there's a lot of politics outside this room that are going to define what [we] have to do."

Finally, transitions yielded specific new agreements on the ultimate directions teams' work should take. Regardless of how much or how little members argued during phase 1, every team that completed its task agreed at transition on plans that formed the basis for the completion of the work. In teams with easy phase

Table 3: Transition Meetings in the Eight Groups

A. Student team A: Day 6 of 11-day span
Team revises first draft of case analysis; plans final draft.

Opening	(1)	I think, what he said today in class — I have . . . lots of criticism on our outline. . . . We've done it — and it's super — but we need to do other things too.
Closing	(2)	The problem is, we're very short on time.

B. Student team B: Day 7 of 15-day span
Team progresses from argument over how its task should be defined to rough outline of case analysis.

Opening	(1)	This is due next Monday, right?
	(2)	Right. Time to roll.
Later	(3)	Not bad! We spent one hour on one topic, and an hour on another! . . . We're moving along here, too. I feel a lot better at this meeting than I have —
	(4)	Well . . . we're also making decisions to be task-oriented, and take the problem at its face value —

C. Student team C: Day 4 of 7-day span
Team progresses from argument over details of competing plans, with no discussion of overall goals, to goal clarification and complete outline of product.

Opening	(1)	This morning I redesigned the whole presentation! I don't know what the content is, but —
Later	(2)	(Surveying blackboard) OK — we've got goals! Those are the U.S. goals for [X topic]. . . . The [outline for the paper is] the lead-in, the goals, and the strategy.
	(3)	That makes sense! . . .
	(4)	I like it!

D. Community fundraising agency committee: Meeting 3 of four preset meetings
Team revises first plan for evaluation procedure; agrees on final plan.

Opening	(1)	Does anyone have any problem with the . . . evaluation draft?
	(2)	Let's be realistic — we don't have the staff time to sit down with each [recipient] agency every year.
	(3)	What are we accomplishing, then? . . . We need to know [X]. Otherwise I say, "don't bother!"
Later	(4)	(Summing up a revised version of the plan) If you tell [member agencies] they *will* be evaluated . . . and these are questions you'll be asked, so — get your baloney swinging . . .! [Laughter from team] OK. Let's move on, otherwise we're going to get behind.

E. Bank task force: Day 17 of 34-day span
Team progresses from "answering questions" to designing complete outline of new bank account.

Opening	(1)	I just hope we don't get stuck, toward the end, without —
	(2)	What are we gonna do — just — answer a lot of questions today? — or —
	(3)	. . . basically, we're gonna lay out the characteristics of the account.
Closing	(4)	Oh, I think that's super!
	(5)	I think we got a good product!

F. Hospital administrators: Week 6 of 12-week span
Team progresses from uncertainty and disagreement about goal to a complete program plan.

Opening	(1)	. . . we need to . . . come up with [something to] bounce off Tom next time.
Closing	(2)	We are making progress! I was afraid we weren't moving fast enough!
	(3)	We've made progress, folks!

Table 3: *Continued*

G. Psychiatrists and social workers: Week 9 of 17-week span		
Leader chooses one of three reorganization plans to break stalemate; dissolves team.		
Opening	(1)	Is [plan A] a reasonable way to go? *That's* the question.
Closing	(2)	We are nearing the completion of *our* task . . . the next step is turning [the work] over [to Dr. C.] . . . There *is* disagreement in here, [but] I think . . . we have to come down . . . [on one plan]. . . . Then we are – dissolved. . . . Thank you.
H. University faculty members and administrators: Week 7 of 14-week span		
Team redefines task; progresses from skepticism to commitment.		
Opening	(1)	. . . the task force reached a crossroads last meeting . . . and decided it [must choose] whether it should [continue with its original task] or consider the overall needs. For that reason, we've asked two people at the vice-presidential level to . . . help us deliberate that question.
Closing	(2)	I think we've . . . reached a conclusion today, and that is, we need to include the administrative end [in our task].
	(3)	Hey, I think we're finally giving Connie some good stuff here! Isn't this typical? You go through, you roll along, and then all of a sudden you say, "What are we doing?" Then we go back and reconstitute ourselves! Anyway, processes are taking place!

1 interaction, the agreeableness itself was not a change. But for teams where phase 1 had been conflictful, transition meetings were high points in collaboration. Indeed, in the one team whose members still disagreed at this point, the leader dissolved the group, chose a plan unilaterally, and moved the work forward by shifting it into other hands (G, 2).

Overall, the changes in teams' work tended to be dialectical. Teams that had started fast, with quick decisions and unhesitating construction of their products, paused at their transitions to evaluate finished work and address shortcomings (A and D). For teams that started slowly, unsure or disagreeing about what to do, transitions were exhilarating periods of structuring, making choices, and pulling together (B, C, E, F, and H). In either case, transitional advances depended on the combination of phase 1 learning and fresh ideas. For example, the bankers' transitional raw materials were ideas generated during phase 1, refined and integrated with the help of expertise newly infused into the team. The hospital administrators, newly open to an alternative format, found use for a theme they had discussed but not developed earlier.

Traditional models of group development do not predict a midpoint transition. They present groups as progressing forward if and whenever they accumulate enough work on specific developmental issues — not at a predictable moment, catalyzed by team members' awareness of time limits. Traditional group development models are silent about team-context relations and the influence of such relations on teams' progress. The findings reported here suggest that there is a predictable time in groups' life cycles when members are particularly influenceable by, and interested in, communication with outsiders. Cases in which task delegators contacted teams at this point suggest this interest might be mutual.

Phase 2

Teams' lives were different after the midpoint transition. In all seven surviving teams, members' approaches toward their tasks clearly changed and advanced (Table 2). All seven executed their transitional plans during this period. Posttransitional changes in teams' internal interaction patterns and approaches toward their outside contexts were not so simple. Transitions did not advance every team in these areas, nor did every team use its transition equally well. Internal troubles that went unaddressed during transition sometimes worsened during phase 2, and teams that were lax in matching their work to outside requirements during the transition showed lasting effects.

The student group, which developed strategic "growth options" for its client in phase 1, spent phase 2 building the organizational design, planned at the transitional meeting, to support those options. As the task approach shifted from strategic planning to organizational design, one element of the team's interaction pattern changed. Jack took over from Rajeev as lead questioner. Other than that, the team continued the easy, orderly agreement of its phase 1 interaction style. The team sustained its new perspective on its context, formed at transition, by maintaining attentiveness to the professor's requirements throughout phase 2.

The bankers spent phase 2 executing the details for the account they had designed at transition; they prepared marketing extras, operational machinery, and documents. With this change in focus, the team deepened its transitional move toward working with the organizational context and also dramatically changed its own interaction pattern. The team did not convene as a group during phase 2 but met individually and in pairs with staff members throughout the bank.

The hospital team's phase 1 uncertainty about the task and discussion of relationships did not recur. A consultant, engaged shortly after the midpoint, took charge of planning the program the team chose at its transition; the team's work for the next four meetings consisted of supplying the consultant with information and arranging menus, invitations, and materials for the retreat.

Though the hospital administrators were like the other teams in using phase 2 to carry out transitional plans in task work, their phase 2 changes in interaction patterns and approach toward outside context were more extreme and less benign. Internally, the team fell apart just after the transitional meeting. Two members, who had engaged in restrained competition through phase 1 but had supported each other at the transition, had a falling out. The same weekend, the team leader and one other member engendered resentment by making some unilateral decisions outside the group, and the interaction in meetings deteriorated. The team's transitional openness toward its context also regressed after the chief appeared at the post-midpoint meeting.

Excerpt 7 (E7)

The following comes from an interview with the hospital team's leader:

> He says "Do what you want. Spend what you want." Then he came to the damn meeting and was worried about money! Giving me mixed signals! That's when I decided, I'm gonna spend what I want and make my own decisions. . . .

By the time the division chief met with the team, the decision to hire the facilitator — the largest expense — had already been made. It was "too late" to be "worried about money," and the team never checked its budget with the chief.

Phase 2 was a second period of inertia in teams' lives, shaped powerfully by the events of their transitions. Teams did not alter their basic approaches toward their tasks within this phase. As one hospital team member stated, "We decided what we were going to do [at the midpoint meeting] . . . and the rest was just mechanics."

Since all teams were doing construction work on their projects during phase 2, similar to "performing" in Tuckman's (1965) synthesis, it was a time when teams were more similar both to each other and to the traditional model than they were in phase 1. However, progress was not so much like traditional models in other respects, since it was not so linear. Some teams started performing earlier than others, without previous conflict; other teams returned to internal conflict after their transition and during phase 2 performance. In every team, transitional work centered explicitly on solving task problems, not on solving internal interaction problems; it is not surprising, then, that some teams' internal processes worsened after the major need for collaborative decision making was past.

Completion

Completion was the phase of teams' lives in which their activities were the most similar to each others'. Three patterns characterized final meetings: (1) groups' task activity changed from generating new materials to editing and preparing existing materials for external use; (2) as part of this preparation, their explicit attention toward outside requirements and expectations rose sharply; and (3) groups expressed more positive or negative feeling about their work and each other. At this point, the major differences among the groups involved not what they were doing but how easily their were doing it. Not surprisingly, groups that had checked outside requirements early on and groups that had paced themselves well all along had easier, shorter final meetings.

The last distinct change in the student team's life occurred the day before the paper was due. This meeting was considerably longer than any other; the team now had to keep working until the case analysis was finished. Members' work activities changed from generating ideas to editing what they had into the form required by the instructor. A sample from that meeting is "I'm not disagreeing with anything you're saying. But I think you got 'em in the wrong section." Though the long hours and the need to edit each other's work made the meeting more difficult than usual, by the time the team was ready to give its presentation, members were expressing their feelings that the project had gone well. The presentation went smoothly and the team received a good grade.

The bank executives' final group meeting marked the "finish [of] all the deliberations" about the design of the account and a shift into activities to educate the public and the branch banks about it: "It's one thing to . . . say we're gonna offer the thing . . . [but now] we've gotta get something out [to the staff] on how to *handle* it." The team went over the account one last time, to get it "written in blood" for the advertising copy, due that day. Then, with two extra staff people, members planned the final approach. After the meeting, everyone rushed off with his own assignment for the new task of getting the whole bank ready for opening day. In interviews later, team members proudly described a memorandum the president had sent congratulating everyone on the success of the account.

By the hospital group's last meeting, its work was mostly done. At this point, the interpersonal tension that had been building during phase 2 erupted in an angry discussion about the handling of the consultant's fee and how to present it to the division chief. But the subject was dropped when a member declared it had been "talked about long enough." The team delegated final responsibilities for the conference and ended the meeting early. On the day of the retreat, half the team members arrived late and left early; otherwise, relations among them appeared smooth. At day's end, the division chief — who had not yet received the bill — toasted the team: "I think this is the best one yet, and I'm looking forward to number five."

In every team, discussion of outsiders' expectations was prominent at the last meeting. As teams anticipated releasing their work into outside hands, they scrutinized it freshly, through outsiders' eyes: "We'll be judged poorly if we . . ."; "You can't promise [X] and then do [Y]." Since phase 2 actions carried out, but did not alter, plans made at transition, teams that entered phase 2 with a poor match between product and requirements had an especially hard time confronting outside expectations at completion. But even teams that discovered in last-day meetings that they had major gaps to fill framed their remaining work as rearranging or fixing what they already had, as these excerpts indicate: "I think our content . . . is good . . . it's just a matter of reorganizing it . . ." (Team B) and "I think we have all the ideas. . . . The main task is how to arrange them" (Team A). Though teams' attention to outside requirements was high at last meetings, completion activities did not undo the basic product revisions established at transition.

Discussion

The traditional paradigm portrays group development as a series of stages or activities through which groups gradually and explicitly get ready to perform, and then perform, their tasks. All groups are expected to follow the same historical path. Proponents of existing models specify neither the mechanisms of change nor the role of a group's environment. In contrast, the paradigm suggested by the current findings indicates that groups develop through the sudden formation, maintenance, and sudden revision of a framework for performance; the developmental process is a punctuated equilibrium. The proposed model highlights the processes through which frameworks are formed and revised and predicts both the timing of progress and when and how in their development groups are likely, or unlikely, to be influenced by their environments. The specific issues and activities that dominate groups' work are left unspecified in the model, since groups' historical paths are expected to vary.

The proposed model works in the following way: A framework of behavioral patterns and assumptions through which a group approaches its project emerges in its first meeting, and the group stays with that framework through the first half of its life. Teams may show little visible progress during this time because members may be unable to perceive a use for the information they are generating until they revise the initial framework. At their calendar midpoints, groups experience transitions — paradigmatic shifts in their approaches to their work — enabling them to capitalize on the gradual learning they have done and make significant advances. The transition is a powerful opportunity for a group to alter the course of its life midstream. But the transition must be used well, for once it is past a team is unlikely to alter its basic plans again. Phase 2, a second period of inertial movement, takes its direction from plans crystallized during the transition. At completion, when a team makes a final effort to satisfy outside expectations, it experiences the positive and negative consequences of past choices.

The components of this model raise an interesting set of theoretical questions. Why do lasting patterns form so early and persist through long periods of inertia? Why do teams' behavior patterns and product designs undergo dramatic change precisely halfway through their project calendars? What is the role of a team's context in its development? This exploratory study did not test or prove any prior hypotheses; nonetheless, it is appropriate to ask whether established theory provides any basis for understanding the observed results, to help formulate hypotheses and questions for future testing.

Early Patterns

Why do lasting patterns form so early and persist through long periods of inertia? The present findings show that lasting patterns can appear as early as the first few seconds of a group's life. This finding was unexpected, but it is not unheard of. Reports from the psychoanalytic literature show the power of the first minutes of a therapeutic interview to predict the central issues of the session (Ginnette, 1986; Pittenger, Hockett, & Danehy, 1960: 22b). Quite recently, Bettenhausen and Murnighan found that "unique norms formed in each [of several bargaining groups], typically during their very first agreements" (1985: 359).

The sheer speed with which recurring patterns appear suggests they are influenced by material established before a group convenes. Such material includes members' expectations about the task, each other, and the context and their repertoires of behavioral routines and performance strategies. The presence of these factors would circumscribe the influence of the interaction process that occurs in the first meeting but not rule it out. Bettenhausen and Murnighan (1985) discussed norm formation in terms of what happens when team members encounter the scripts (Abelson, 1976) each has brought to a group's first meeting. Pittenger, Hockett, and Danehy (1960: 16–24) described the opening of a therapeutic interview as the interaction of "rehearsed" material brought in by the patient with the therapist's opening gambit. This construction of first meetings suggests that peoples' earliest responses to each other set lasting precedents about how a team is going to handle the issues, ideas, questions, and performance strategies that members have brought in.

In phase 1, groups define most of the parameters of their situation quickly and examine them no further, concentrating their work and attention on only a few factors. The contrast between this model and the traditional idea that groups take time to generate, evaluate, and choose alternative views before getting to work parallels Simon's (1976) contrast between bounded and perfect rationality, and it may be understood through his argument that people must make simplifying assumptions in order to take any action at all.

The Halfway Point

Why do teams' behavior patterns and product designs undergo dramatic change exactly halfway through their project calendars? The transition can be understood through a combination of two concepts: problemistic search (March & Simon, 1958) and pacing. The idea of problemistic search simply extends the theory of bounded rationality. Its proponents posit that innovation is the result of search and that people do not initiate search unless they believe they have a problem. New perspectives appear to enter a group at transition because team members find old perspectives are no longer viable and initiate a fresh search for ideas.

The problem that stimulates search and stimulates it at a consistent moment in groups' calendars may be explained with the construct of pacing. Groups must pace their use of a limited resource, time, in order to finish by their deadlines. The midpoint appears to work like an alarm clock, heightening members' awareness that their time is limited, stimulating them to compare where they are with where they need to be and to adjust their progress accordingly: it is "time to roll." Since the groups in this research are charged with creating novel products, perspectives created quickly at the first meeting are likely to be found wanting in some way. For example, it may be perfectly suitable to begin with the approach "we're mapping out the task," but that approach must change at some time if there is to be a product. Even groups that started with a plan they liked learned by working on it to see flaws that were not visible when the plan was just an idea.

This model has some important qualifications. If the midpoint is primarily a moment of alarm, when groups feel "we need to move forward *now*," then the transition is an opportunity for, not a guarantee of, progress. This allows for the possibility that a group, like an individual, might feel strongly that it is time to move ahead, yet be unable to do so. Similarly, to hypothesize that transitions are catalyzed by groups' comparison of their actual progress with their desired progress leaves room for the chance that a group may — correctly or incorrectly — be largely satisfied and proceed with little visible change. These qualifications are consistent with the observation that groups' historical paths vary, and they provoke further research by posing the question, what factors affect the success of groups' transitions?

Why the consistent midpoint timing? Halfway is a natural milestone, since teams have the same amount of time remaining as they have already used, and they can readily calibrate their progress. Adult development research offers analogous findings. At midlife, people shift their focus from how much time has passed to how much time is left (Jaques, 1955). Levinson found a major transition at midlife, characterized by "a heightened awareness of mortality and a desire to use the remaining time more wisely" (1978: 192). Nonetheless, it would be premature to base the entire weight of these findings on the midpoint timing of the transition. Some groups may work on schedules that make times other than the midpoint highly salient. Ultimately, the midpoint itself is not as important as the finding that groups use temporal milestones to pace their work and that the event of reaching those mile-stones pushes groups into transitions. This study raises, but cannot answer, the question of what sets the alarm to go off when it does and precisely how it works in groups.

Context

What is the role of a team's context in its development? Traditional group development theory leaves little room for environmental influence on the course of development; all groups are predicted to go through the same steps, and all are predicted to suspend opinions of what they are about until they have thrashed that issue out through their own internal processes. Neither do these theories comment about development-linked changes in interaction between a group and its context. In contrast, the current findings suggest that the outside context may play a particularly important role in a group's developmental path at three points: the design of the group and two well-defined critical periods.

As noted, the speed with which distinctive patterns appear suggests the influence of materials imported into the group. The finding is congruent with, but does not test, a viewpoint from the group-performance research tradition. In that view, the *design* of a group — the composition of the team, the structure of the task, the contextual supports and circumstances under which the team is formed — precedes and conditions the interaction that transpires among members (Hackman, 1986). In terms of the current model, the pool of materials from which a team fashions its first framework is set by the design and designer of the group.

A critical period is a time in an organism's life within which a particular formative experience will take and after which it will not (Etkin, 1967). Though the analogy is imperfect, there appear to be two critical periods when groups are much more open to fundamental influence than they are at other times. The first is the initial meeting. As a time when the interaction in the group sets lasting precedents, it holds special potential to influence a team's basic approach toward its project.

The transition is the second chance. Not only did teams open up to outside influence at this point — they actively used outside resources and requirements as a basis for recharting the course of their work. The transition appears to be a unique time in groups' lives. It is the only period when the following three conditions are true at once: members are experienced enough with the work to understand the meaning of contextual requirements and resources, have used up enough of their time that they feel they must get on with the task, and still have enough time left that they can make significant changes in the design of their products.

In contrast, teams did not make fundamental changes of course in response to information from their contexts during phase 1 and phase 2, when ideas that did not fit with their approach to the task did not appear to register. That observation does not suggest that teams universally ignore or cut off environmental communication during phases 1 and 2, but it suggests that outsiders are unlikely to turn teams around during those times.

The three example teams showed how groups may insulate themselves from environmental input at some times yet seek it during transitions — partly to get help limiting their own choices and moving forward, partly to increase the chances that their product will succeed in their environment. That pattern has interesting implications for the theoretical debate between population ecologists, who argue that environments "select," and advocates of resource dependency, who argue that systems "adapt." Researchers have already observed that organizations change through alternating periods of momentum and revolution (Miller & Friesen, 1984; Tushman & Romanelli, 1985). Further, organizations commonly construct time-related goals for productivity and growth, such as monthly, annual, and five-year plans, as well as possibly much longer-term objectives for their ultimate growth schedules. It appears worth investigating (1) whether pacing or life cycle issues affect the timing or success of organizational revolutions and (2) how organization-environment communication, or lack of it, during revolutionary periods particularly affects outcomes. Interaction with an environment may be very likely to foster and shape adaptation at certain predictable times in a system's life cycle and unlikely to do so at other times. If its environment changes dramatically when an organization is also entering a change phase, that organization may be more likely to adapt. Organizations that are instead in a phase of inertia will be less able to respond and may be selected out. Since this study did not include interviews with external stakeholders or observation of them outside teams' meetings, more research is needed to study the effects of environmental influence attempts during phases 1 and 2, versus during transition.[10]

Limitations of the Study

This study must be interpreted with caution. It was hypothesis-generating, not hypothesis-testing; the model is expressly provisional. One person conducted the analysis. As Donnellon, Gray, and Bougon (1986: 54) pointed out, the use of a single judge is important in discourse analysis, where the goal is to create an in-depth understanding of a whole event, but it increases the need for further research. There are also limits on the type of group to which the findings might apply. The transition involves groups' revising their understanding of and approach to their work in response to time limits. Accordingly, results should apply only to groups that have some leeway to modify their work processes and must orient themselves to a time limit. The length of the time span should not matter, though that is a question for empirical research.

Comparison with Past Findings

Why did this study result in findings so different from the findings of previous group development research? An important possibility is that the paradigm of unitary stage theory directed previous researchers' attention away from phenomena of special interest here. The developmental stage paradigm naturally focuses on the stages themselves, not on the process of change, since all systems are assumed to progress through the same stages in a forward direction. Such events as T-groups' characteristic revolt against the leader may be midpoint transitions, but past researchers did not note their timing or think in such terms. The theoretical prominence of the environment is also limited in the traditional models because is does not alter the basic sequence of stages. In contrast, punctuated equilibrium paradigms direct attention to periods of stability and to change processes, provoking questions about what happens within a team and between a team and its context during the short periods of time when systems are especially plastic and labile. Finally, the traditional paradigm raised different questions about group process. Many past studies conceptualized and examined group process at the microanalytic level of members' sentence-by-sentence rhetoric and speech patterns, whereas this study encompassed the more macroanalytic level of group actions, such as revising plans and contacting outside supervisors. Such actions would be undetectable to traditional coding schemes, as would one of the most important clues in the study, the one-shot comments about time that group members made as they began their transitions.

The work of Poole (1983a, 1983b) suggests another possibility. He found that groups developed decisions within single meetings in multiple, not unitary, sequences, and proposed that past research did identify the key components of the development of group decisions, but that outside the controlled conditions and broad category systems of past laboratory research, it is possible to see that groups treat those components as blank spaces on an outline. They may fill in the blank spaces in a variety of sequences, depending on a host of task-related variables. Finally, the nature of its task affects the development of a group (Poole, 1983b). Past research has concentrated on a few types of group and tasks, with little attention to naturally-occurring groups responsible for creating concrete products for outside use and evaluation.

Implications for Action

The results reported here have many implications for managers working with groups. Although traditional theory implies that group leaders have plenty of time at a project's beginning before the group will choose its norms and get to work, this model implies that a group's first meeting will set lasting precedents for how the group will use the first half of its time. That finding suggests that group leaders prepare carefully for the first meeting, and it identifies a key point of intersection between group development and group-effectiveness research on team design. According to traditional theory, a group must also expect an inevitable storming stage. In contrast, the proposed model suggests that groups use the first meeting to diagnose the unique issues that will preoccupy them during phase 1.

The proposed model also suggests that a group does not necessarily need to make visible progress with a steady stream of decisions during phase 1 but does need to generate the raw material to make a successful transition. For example, groups that begin with a clear plan may do best to use phase 1 to flesh out a draft of that plan fully enough to see its strong and weak points at the transition. Groups that begin with a deep disagreement may do best to pursue the argument fully enough to understand by transition what is and is not negotiable for compromise. A leader who discovers at the first meeting that the group adamantly opposes the task may do best to decide whether to restart the project or help the group use phase 1 to explore the issues enough to determine, at transition, whether it can reach an acceptable formulation of the task. In such a case, a leader might want to redefine a group's task as a preliminary diagnostic project, with a shorter deadline. Once past the first meeting, phase 1 interventions aimed at fundamentally altering a group, rather than at helping it pursue its first framework more productively, may be unsuccessful because of members' resistance to perceiving truly different approaches as relevant to the concerns that preoccupy them.

The next new implication of the present model is that the midpoint is a particularly important opportunity for groups and external managers to renew communication. Again, note that the teams and supervisors studied did not all automatically do this or do it uniformly well. The special challenge of the transition is to use a group's increased information, together with fresh input from its environment, to revise its framework knowledgeably and to adjust the match between its work and environmental resources and requirements. This is another point of special intersection between group development and group-effectiveness research, since that research should be especially helpful in evaluating and revising a group's situation (Hackman & Walton, 1986). Further research is needed to explore ways to manage the transition process productively.

Once the transition is past, the major outlines of a group's project design are likely to be set; the most helpful interventions are likely to be aimed at helping the group execute its work smoothly. For external managers, this may be an especially important time to insure a group's access to needed resources.

Conclusions

The concepts highlighted here center around the broad theme of change over time in groups' lives. This kind of knowledge about groups is particularly needed now, given the increasing importance of groups in high-commitment organizations (Walton & Hackman, 1986) and in young, high-technology industries (Mintzberg, 1981).

The pattern of continuity and change, observed directly in eight groups, also matches a punctuated equilibrium pattern that others have postulated at different levels of analysis. These formulations range from Kuhn's (1962) concept of normal science versus scientific revolution, through Abernathy and Utterback's (1982) description of radical versus evolutionary innovation in industries and Miller and Friesen's (1984) model of momentum and revolution in organizations, to Levinson's (1978) theory of adult development as alternating periods of stability and transition. Findings about small groups cannot be generalized directly to individual lives, growing organizations, or developing industries; nevertheless, knowledge about group development should stimulate and enrich our learning about inertia and change in human systems across those levels of analysis.

Notes

1. This work was called to my attention by a reviewer.
2. This paper uses the dictionary definition of inertia as the tendency of a body to remain in a condition: if standing still, to remain so; if moving, to keep moving on the same course.
3. Three dimensions of a group's stance on a task emerged from the data. Members may accept or object to an assignment, may be certain or uncertain what to do about a task, and may converge or diverge with each other about these issues. The dimensions may be arrayed in a $2 \times 2 \times 2$ matrix to suggest a number of potential answers to the question "Where do we stand?" The three dimensions are primarily concerned with members' approaches toward context, task, and internal interaction, yet they are closely intertwined.
4. All names used in this report are pseudonyms.
5. The notation "E2, 1" identifies the excerpt (E2) and the line or lines (1) of dialogue.
6. The authority designation comes from Mann, Gibbard, and Hartman (1967).
7. Two additional indicators of transition, a pretransition low point and a change in groups' routines, are not covered here because of space limitations. A discussion of all seven indicators is available in Gersick (1984).
8. In the discussion of indicators, letters identify teams, and numbers identify lines of dialogue in Table 3.
9. For example, team H decided to schedule a special meeting to confront top administrators about its mission. Just after that, the leader received two independent requests from administrators to change the team's direction.
10. Gersick (1983) does include and discuss additional evidence of teams dismissing or not understanding outside requirements during phases 1 and 2.

References

Abelson, R. P. 1976. Script processing in attitude formation and decision making. In J. Carroll & J. Payne (Eds.), *Cognition and social behavior*: 33–45. Hillsdale, N.J.: Lawrence Erlbaum Associates.

Abernathy, W., & Utterback, J. 1982. Patterns of industrial innovation. In M. Tushman & W. Moore (Eds.), *Readings in the management* of innovation: 97–108. Boston, Mass.: Pitman Publishing.
Alderfer, C. P. 1977. Group and intergroup relations. In J. R. Hackman & J. L. Suttle (Eds.), *Improving life at work*: 227–296. Santa Monica, Calif.: Goodyear Publishing.
Bales, R. F., & Strodtbeck, F. L. 1951. Phases in group problem solving. *Journal of Abnormal and Social Psychology*, 46: 485–495.
Bell, M. A. 1982. Phases in group problem solving. *Small Group Behavior*, 13: 475–495.
Bennis, W., & Shepard, H. 1956. A theory of group development. *Human Relations*, 9: 415–437.
Bettenhausen, K., & Murnighan, J. K. 1985. The emergence of norms in competitive decision-making groups. *Administrative Science Quarterly*, 30: 350–372.
Bion, W. R. 1961. *Experiences in groups.* New York: Basic Books.
Donnellon, A., Gray, B., & Bougon, M. 1986. Communication, meaning, and organized action. *Administrative Science Quarterly*, 31: 43–55.
Eldrege, N., & Gould, S. J. 1972. Punctuated equilibria: An alternative to phyletic gradualism. In T. J. Schopf (Ed.), *Models in paleobiology*: 82–115. San Francisco: Freeman, Cooper and Co,
Etkin, W. 1967. *Social behavior from fish to man.* London: University of Chicago Press.
Fisher, B. A. 1970. Decision emergence: Phases in group decision-making. *Speech Monographs*, 37: 53–66.
Gersick, C. G. 1982. Manual for group observations. In J. R. Hackman (Ed.), *A set of methodologies for research on task performing groups.* Technical report no. 1, Research Program on Group Effectiveness, Yale School of Organization and Management, New Haven, Connecticut.
Gersick, C. G. 1983. *Life cycles of ad hoc task groups.* Technical report no. 3, Research Program on Group Effectiveness, Yale School of Organization and Management, New Haven, Connecticut.
Gersick, C. G. 1984. *The life cycles of ad hoc task groups: Time, transitions, and learning in teams.* Unpublished doctoral dissertation, Yale University, New Haven, Connecticut.
Ginette, R. 1986. *OK, let's brief real quick.* Paper presented at the 1986 meeting of the Academy of Management, Chicago, Ill.
Gladstein, D. 1984. Groups in context: A model of task group effectiveness. *Administrative Science Quarterly*, 29: 499–517.
Glaser, B., & Strauss, A, 1967. *The discovery of grounded theory: Strategies for qualitative research.* London: Wiedenfeld and Nicholson.
Goodstein, L. D., & Dovico, M. 1979. The decline and fall of the small group. *Journal of Applied Behavioral Science*, 15: 320–328.
Hackman, J. R. 1985. Doing research that makes a difference. In E. Lawler, A. Mohrman, S. Mohrman, G. Ledford, & T. Cummings (Eds.), *Doing research that is useful for theory and practice*: 126–148. San Francisco: Jossey-Bass.
Hackman, J. R. 1986. The design of work teams. In J. Lorsch (Ed.), *Handbook of organizational behavior*: 315–342. Englewood Cliffs, N.J.: Prentice-Hall.
Hackman, J. R. (Ed.). *Groups that work.* San Francisco: Jossey-Bass. Forthcoming.
Hackman, J. R., & Walton, R. E. 1986. Leading groups in organizations. In P. S. Goodman & Associates (Eds.), *Designing effective work groups*: 72–119. San Francisco: Jossey-Bass.
Hare, A. P. 1976. *Handbook of small group research* (2nd ed.). New York: Free Press.
Harris, S., & Sutton, R. 1986. Functions of parting ceremonies in dying organizations. *Academy of Management Journal*, 29: 5–30.
Hellriegel, D., Slocum, J., & Woodman, R. 1986. *Organizational behavior* (4th ed.). St. Paul: West Publishing Co.
Huse, E., & Cummings, T. 1985. *Organization development and change* (3rd ed.). St. Paul: West Publishing Co.
Jaques, E. 1955. Death and the mid-life crisis. *International Journal of Psychoanalysis*, 46: 502–514.
Katz, R. 1982. The effects of group longevity on project communication and performance. *Administrative Science Quarterly*, 27: 81–104.

Katz, R., & Tushman, M. 1979. Communication patterns, project performance, and task characteristics: An empirical evaluation and integration in an R & D setting. *Organizational Behavior and Human Performance*, 23: 139–162.

Kuhn, T. S. 1962. *The structure of scientific revolutions.* Chicago: University of Chicago Press.

Labov, W., & Fanshel, D. 1977. *Therapeutic discourse.* New York: Academic Press.

LaCoursiere, R. B. 1980. *The life cycle of groups: Group developmental stage theory.* New York: Human Sciences Press.

Levinson, D. J. 1978. *The seasons of a man's life.* New York: Alfred A. Knopf.

Levinson, D. J. 1986. A conception of adult development. *American Psychologist*, 41: 3–14.

Mann, R., Gibbard, G., & Hartman, J. 1967. *Interpersonal styles and group development.* New York: John Wiley & Sons.

March, J., & Simon, H. 1978. *Organizations.* New York: John Wiley & Sons.

McGrath, J. E. 1984. *Groups: Interaction and performance.* Englewood Cliffs, N.J.: Prentice Hall.

McGrath, J. E. 1986. Studying groups at work: Ten critical needs for theory and practice. In P. S. Goodman & Associates (Eds.), *Designing effective work groups*: 363–392. San Francisco: Jossey-Bass.

Miller, D., & Friesen, P. 1984. *Organizations: A quantum view.* Englewood Cliffs, N.J.: Prentice-Hall.

Mills, T. 1979. Changing paradigms for studying human groups. *Journal of Applied Behavioral Science*, 15: 407–423.

Mintzberg, H. 1981. Organization design, fashion or fit? *Harvard Business Review*, 59(1): 103–116.

Pittenger, R., Hockett, C., & Danehy, J. 1960. *The first five minutes: A sample of microscopic interview analysis.* Ithaca, N.Y.: Paul Martineau.

Poole, M. S. 1981. Decision development in small groups 1: A comparison of two models. *Communication Monographs*, 48: 1–24.

Poole, M. S. 1983a. Decision development in small groups II: A study of multiple sequences of decision making. *Communication Monographs*, 50: 206–232.

Poole, M. S. 1983b. Decision development in small groups III: A multiple sequence model of group decision development. *Communication Monographs*, 50: 321–341.

Scheidel, T., & Crowell, L. 1964. Idea development in small discussion groups. *Quarterly Journal of Speech*, 50: 140–145.

Schutz, W. C. 1958. *FIRO: A three-dimensional theory of interpersonal behavior.* New York: Rinehart & Winston.

Seeger, J. A. 1983. No innate phases in group problem solving. *Academy of Management Review*, 8: 683–689.

Simon, H. A. 1976. *Administrative behavior* (3rd ed.). New York: Free Press.

Slater, P. E. 1966. *Microcosm: Structural, psychological, and religious evolution in groups.* New York: John Wiley & Sons.

Szilagy, A., & Wallace, M. 1987. *Organizational behavior and performance* (4th ed.). Glenview, Ill.: Scott, Foresman & Co.

Tosi, H., Rizzo, J., & Carroll, S. 1986. *Managing organizational behavior.* Marshfield, Mass.: Pitman Publishing.

Tuckman, B. 1965. Developmental sequence in small groups. *Psychological Bulletin*, 63: 384–399.

Tuckman, B., & Jensen, M. 1977. Stages of small-group development. *Group and Organizational Studies*, 2: 419–427.

Tushman, M. L., & Romanelli, E. 1985. Organizational evolution: A metamorphosis model of convergence and reorientation. In L. Cummings & B. Staw (Eds.), *Research in organizational behavior*, vol. 7: 171–222. Greenwich, Conn.: JAI Press.

Walton, R. E., & Hackman, J. R. 1986. Groups under contrasting management strategies. In P. Goodman & Associates (Eds.), *Designing effective work groups*: 168–201. San Francisco: Jossey-Bass.

50

Relations Between Work Group Characteristics and Effectiveness: Implications for Designing Effective Work Groups

Michael A. Campion, Gina J. Medsker and A. Catherine Higgs

Source: *Personnel Psychology* 46 (4) (1993): 823–850.

The use of work groups in organizations is graining substantial popularity (e.g., Banas, 1988; Goodman, Ravlin, & Schminke, 1987; Guzzo & Shea, 1992; Magjuka & Baldwin, 1991; Majchrzak, 1988). The difficulty with groups is that sometimes they lead to negative outcomes, such as low productivity (Whyte, 1955), poor decisions (Janis, 1972), and conflict (Alderfer, 1977). However, according to some current models (e.g., Gladstein, 1984; Hackman, 1987) and reviews (e.g., Goodman, Devadas, & Hughson, 1988; Katzell & Guzzo, 1983), groups hold the potential for simultaneously increasing both productivity and employee satisfaction. This is very important. From a work design point of view, the establishment of groups is consistent with a psychological approach, and is thus intended to increase satisfaction and related outcomes. But psychological approaches to work design have been historically, theoretically, and empirically in conflict with traditional engineering approaches (e.g., specialization, assembly lines, etc.) which are intended to increase efficiency and related outcomes (Campion, 1988; Campion & McClelland, 1991; Campion & Thayer, 1985). Therefore, if work groups are truly related to both productivity and satisfaction, they may be the key to avoiding the production-satisfaction trade-off previously presumed to be inherent in work design. In summary, work groups are gaining importance in many organizations and they present many potential risks and opportunities, so there is a need to understand the characteristics of effective work groups.

The Present Study

This study adopts a work design perspective on groups. In that tradition, it attempts to examine relationships between design characteristics and various outcomes. It is recognized that other perspectives on groups exist (e.g., organizational design perspective), and that they might conceptualize the issues differently (e.g., regarding trade-offs) and examine different variables (e.g., centralization, formalization, etc.).

Specifically, the study tries to make three contributions. First, it reviews a wide range of literature and derives five common themes or clusters of work group characteristics that may be related to effectiveness. The review includes social psychology (e.g., McGrath, 1984; Steiner, 1972), socio-technical theory (e.g., Cummings, 1978; Pasmore, Francis, & Haldeman, 1982), industrial engineering (e.g., Davis & Wacker, 1987; Majchrzak, 1988), and, in particular, organizational psychology (e.g., Gladstein, 1984; Hackman, 1987; Guzzo & Shea, 1992; Sundstrom, De Meuse, & Futrell, 1990). It also delineates an extensive set of 19 characteristics within these themes, and then develops a measure.

Second, this study relates these characteristics to effectiveness criteria in a field setting with natural work groups. Most group research has involved concocted groups in the laboratory, and it is not absolutely certain that inferences can be made about natural groups based on this research (Guzzo & Shea, 1992). More empirical research is needed to confirm the generalizability of findings from laboratory studies to actual work settings. This study answers the frequent call in recent reviews for more field research on groups (e.g., Levine & Moreland, 1990; McGrath, 1986; Shea & Guzzo, 1987).

Third, this study is more methodologically rigorous than many previous efforts. Consistent with most theories (e.g., Gladstein, 1984; Hackman, 1987; Sundstrom et at., 1990) and some previous studies (e.g., Gladstein, 1984; Goodman, 1979; Wall, Kemp, Jackson, & Clegg, 1986; Walton, 1972), work group effectiveness is defined in terms of both productivity and employee satisfaction. The inclusion of productivity criteria enhances the objectivity of the effectiveness evaluation, and it avoids the sole reliance on affective outcomes which typifies much of the research in the area. The other criteria examined in this study — employee satisfaction and manager judgments of effectiveness — are measured using methods which minimize common method variance. Finally, large samples and multiple sources of respondents are also used to enhance the rigor of the empirical evaluation.

Work Group Characteristics Related to Effectiveness

The five themes below are summaries of key components of previous theories. Together, the themes depict a hybrid conceptual framework (Fig. 1) based on the models of Gladstein (1984); Hackman (1987); Guzzo and Shea (1992); and Tannenbaum, Beard, and Salas (1992).

Job Design

This theme is most closely linked to the work of Hackman (1987), but is also reflected in the group structure component of Gladstein's (1984) model, the group task school of thought in Guzzo and Shea's (1992) review, and the task characteristics and work structure components of Tannenbaum et al.'s (1992) model. This theme contains work group characteristics that derive directly from theories of motivational job design. The main distinction is in terms of level of application rather than content (Campion & Medsker, 1992; Shea & Guzzo,

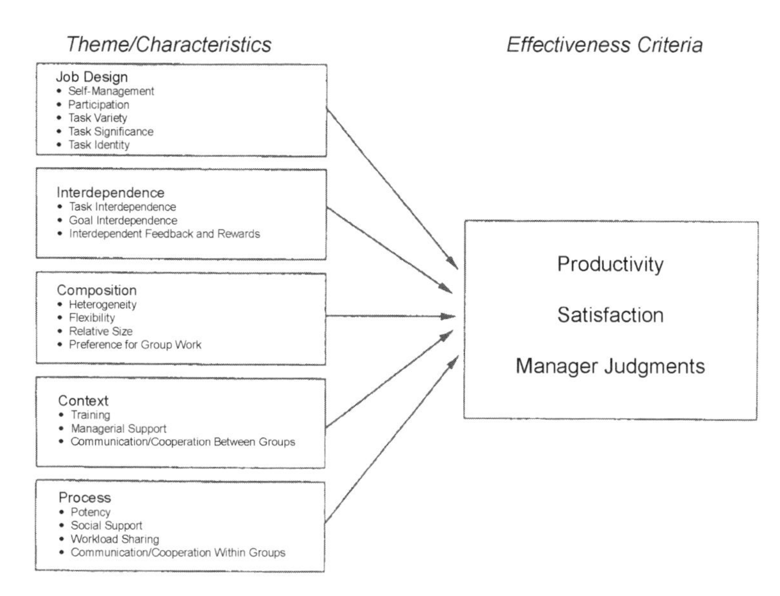

Figure 1: Themes and characteristics related to work group effectiveness.

1987; Wall et al., 1986). All the job characteristics of Hackman and colleagues (e.g., Hackman & Lawler, 1971; Hackman & Oldham, 1980) can be applied to groups, even though there have been few tests at the group level.

One characteristic in this theme is *self-management*, which is the group level analogy to autonomy at the individual job level. It is central to many definitions of effective work groups (e.g., Cummings, 1978, 1981; Hackman, 1987; Pearce & Ravlin, 1987) and part of most interventions (e.g., Cordery, Mueller, & Smith, 1991; Goodman, 1979; Goodman et al., 1988; Pasmore et al., 1982; Wall et al., 1986; Walton, 1972). A related characteristic is *participation*. Regardless of management involvement in decision making, work groups can still be distinguished in terms of the degree to which all members are allowed to participate in decisions (McGrath, 1984; Porter, Lawler, & Hackman, 1987). Self-management and participation are presumed to enhance group effectiveness by increasing members' sense of responsibility and ownership of the work. These characteristics may also enhance decision quality by increasing relevant information and by putting decisions as near as possible to the point of operational problems and uncertainties.

Another characteristic is *task variety*, or giving each member the chance to perform a number of the group's tasks. Variety motivates by allowing members to use different skills (Hackman, 1987) and by allowing both interesting and dull tasks to be shared among members (Davis & Wacker, 1987; Walton, 1972). *Task significance* is also applicable. Members should believe that their group's work has significant consequences, either for others inside the organization or its customers (Hackman, 1987). Finally, group work should have *task identity* (Hackman, 1987) or task differentiation (Cummings, 1978), which is the degree to which the group completes a whole and separate piece of work. Identity may increase motivation because it increases a group's sense of responsibility for a meaningful piece of work (Hackman, 1987), and it may facilitate cooperation within a group and increase control over sources of disturbance from goal attainment (i.e., technical variances) by keeping those sources within group boundaries (Cummings, 1978).

Interdependence

This theme comes primarily from the work of Guzzo and Shea (1992; Shea & Guzzo, 1987), although it is implicit in all the models. Interdependence is often the reason groups are formed in the first place (Mintzberg, 1979), and it is a defining characteristic of groups (Salas, Dickinson, Converse, & Tannenbaum, 1992; Tannenbaum et al., 1992; Wall et al., 1986). Interdependence may increase the motivational properties of work or the efficiencies with which the work is done, and thus may be related to effectiveness.

One form of interdependence is *task interdependence.* Group members interact and depend on one another to accomplish the work. Interdependence may vary across groups, increasing as work flow goes from pooled to sequential to reciprocal (Thompson, 1967). There has been little research at the group level, but interdependence among tasks in the same job (Wong & Campion, 1991) or between jobs (Kiggundu, 1983) has been related to increased motivation. It may also increase group effectiveness because it enhances the sense of responsibility for others' work (Kiggundu, 1983) or because it enhances the reward value of group accomplishments (Shea & Guzzo, 1987).

Another form of interdependence is *goal interdependence.* Goal setting is a well documented individual level performance improvement technique (Locke & Latham, 1990). There is less evidence at the group level, but a clearly defined mission or purpose is thought to be critical to group effectiveness (Davis & Wacker, 1987; Gladstein, 1984; Guzzo & Shea, 1992; Hackman, 1987; Hackman & Walton, 1986; Sundstrom et al., 1990). Its importance has also been shown in some empirical studies on groups (e.g., Buller & Bell, 1986; Koch, 1979; Pearson, 1987; Pritchard, Jones, Roth, Stuebing, & Ekeberg, 1988; Woodman & Sherwood, 1980). Not only should goals exist for groups, but individual members' goals must be linked to the groups' goals to be maximally effective.

Finally, *interdependent feedback and rewards*, or what Guzzo and Shea (1992) call outcome interdependence, is also important to group effectiveness. Individual feedback and rewards should be linked to the group's performance in order to motivate group-oriented behavior. This characteristic is recognized in many other theoretical treatments (e.g., Gladstein, 1984; Hackman, 1987; Hackman & Walton, 1986; Kolodny & Kiggundu, 1980; Leventhal, 1976; Pearce & Ravlin, 1987; Steiner, 1972; Sundstrom et al., 1990) and some research studies (e.g., Koch, 1979; Pasmore et al., 1982; Pritchard et al., 1988; Wall et al., 1986). Most of what is known about the effects of feedback and rewards on performance has been from research at the individual level, however, and it is uncertain how well the findings generalize to the group level (Shea & Guzzo, 1987). Feedback is one of the motivating job characteristics discussed by Hackman (Hackman & Oldman, 1980), but it is included here because of the need for interdependence of feedback for group members.

Composition

The composition of the work group is a theme in all the models of effectiveness. Gladstein (1984) and Guzzo and Shea (1992) refer to it directly as group composition, while Hackman (1987) refers to it under group design, and Tannenbaum et al. (1992) refer to it under team characteristics. The importance of composition has not been widely investigated for its impact on task performance, however, and the evidence has been mixed when it has been investigated (Guzzo & Shea, 1992).

Several aspects of composition may influence effectiveness. For one, membership *heterogeneity* in terms of abilities and experiences has been found to have a positive effect on performance. This is especially the case when tasks assigned to the group are diverse, because a wide range of competencies are needed (Gladstein, 1984; Goodman, Ravlin, & Argote, 1986; Hackman, 1987; Pearce & Ravlin, 1987; Shaw, 1983; Wall et al., 1986), and when tasks are disjunctive, because performance is determined by the most competent member (Steiner, 1972). Heterogeneity may also increase effectiveness because employees can learn from each other. On the other hand, the beneficial effects are unclear because most data are based on problem solving and creativity outcomes. Homogeneity may lead to better outcomes when satisfaction, conflict, communication (Pearce & Ravlin, 1987), and turnover (Jackson et al., 1991) are considered. Thus, heterogeneity is expected to have a positive effect in the present study, but the prediction is made with caution.

Another composition characteristic of effective groups is whether members have *flexibility* in terms of job assignments (Goodman, 1979; Poza & Markus, 1980; Sundstrom et al., 1990; Walton, 1972). If members can perform each other's jobs, effectiveness is enhanced because they can fill in as needed. *Relative size* is another aspect of composition. Groups need to be large enough to accomplish work assigned to them, but when too large, groups may be dysfunctional due to heightened coordination needs (Gladstein, 1984; O'Reilly & Roberts, 1977; Steiner, 1972) or reduced involvement (McGrath, 1984; Wicker, Kirmeyer, Hanson, & Alexander, 1976). Thus, groups should be staffed to the smallest number needed to do the work (Goodman et al., 1986; Hackman, 1987; Hackman & Walton, 1986; Sundstrom et al., 1990). However, most previous research on size has been in the laboratory (Sundstrom et al., 1990), so it is unclear if these findings generalize to actual work groups.

A final characteristic is employee *preference for group work*. Employees who prefer to work in groups may be more satisfied and effective in groups (Cummings, 1981; Hackman & Oldham, 1980). This preference is somewhat similar to cohesiveness (Cartwright, 1968; Goodman et al., 1987; Zander, 1979). It differs in that cohesiveness refers to attraction to and the desire to remain in a particular group, while preference for group work is not specific to a particular group, but refers to a general preference for working in groups. Research supports the notion that employee preferences may influence their reactions to their jobs (Fried & Ferris, 1987; Hackman & Oldham, 1980; Hulin & Blood, 1968), but little research has focused on this issue at the group level.

Context

Organizational context and resources are considered in all recent models of work group effectiveness. Gladstein (1984) considers organizational level resources, Hackman (1987) considers organizational context, Guzzo and Shea (1992) consider contextual influences, and Tannenbaum et al. (1992) consider organizational and situational characteristics.

One resource that groups need is adequate *training*. Training is an extensively researched determinant of group performance (for reviews see Dyer, 1984; Salas et al., 1992), and training is included in most interventions (e.g., Cordery et al., 1991; Goodman, 1979; Pasmore et al., 1982; Tannenbaum et al., 1992; Wall et al., 1986; Walton, 1972). Training content often includes team philosophy, group decision making, and interpersonal skills, as well as technical knowledge. It was recently shown that group member familiarity with the work and environment is related to productivity (Goodman & Leyden, 1991). Yet, the overall evidence in support of team training is mixed, methodologies of most studies have been weak, and most studies have focused on process outcomes rather than effectiveness (Baker, Dickinson, & Salas, 1991; De Meuse & Liebowitz, 1981; Shea & Guzzo, 1987).

Managerial support is another contextual characteristic. Management controls resources (e.g., material and information) required to make group functioning possible (Shea & Guzzo, 1987), and an organization's culture and top management must support the use of groups (Sundstrom et al., 1990). Teaching facilitative leadership to management is often a feature of work group interventions (Pasmore et al., 1982). Although managerial support seems logically related to group effectiveness, there has been little prior research examining its influence.

Finally, *communication and cooperation between groups* is a contextual characteristic because it is often the responsibility of the management. Supervising group boundaries (Brett & Rognes, 1986; Cummings, 1978) and externally integrating the group with the rest of the organization (Sundstrom et al., 1990) enhances effectiveness. However, research has not extensively tested this, and there is little data on the link between intergroup relations and group effectiveness (Guzzo & Shea, 1992).

Process

Originally proposed by McGrath (1964), an input-process-output perspective is probably the dominant view of groups historically (Guzzo & Shea, 1992). The four themes above deal with the inputs to the group. Process describes those things that go on in the group that influence effectiveness. Gladstein's (1984) and Tannenbaum et al.'s (1992) models refer directly to group process, while Hackman (1987) refers to process criteria of effectiveness, and Guzzo and Shea (1992) refer to the social interaction process.

One process characteristic is *potency*, or the belief by a group that it can be effective (Guzzo & Shea, 1992; Guzzo, Yost, Campbell, & Shea, 1993; Shea & Guzzo, 1987). It is similar to the lay-term of "team spirit" and the notions of self-efficacy (Bandura, 1982) and high expectancy (Vroom, 1964). Hackman (1987) argues that groups with team spirit (potency) are more committed and willing to work hard for the group, but there has been little research on potency thus far (Guzzo et al., 1993).

Another process characteristic is *social support*. Effectiveness may be enhanced when members help each other and have positive social interactions. Gladstein (1984) describes supportiveness as a group maintenance behavior. Like social facilitation (Harkins, 1987; Zajonc, 1965), social support is arousing and may enhance effectiveness by sustaining effort on mundane tasks.

Another process characteristic is *workload sharing*, which enhances effectiveness by preventing social-loafing or free-riding (Albanese & Van Fleet, 1985; Harkins, 1987; Latane, Williams, & Harkins, 1979). To enhance sharing, group members should believe their individual performance can be distinguished from the group's, and that there is a link between their performance and outcomes. Most research has been conducted in laboratory settings, however (Sundstrom et al., 1990). It is assumed to relate to greater productivity, but the actual connection to productivity has not been tested.

Finally, *communication and cooperation within the work group* are also important to effectiveness. They have long been shown to influence effectiveness in laboratory studies (Deutsch, 1949; Leavitt, 1951), and they are considered in many current models (e.g., Gladstein, 1984; Pearce & Ravlin, 1987), but they have not been extensively field tested.

In the sections below, these characteristics are examined empirically in terms of their ability to predict several effectiveness criteria.

Method

Setting

The study was conducted in 5 geographic units of a large financial services company. Each unit supported 1 to 3 geographic territories ($M = 1.80$, $SD = .84$), for a total of 9 territories. Each territory was divided into 5 to 14 subterritories ($M = 8.89$, $SD = 2.67$), for a total of 80 subterritories. Each subterritory was staffed with a single work group and manager. The groups ranged in size from 6 to 30 ($M = 14.93$, $SD = 4.88$), but were more comparable in size within a territory. They were formal groups in that employees were permanently assigned; viewed themselves and were seen by others as groups; and interacted and shared resources to accomplish mutual tasks, responsibilities, and goals (Shea & Guzzo, 1987; Sundstrom et al., 1990).

Jobs were clerical and involved processing paperwork for other units that sold the products. Tasks included sorting, coding, computer keying, quality checking, answering customer inquiries, and related activities. Each group performed the same set of tasks. Jobs were interdependent in several ways. In addition to shared resources and responsibilities, work was often sequentially interdependent in that products flowed from some employees to others, and it was often reciprocally interdependent in that products flowed back and forth between employees. They were also interdependent in that members depended on each other for their knowledge of different products. Thus, the groups were teams and were referred to as such by the organization; they were not simply collections of individual workers (Salas et al., 1992).

Aside from performing the same work, the groups were similar in many other ways. Because they were in the same division of the same company, they were managed under the same policies and practices. Physical settings were very similar; furniture was identical and buildings were *very* comparable. Employees were similar in many ways (e.g., sex, education) as were managers (e.g., education, tenure) as described below. As a check, the measures were correlated with the demographics (e.g., age, tenure, sex, and education), and only trivial or nonsignificant relationships were observed.

Sample

Because the unit of analysis in group research is the group (McGrath, 1986; Shea & Guzzo, 1987), a sufficiently large and randomly selected sample of employees had to be included from each group so that the data accurately estimated the population parameters (i.e., values that would have been obtained had all the employees in each group been included). Using standard sampling accuracy formulas (e.g., Warwick & Lininger, 1975) and assuming an average variance on the measures of .50 ($SD = .71$; based on previous research and confirmed post hoc), an average 95% confidence interval of plus or minus 15% on the measures (i.e., .6 on the 1–5 scales) would require sampling no more than 5 members per group for the range of group sizes.

Thus, 5 employees were sampled from each of the 80 groups for a total of approximately 400 (usable sample = 391). Managers were also included for 77 of the 80 groups (7 managed two groups and provided data on both). Sample sizes below vary from 75 to 79 groups due to incomplete data and are indicated in the tables. Statistical power was 93% to detect an $r = .30$ and 70% for an $r = .20$ ($p < .10$, one-tailed; Cohen, 1977). To balance Type I and II errors, both the $p < .05$ and $p < .10$ significance levels were interpreted.

Employees were nearly all female (96.1%). Average age was 32.9 years ($SD = 9.9$), with half being 30 years old or younger. Average tenure was 6.0 years ($SD = 6.3$), with half having 3 years or less. Almost half (44.2%) had a high school education only, 51.9% had some additional education, but only 1.6% had a 4-year degree or more. Half the managers were female (51.5%). Average age was 29.3 years ($SD = 3.6$), with 69.1% under 30 years. Average tenure was 3.9 years ($SD = 2.5$), with 63.2% having 2 or 3 years. Nearly all had a 4-year college degree or more (92.6%).

Measurement Overview

Three objectives guided measurement based on McGrath's (1986) recommendations for studying work groups. First, multiple constructs of both characteristics and effectiveness were assessed, and data were collected from multiple sources for each. Characteristics were obtained from employees and managers; effectiveness was obtained from employees, managers, and records. Thus, self-perceptions, observer perceptions, and objective measures were used.

Second, common method variance between characteristics and effectiveness measures was minimized. Methodological separation was accomplished by using different data sources or time frames, by including respondents who only provided one set of measures but not both, or by using objective records.

Third, the group was the level of analysis. For some measures, data were collected at the group level; for others, data were collected from individuals and aggregated to the group level. Aggregation is a controversial issue, but several recommendations have emerged (e.g., Goodman et al., 1987; James, 1982; Roberts, Hulin, & Rousseau, 1978; Van de Ven & Ferry, 1980).

One recommendation is that there should be a strong rationale or "composition" theory to justify aggregation (Roberts et al., 1978, p. 84). As in aggregation in climate research (James, 1982, p. 219), this study views the characteristics as "macro perceptions" or shared views of the group. Another rationale (Van de Ven & Ferry, 1980) is that the meaning of the characteristics do not change from the individual to the group perspective. Further, in the work design literature it is not unusual to conceptualize and measure design at the incumbent level when examining individual positions, and then aggregate to the job level when examining positions held by multiple people (e.g., Algera, 1983; Campion, 1988; Campion & McClelland, 1991).

Another recommendation is that measures refer to the level of interest (Van de Ven & Ferry, 1980). In this study, most items refer to the group. Those referring to the individual are in the context of group membership. Lastly, the study performs the recommended check of demonstrating that the ratio of within- to between-group variance is statistically significant before aggregation (Goodman et al., 1987; James, 1982; Roberts et al., 1978).

Measures of Work Group Design

A questionnaire was developed to assess the 19 characteristics. It was completed individually by five randomly selected employees and the manager of each group. Based on the literature, three items were written for nearly all characteristics to obtain minimally adequate internal consistency yet limit length. Each characteristic's items were grouped under a descriptive label to help respondents clearly understand their meaning (with minor changes to labels to clarify meanings to laypersons). "Team" was used to refer to the group. A 5-point response format was used ranging from 5 = "strongly agree" to 1 = "strongly disagree." Items were averaged to form a scale for each characteristic. A copy of the 54-item questionnaire is in the Appendix.[1]

The 54 items were too many to use in confirmatory factor analysis (Bentler & Chou, 1987), so exploratory factor analysis was used to examine the acceptability of maintaining the 19 characteristics as separate scales ($n = 8.7$ per item). Common factor analysis was used because the factors were presumed to represent underlying attributes (Ford, MacCallum, & Tait, 1986). Although simpler solutions could be derived, it is noteworthy that each of the 19 characteristics loaded on its own factor when a 19-factor solution was imposed. That is, all items for each characteristic had their highest loadings on the same factor, separate from items of other characteristics. And there were only a few cross-loadings of .30 or above. This was found with both orthogonal and oblique rotations. Principal components analysis produced fairly similar results. The 19 factors explained 73% of the total variance, and 17 of the 19 characteristics loaded on their own factors with either orthogonal or oblique rotations. As an additional assessment of the independence of characteristics, intercorrelations among scales were examined (Table 1). With exceptions, intercorrelations were generally low (average r using z transformation = .22). Based mainly on conceptual distinctions between characteristics, but bolstered by these analyses, the 19 characteristics are kept separate in analyses below.

Several types of reliability were examined (Table 1). First, internal consistency reliabilities assessed unidimensionality. Only one was much below .60. Second, intraclass correlations assessed interrater reliability of the aggregate responses across the five employees in each group (Cronbach, Gleser, Nanda, & Rajaratnam, 1972). With five exceptions, all were significant although some were modest in size. Third, interrater agreement was assessed using the James, Demaree, and Wolf (1984) procedure (see Kozlowski & Hattrup, 1992), which compares observed variance of the ratings with the null variance expected with slight positive leniency. With three exceptions, agreement was moderate (.50) to high. Fourth, manager ratings were correlated with average employee ratings. Only 11 of 19 were significant, and most were small. In summary, each analysis showed some scales had low reliability. But each analysis suggested the scales were reliable as a set, all 19 scales showed reliability in one or more analyses, and 15 of 19 showed reliability in two or more analyses. All scales are tested below, but results for scales with low reliabilities should be interpreted cautiously. Further, employees and managers converged only moderately, and thus are tested separately.

Measures of Work Group Effectiveness

Three measures of work group effectiveness were collected.

Productivity

Interviews with managers and employees were conducted to determine the productivity measures collected, the degree to which they were contaminated or deficient as criteria, and the extent to which they were used to evaluate

Table 1: Means, Standard Deviations, Reliabilities, and Intercorrelations Among the Work Group Characteristics

Themes/Characteristics	*M*[a]	*SD*	*r*[b]	*r*[c]	*r*[d]	*r*[e]	1[f]	2	3	4	5	6	7	8	9	10	11	12	13	14	15	16	17	18
Job design																								
(1) Self-management	3.33	.79	.76	.41**	.58	.24**																		
(2) Participation	3.44	.81	.88	.54**	.66	.20*	.55																	
(3) Task variety	3.14	.76	.71	.43**	.58	−.03	.34	.44																
(4) Task significance	4.24	.56	.74	.26*	.55	.07	.21	.33	.27															
(5) Task identity	4.03	.56	.71	.24*	.83	−.03	.20	.24	.25	.35														
Interdependence																								
(6) Task interdependence	3.54	.66	.61	.04	.68	.05	.06	.17	.19	.15	.09													
(7) Goal interdependence	3.37	.77	.68	.03	.56	.12	.28	.33	.31	.34	.23	.12												
(8) Inter. feedback & rewards	3.05	.75	.59	.16	.50	.11	.20	.26	.28	.14	.10	.11	.29											
Composition																								
(9) Heterogeneity	4.05	.51	.74	.04	.87	.15*	.09	.08	.04	.24	.14	.11	.04	.09										
(10) Flexibility	3.05	.79	.66	.33**	.46	.18	.21	.25	.37	.17	.21	.04	.14	.16	.05									
(11) Relative size	2.82	1.05	—	.42**	.18	.15*	.13	.11	.14	.10	.07	.00	.03	−.01	−.07	.15								
(12) Preference for group work	3.84	.78	.90	.53**	.71	.30**	.12	.37	.32	.40	.24	.21	.29	.19	.16	.22	.03							
Context																								
(13) Training	3.39	.81	.81	.39**	.59	.30**	.19	.28	.29	.27	.24	.07	.18	.11	.03	.28	.18	.32						
(14) Managerial support	4.00	.62	.74	.14	.78	.18*	.27	.30	.21	.42	.30	.07	.26	.16	.27	.19	.07	.29	.34					
(15) Comm./Coop. bet. groups	3.46	.67	.47	.21*	.57	.15*	.13	.13	.14	.20	.08	.03	.10	.08	.05	.20	.09	.16	.25	.17				
Process																								
(16) Potency	3.67	.77	.80	.66**	.65	.53**	.27	.39	.38	.43	.23	.10	.22	.18	.05	35	.16	.50	.34	.28	.18			
(17) Social support	3.85	.64	.78	.44**	.79	.14	.22	.45	.37	.50	.25	.20	.25	.14	.21	.28	.10	.51	.34	.38	.22	.55		
(18) Workload sharing	3.22	.92	.84	.58**	.36	.37**	.19	.26	.24	.23	.10	.05	.16	.17	−.04	.32	.05	.29	.34	.23	.14	.45	.38	
(19) Comm./Coop. within group	3.87	.65	.81	.57**	.80	.20*	.18	.40	.29	.42	.18	.16	.28	.15	.15	.28	.07	.46	.32	.35	.20	.57	.62	.48

[a] $n = 391$ employees and 77 managers.
[b] Internal consistency reliabilty.
[c] Interrater reliability (intraclass correlation).
[d] Interrater agreement (James et al., 1984).
[e] Correlation between employees and managers.
[f] Intercorrelations of .08 significant at $p < .05$, one-tailed.
* $p < .10$; ** $p < .05$, one-tailed.

effectiveness. Indications were that the measures most carefully collected and closely monitored were indicators of the amount of work not finished on a weekly basis which was received by the group from the subterritory it supported. That is, the groups' goals were not to reach the highest productivity per se, but to complete all the work that came in each week. Most territories did not even record the amount of work completed, but they did record most of these six measures related to unfinished work per week: (1) New Work Unfinished — number of new pieces of work not finished, (2) Percentage of New Work Unfinished — amount of new work unfinished as a percentage of new work received, (3) Revisions Unfinished — number of revisions to existing pieces of work not finished, (4) Percentage of Revisions Unfinished — number of revisions unfinished as a percentage of revisions received, (5) Calls Not Answered — number of phone calls to members of the group not answered, and (6) Percentage of Calls Not Answered — number of calls not answered as a percentage of calls received.

Each piece of work required the same set of tasks (e.g., coding, computer keying, quality checking, etc.). Although pieces of work varied somewhat in difficulty, distribution of difficulty was considered equivalent across groups in a given territory. Group size was used to adjust for differences in workload generated by the subterritories or skills among employees. Groups with higher workloads or fewer trained employees were assigned more employees. Group size did not change frequently because workload was fairly stable. Thus, groups were comparable *within* a territory, even though they differed in number of employees, and there was no need to standardize productivity data based on group size. There were differences across territories, however, such as complexity of the work and average group size. Therefore, productivity measures were standardized across territories using *z*-scores.

Although productivity is often stable (e.g., Deadrick & Madigan, 1990), the range of jobs studied has been limited. Thus, productivity data was collected and aggregated for each group over a long period ($M = 27.89$ weeks per group, $SD = 3.88$). To avoid temporal influences, the time period was the same for each group, from 3 months before to 3 months after the collection of the characteristics data. Intraclass correlations were used to assess reliability, or the degree of variance in productivity across weeks within a group compared to between groups. They can be interpreted as the correlations between the mean of this 30 weeks of productivity and the mean of another (hypothetical) 30 weeks. Average intraclass correlations ranged from .77 to .95 ($p < .05$), thus suggesting substantial reliability.

The six measures were intercorrelated, so they were averaged into a composite ($M = .00$, $SD = .42$, internal consistency $= .74$). All measures were not available for all groups (range from 46–79), so the composite was based on the available data for each group. Analyses with measures having the least missing data were similar, so only data for the composite are presented.[2] The signs on the correlations were reversed so that positive numbers indicate relationships with *higher* productivity (i.e., less work not finished).

Employee Satisfaction

To avoid common method variance, the organization's opinion survey was used as the measure of satisfaction rather than adding a scale to the questionnaire. That is, it was administered at a different point in time (3 months earlier) and for an unrelated purpose, thus mitigating any consistency or priming effects. Data were obtained from all employees (total $n = 1{,}175$), not just the 5 who provided other measures. This gave the maximum data for each group ($M = 14.87$ employees per group, $SD = 5.52$), enhanced interrater reliability, and further reduced common method variance because satisfaction data were included from many additional employees who did not provide characteristics data.

The aggregate data from all employees in each group was used as the satisfaction measure. Such aggregation of satisfaction data is common, and may be somewhat justified by the definition of morale as referring to either the individual or group (*Webster's*, 1965), even though the practice is not without criticism (Roberts et al., 1978).

The survey consisted of 71 items on a range of topics. Five-point response formats were used, usually ranging from 5 = "very satisfied" or "strongly agree" to 1 = "very dissatisfied" or "strongly disagree." A principal components analysis revealed 12 factors explaining 61% of total variance: supervision, job, quality of service, upper management, career development, rewards, management's commitment to quality, employee relations, communications, co-workers, recognition, and workload. Scales were formed for each factor. They showed good reliability and a pattern of relationships very similar to a single average composite combining all items ($M = 3.54$, $SD = .55$). Thus, only results for the composite are presented.[3] Internal consistency was .97, interrater reliability was .79 ($p < .05$), and interrater agreement was .96. Satisfaction was fairly independent of productivity ($r = .15$, $p < .10$).

Manager Judgments of Effectiveness

Managers evaluated all groups in their territories on four items in their questionnaire: (1) Quality of Work, (2) Customer Service, (3) Satisfaction of the Members, and (4) Productivity. These items reflected the company's definition of effectiveness. Having both productivity and satisfaction is also consistent with effectiveness definitions in the literature (Gladstein, 1984; Hackman, 1987; Sundstrom et al., 1990). A 5-point response format was used ranging from 5 = "well above" to 1 = "well below" the average in the territory. Reliability was increased, and common method variance decreased, by collecting judgments on each group from all managers in each territory ($M = 6.18$ managers judging each group, $SD = 1.52$) as opposed to just collecting judgments from the manager providing characteristics data.

Principal components analysis revealed one factor (explaining 64% of total variance), thus items were averaged into a composite ($M = 3.31$, $SD = .46$). Relationships with characteristics were highly comparable between items and the composite.[4] Internal consistency was .82, interrater reliability was .75 ($p < .05$), and interrater agreement was .77. Judgments were related to productivity ($r = .56$, $p < .05$), but more independent of satisfaction ($r = .29$, $p < .05$).

Procedures

Researchers visited two sites prior to data collection to qualitatively evaluate the conceptual framework and degree to which the characteristics would capture the differences between groups (cf. Strauss & Corbin, 1990), as well as to identify effectiveness criteria. Discussions were held with intact work groups and the managers at each site.

Selection of employees began by including those involved in a study of job design 2 years before (Campion & McClelland, 1991) who were still with the company and assigned to a group ($n = 126$). Additional employees were randomly sampled using a random number table and alphabetical listings so that 5 employees were included from each group (additional $n = 265$). Using employees who were in the prior study did not substantially decrease randomness because they were originally randomly sampled. All employees agreed to participate if available. Those unable to do so due to absenteeism or scheduling problems ($n = 51$) were replaced by randomly chosen alternates. Questionnaires were completed at individual work stations on company time.

Managers of all 80 groups agreed to participate, but 3 were unavailable. They were instructed to respond to group characteristics questions based on their perceptions of employees' views of the group. Productivity, opinion survey, and demographic data were obtained from records. Productivity data required considerable study and communication with personnel from each site to ensure comparability across sites and minimize contamination and deficiency.

Results

Primary Analyses

Primary analyses correlated the five sets of work group characteristics with the three effectiveness criteria (Table 2). Job design characteristics were related to all criteria, with half the relationships significant and in the positive direction. Self-management and participation related to effectiveness in five of six analyses. Variety and significance showed three positive relationships each. Task identity was unrelated to any of the criteria.

To examine the predictiveness of all job design characteristics together, and control for experiment-wise error rate, characteristics were averaged to a composite and correlated with the criteria. Unit-weighted averages were more robust than differentially weighted regressions (Wainer, 1976), and unlike regressions, unit-weighted averages did not lose statistical power. Five of the six correlations were significant (Table 2), although modest in size.

Interdependence characteristics were related to all three criteria, but proportionately fewer of the correlations were significant. Each characteristic showed

Table 2: Correlations of Work Group Characteristics Reported by Employees and Managers with Productivity, Employee Satisfaction, and Manager Judgments of Effectiveness

Themes/ characteristics	Productivity		Employee satisfac.		Manager judge.	
	Empl. data ($n=78$)	Manag. data ($n=75$)	Empl. data ($n=78$)	Manag. data ($n=75$)	Empl. data ($n=79$)	Manag. data ($n=76$)
Job design						
Self-management	.23**	.18*	.13	.16*	.28**	.16*
Participation	.15*	.22*	.34**	.11	.16*	.16*
Task variety	.10	.20**	.23**	.09	.19**	.12
Task significance	.14	.10	.20**	−.22**	.21**	.19**
Task identity	.06	.07	.07	−.06	.06	.12
Composite	.19**	.25**	.28**	.07	.25**	.25**
Interdependence						
Task interdep.	.14*	.06	.01	.06	.05	−.14
Goal interdep.	.13	.01	.11	.10	.18*	−.02
Inter, feedback and rewards	.13	.08	.27**	.16*	.13	.06
Composite	.20**	.08	.20**	.18*	.18*	−.04
Composition						
Heterogeneity	−.05	−.15*	−.05	−.04	.03	−.14
Flexibility	.12	−.02	−.02	−.01	.35**	.19**
Relative size	.23**	.19**	.23**	.25**	.29**	.24**
Preference for group work	.10	.01	.18*	.05	.13	−.10
Composite	.21**	.08	.19**	.17*	.36**	.20**
Context						
Training	.11	.08	.18**	.08	.19**	.15*
Managerial support	.14	.16*	.28**	.20**	.09	.18*
Comm./Coop. between groups	.04	−.06	−.03	−.12	−.01	.02
Composite	.13	.08	.20**	.06	.14	.16*
Process						
Potency	.29**	.22**	.20**	.27**	.38**	.28**
Social support	.20**	.12	.03	−.06	.13	.14
Workload sharing	.21**	.22**	.06	.07	.20**	.23**
Comm./Coop. within the group	.18**	.20**	.08	−.01	.18*	.13
Composite	.26**	.25**	.11	.11	.27**	.27**

*$p<.10$; **$p<.05$

one or two positive significant relationships, but the composite was significantly related to effectiveness in four of six analyses.

Composition characteristics were related to all three criteria, especially manager judgments. A third of the correlations were positive and significant. Relative size was related to effectiveness in all six analyses, flexibility had two positive relationships, preference for group work had one positive relationship, but heterogeneity only showed a reversal. The composite was significant in five of six analyses.

Context characteristics related mostly to satisfaction and manager judgments criteria, with a third of the correlations positive and significant. Managerial support had four relationships, training had three, and communication/cooperation between groups had none. Two of six correlations with the composite were significant.

Finally, process characteristics related mostly to productivity and manager judgments criteria, with over half the correlations positive and significant. Potency was related in all six analyses. Workload sharing had four relationships, communication/cooperation within the group had three, and social support had one. The composite was significant in four of six analyses.

In summary, many relationships were observed between group characteristics and effectiveness, even though small in size. The manager judgments criterion was most predictable, followed by productivity, and then satisfaction. All five themes predicted some of the effectiveness criteria. Job design and process characteristics were slightly more predictive than interdependence, composition, and context characteristics. Characteristics data provided by employees and managers were somewhat similar (i.e., both significant in about a third of the cases, both non-significant in about a third, and one significant and the other not in about a third), thus modestly strengthening the findings.

Supplementary Analysis

The practical significance of the findings was examined because results in Table 2 suggested small effects. Based on the average of the 19 characteristics, the best (top ranked one third) and worst (bottom ranked one third) groups were identified and compared on two effect size indicators: standard deviation differences and differences expressed as percentages of the means. For employee data, the standard deviation (percentage of mean) differences were .66 (33%) for productivity, .52 (5%) for satisfaction, and .70 (12%) for manager judgments. Differences for manager data were slightly smaller, with differences of .26 (12%) for productivity, .22 (2%) for satisfaction, and .43 (7%) for manager judgments. Thus, differences between best and worst groups were practically important, especially in terms of productivity.

Discussion

Summary and Conclusions

Based on a review of the work group effectiveness literature, 5 themes and 19 characteristics were delineated. They were then evaluated against both objective and subjective criteria of effectiveness for 80 work groups.

Job design characteristics were very useful in predicting effectiveness. They related to all three criteria. Except for task identity, all the characteristics showed positive relationships with most criteria. Self-management and participation were the most predictive, perhaps partly because they were the more readily observable characteristics of effective work groups (cf. higher correlations between employees and managers in Table 1).

Theoretically, the findings suggest that the model validated so many times at the job level may also be valid at the group level. The motivational value of group work may come in part because such work designs, especially self-managed groups, enhance the motivational quality of members' jobs.

Interdependence characteristics, which are much more recent in the literature (Shea & Guzzo, 1987) and relatively untested, may also have some value. They showed several relationships with effectiveness criteria. In particular, interdependent feedback and rewards were related to employee satisfaction in both samples.

Composition characteristics showed relationships with all three criteria, but mainly with manager judgments. This may be because composition is determined by staffing, which is an important responsibility of managers. Relative size was related to all criteria in both samples, with larger groups more effective. Relationships were also observed for flexibility, with groups having flexible members viewed as more effective by managers. And preference for group work also showed one positive relationship with satisfaction.

Heterogeneity showed no positive relationships with effectiveness. This could have been partly due to the lack of heterogeneity in the sample (e.g., nearly all female and similar levels of education), but this may have also been due to improper construct operationalization. The literature may have recommended that a variety of different skills be present in the group (Cordery et al., 1991; Gladstein, 1984; Goodman et al., 1986; Pearce & Ravlin, 1987; Shaw, 1983; Wall et al., 1986). That is, all members must be skilled, but in different areas. Whereas, the measures in this study assessed the variation of skill levels in the group, perhaps implying that some members were skilled and others were not. The scale in the Appendix has been modified for future research to be more consistent with the former meaning of heterogeneity.

Context characteristics related mainly to satisfaction and manager judgments, but characteristics relevant to each were different. Management support was more predictive of employee satisfaction, while training was more predictive of manager judgments. This may represent the inputs to the group perceived as most valuable by each party. Employees view manager support as most critical, while managers view observable contributions like training as most critical. Communication and cooperation between groups was not related to effectiveness probably because the groups were very independent in this sample.

It is a recent trend to recognize the importance of context and resources (Guzzo & Shea, 1992). These characteristics showed somewhat fewer relationships than some others, but the results suggest they do add to our understanding of potential determinants of effectiveness.

Process characteristics related mainly to productivity and manager judgments. Potency related to all three criteria in both samples. It was the strongest predictor of all characteristics, thus supporting assertions as to the importance of the construct (Shea & Guzzo, 1987; Guzzo et al., 1993). Workload sharing was also very predictive, and social support and communication and cooperation within the group showed several relationships. These results highlight the importance of proper group processes to the functioning of effective work groups (Gladstein, 1984; Hackman, 1987; McGrath, 1964).

Implications for Work Group Design

The implication of relating work group characteristics to effectiveness is that such information might be used to design more effective work groups. As such, this study may make practical contributions. First, it focuses attention on characteristics management can influence. The degree to which they can be controlled or designed into groups by management varies, however. Input characteristics (i.e., job design, interdependence, composition, and context) are more directly controllable than process characteristics. Process may be only indirectly affected by management through encouragement, modeling, and reinforcement. Nevertheless, identifying and validating these characteristics is a first step in learning how to design effective work groups. Second, the study provides 19 characteristics and a 54-item measure (Appendix) as practical tools for designing work groups. Each characteristic can be viewed as a group design recommendation. With due consideration of the limits of the study, it is cautiously recommended that groups be designed to have higher levels of each characteristic (e.g., higher levels of self-management, interdependence, managerial support, etc.). The Appendix could even be converted into a work design checklist for enhancing team effectiveness. Third, it illustrates the potential importance of proper design in terms of productivity and satisfaction differences associated with groups that are high or low on the characteristics.

Practical implications could also be recognized by conceptualizing these work group characteristics within a human resources (HR) management framework. That is, many characteristics relate to HR activities line managers perform (e.g., staffing, training, assigning work, appraising performance, allocating rewards, etc.). Linking characteristics to HR activities has several advantages. First, it helps line managers understand how they can create and maintain effective work groups as part of their HR responsibilities. Second, by linking to HR activities organizations understand, it provides focal points for work group interventions. Note that reviews of interventions by Katzell and Guzzo (1983) and Guzzo, Jette, and Katzell (1985) were also organized by HR activities. Third, it might enhance awareness in HR departments of their responsibilities regarding work groups (e.g., advising management how to staff, train, appraise, and reward groups), in addition to traditional concerns for individual employees (Shea & Guzzo, 1987). Finally, integration of work group design with HR activities may identify important interactions not recognized previously (e.g., between job design and compensation, Campion & Berger, 1990).

Limitations and Future Research

The implications of the study should perhaps be viewed as propositions for future research given the study's limitations. Some ideas for future research derive from methodological limitations. First, statistical power was only moderate for small effect sizes. Group research is susceptible to this problem because of the group level of analysis. Objective criteria like productivity exacerbate the problem because of smaller effect sizes than subjective measures with common method variance. Second, reliabilities of some scales were low. Future studies might lengthen or purify some scales, include more than five employees per group, and examine perceptual differences between employees and managers. Third, some data were collected from individuals and then aggregated to the group. Future research might use a group level of measurement (e.g., have groups give consensus ratings). Fourth, passive observation research does not allow causal inferences, and thus causation could be reversed (e.g., employees were aware of their effectiveness and described the groups accordingly). However, there is substantial laboratory experimental evidence that many of these characteristics cause the outcomes (e.g., Cartwright & Zander, 1968; Levine & Moreland, 1990; McGrath, 1984; Steiner, 1972; Zander, 1979), and the present study complements this research by assessing generalizability to the field. Nevertheless, field experiments should be conducted. Fifth, static research does not allow an examination of change over time, as is likely with work design (Campion & McClelland, 1993; Campion & Medsker, 1992; Griffin, 1991).

Other ideas for future research are more theoretical. First, tasks and technologies may be moderators of design-outcome relationships (Fry & Slocum, 1984; Gladstein, 1984). For example, heterogeneity may relate to productivity in creative tasks, and communication between groups might relate to productivity in groups with highly interdependent tasks. Second, future research might combine and test the themes in an integrated input-process-output model. It would be useful to know which inputs enhance key process variables, like potency, and whether these process variables mediate the influence of input variables on the outcomes. Third, other potentially important design characteristics could be examined in future research. For example, leadership and employee abilities have been shown to be highly influential in other areas of personnel research and most certainly play a role in determining group effectiveness. Finally, from a practical perspective, more needs to be known about how managers can actually affect these design characteristics when implementing group work design in organizations.

Notes

1. The heterogeneity scale was slightly modified. See the Discussion for explanation. Original items are available from the authors.
2. Results for relationships between the characteristics and individual productivity measures are available from the authors.
3. Results for relationships between the characteristics and individual satisfaction subscales are available from the authors.
4. Results for relationships between the characteristics and individual items of the managers' effectiveness judgments are available from the authors.

References

Albanese R, Van Fleet DD. (1985). Rational behavior in groups: The free-riding tendency. *Academy of Management Review*, 10, 244–255.

Alderfer CP. (1977). Group and intergroup relations. In Hackman JR, Suttle IL (Eds.), *Improving life at work: Behavioral science approaches to organizational change* (pp. 227–246). Santa Monica, CA: Goodyear.

Algera JA. (1983). Objective and perceived task characteristics as a determinant of reactions by task performers. *Journal of Occupational Psychology*, 56, 95–107.

Baker TA, Dickinson TL, Salas E. (1991, April). *The influence of team and individual training in interdependent tasks.* Paper presented at the Annual Conference of the Society of Industrial and Organizational Psychology, St. Louis, MO.

Banas PA. (1988). Employee involvement: A sustained labor/management initiative at the Ford Motor Company. In Campbell JP, Campbell RJ & Associates (Eds.), *Productivity in organizations* (pp. 255–294). San Francisco: Jossey-Bass.

Bandura A. (1982). Self-efficacy mechanism in human agency. *American Psychologist*, 37, 122–147.

Bentler PM, Chou CR (1987). Practical issues in structural modeling. *Sociological Methods & Research*, 15, 78–117.

Brett J, Rognes J. (1986). Intergroup relations in organizations. In Goodman PS (Ed.), *Designing effective work groups* (pp. 202–236). San Francisco: Jossey-Bass.

Buller PF, Bell CH Jr. (1986). Effects of team building and goal setting on productivity: A field experiment. *Academy of Management Journal*, 29, 305–328.

Campion MA. (1988). Interdisciplinary approaches to job design: A constructive replication with extensions. *Journal of Applied Psychology*, 73, 464–481.

Campion MA, Berger CL. (1990). Conceptual integration and empirical test of job design and compensation relationships. *Personnel Psychology*, 43, 525–554.

Campion MA, McClelland CL. (1991). Interdisciplinary examination of the costs and benefits of enlarged jobs: A job design quasi-experiment. *Journal of Applied Psychology*, 76, 186–198.

Campion MA, McClelland CL. (1993). Follow-up and extension of the interdisciplinary costs and benefits of enlarged jobs. *Journal of Applied Psychology*, 78, 339–351.

Campion MA, Medsker GJ. (1992). Job design. In Salvendy G (Ed.), *Handbook of industrial engineering* (2nd ed., pp. 845–881). New York: Wiley.

Campion MA, Thayer PW. (1985). Development and field evaluation of an interdisciplinary measure of job design. *Journal of Applied Psychology*, 70, 29–43.

Cartwright D. (1968). The nature of group cohesiveness. In Cartwright D, Zander A (Eds.), *Group dynamics: Research and theory*. New York: Harper & Row.

Cartwright D, Zander A (Eds.). (1968). *Group dynamics: Research and theory*. New York: Academic Press.

Cohen J. (1977). *Statistical power analysis for the behavioral sciences* (Rev. ed.). New York: Academic Press.

Cordery JL, Mueller WS, Smith LM. (1991). Attitudinal and behavioral effects of autonomous group working: A longitudinal field study. *Academy of Management Journal*, 34, 464–476.

Cronbach LJ, Gleser GC, Nanda H, Rajaratnam N. (1972). *The dependability of behavioral measurements: Theory of generalizability for scores and profiles.* New York: Wiley.

Cummings TG. (1978). Self-regulating work groups: A socio-technical synthesis. *Academy of Management Review*, 3, 625–634.

Cummings TG. (1981). Designing effective work groups. In Nystrom PC, Starbuck WH (Eds.), *Handbook of organization design* (Vol. 2, pp. 250–271). New York: Oxford University Press.

Davis LE, Wacker GL. (1987). Job design. In Salvendy G (Ed.), *Handbook of human factors* (pp. 431–452). New York: Wiley.

Deadrick DL, Madigan RM. (1990). Dynamic criteria revisited: A longitudinal study of performance stability and predictive validity. *Personnel Psychology*, 43, 717–744.

DeMeuse KP, Liebowitz SJ. (1981). An empirical analysis of team building research. *Group and Organization Studies*, 6, 357–378.

Deutsch M. (1949). An experimental study of the effects of cooperation and competition upon group process. *Human Relations*, 2, 199–231.

Dyer J. (1984). Team research and team training: A state-of-the-art review. In Muckler FA (Ed.), *Human factors review* (pp. 285–323). Santa Monica, CA: Human Factors Society.

Ford JK, MacCallum RC, Tait M. (1986). The application of exploratory factor analysis in applied psychology: A critical review and analysis. *Personnel Psychology*, 39, 291–314.
Fried Y, Ferris GR. (1987). The validity of the job characteristics model. *Personnel Psychology* 40, 287–322.
Fry LW, Slocum JW. (1984). Technology, structure, and workgroup effectiveness: A test of a contingency model. *Academy of Management Journal*, 27, 221–246.
Gladstein DL. (1984). Groups in context: A model of task group effectiveness. *Administrative Science Quarterly*, 29, 499–517.
Goodman PS. (1979). *Assessing organizational change: The Rushton quality of work experiment.* New York: Wiley.
Goodman PS, Devadas R, Hughson TLG. (1988). Groups and productivity: Analyzing the effectiveness of self-managing teams. In Campbell JP, Campbell RJ (Eds.), *Productivity in organizations* (pp. 295–327). San Francisco: Jossey-Bass.
Goodman PS, Leyden DP. (1991). Familiarity and group productivity. *Journal of Applied Psychology*, 76, 578–586.
Goodman PS, Ravlin EC, Argote L. (1986). Current thinking about groups: Setting the stage for new ideas. In Goodman PS (Ed.), *Designing effective work groups* (pp. 1–27). San Francisco: Jossey-Bass.
Goodman PS, Ravlin EC, Schminke M. (1987). Understanding groups in organizations. In Staw BM, Cummings LL (Eds.), *Research in organizational behavior* (Vol. 9, pp. 121–173). Greenwich, CT: JAI Press.
Griffin RW. (1991). Effects of work redesign on employee perceptions, attitudes, and behaviors: A long-term investigation. *Academy of Management Journal*, 34, 425–435.
Guzzo RA, Jette RD, Katzell RA. (1985). The effects of psychologically based intervention programs on work productivity: A meta-analysis. *Personnel Psychology*, 38, 275–291.
Guzzo RA, Shea GP. (1992). Group performance and intergroup relations in organizations. In Dunnette MD, Hough LM (Eds.), *Handbook of industrial and organizational psychology* (Vol. 3, pp. 269–313). Palo Alto: Consulting Psychologists Press.
Guzzo RA, Yost PR, Campbell RJ, Shea GP. (1993). Potency in groups: Articulating a construct. *British Journal of Social Psychology*, 32, 87–106.
Hackman JR. (1987). The design of work teams. In Lorsch JW (Ed.), *Handbook of organizational behavior* (pp. 315–342). Englewood Cliffs, NJ: Prentice-Hall.
Hackman JR, Lawler EE III. (1971). Employee reactions to job characteristics [Monograph]. *Journal of Applied Psychology*, 55, 259–286.
Hackman JR, Oldham GR. (1980). *Work redesign.* Reading, MA: Addison-Wesley.
Hackman JR, Walton RE. (1986). Leading groups in organizations. In Goodman PS (Ed.), *Designing effective work groups* (pp. 72–119). San Francisco: Jossey-Bass.
Harkins SG. (1987). Social loafing and social facilitation. *Journal of Experimental Social Psychology*, 23, 1–18.
Hulin CL, Blood MR. (1968). Job enlargement, individual differences, and worker responses. *Psychological Bulletin*, 69, 41–55.
Jackson SE, Brett JF, Sessa VI, Cooper DM, Julin JA, Peyronnin K. (1991). Some differences make a difference: Individual dissimilarity and group heterogeneity as correlates of recruitment, promotions, and turnover. *Journal of Applied Psychology*, 76, 675–689.
James LR. (1982). Aggregation bias in estimates of perceptual agreement. *Journal of Applied Psychology*, 67, 219–229.
James LR, Demaree RG, Wolf G. (1984). Estimating within-group interrater reliability with and without response bias. *Journal of Applied Psychology*, 69, 85–98.
Janis IL. (1972). *Victims of groupthink: A psychological study of foreign policy decisions and fiascos.* Boston: Houghton Mifflin.
Katzell RA, Guzzo RA. (1983). Psychological approaches to productivity improvement. *American Psychologist*, 38, 468–472.
Kiggundu MN. (1983). Task interdependence and job design: Test of a theory. *Organizational Behavior and Human Performance*, 31, 145–172.
Koch JL. (1979). Effects of goal specificity and performance feedback to work groups on peer leadership, performance, and attitudes. *Human Relations*, 33, 819–840.

Kolodny HF, Kiggundu MN. (1980). Towards the development of a sociotechnical systems model in woodlands mechanical harvesting. *Human Relations*, 33, 623–645.

Kozlowski SWJ, Hattrup K. (1992). A disagreement about within-group agreement: Disentangling issues of consistency versus consensus. *Journal of Applied Psychology*, 77, 161–167.

Latané B, Williams K, Harkins S. (1979). Many hands make light the work: The causes and consequences of social loafing. *Journal of Personality and Social Psychology*, 37, 822–832.

Leavitt HJ. (1951). Some effects of certain communication patterns on group performance. *Journal of Abnormal and Social Psychology*, 46, 38–50.

Leventhal GS. (1976). The distribution of rewards and resources in groups and organizations. In Berkowitz L, Walster E (Eds.), *Advances in experimental social psychology* (Vol. 9, pp. 91–131). New York: Academic Press.

Levine JM, Moreland RL. (1990). Progress in small group research. *Annual Review of Psychology*, 41, 585–634.

Locke EA, Latham GP. (1990). *A theory of goal setting and task performance.* Englewood Cliffs, NJ: Prentice Hall.

Magjuka RJ, Baldwin TT. (1991). Team-based employee involvement programs for continuous organizational improvement: Effects of design and administration. *Personnel Psychology*, 44, 793–812.

Majchrzak A. (1988). *The human side of factory automation.* San Francisco: Jossey-Bass.

McGrath JE. (1964). *Social psychology: A brief introduction.* Holt, Rinehart, and Winston.

McGrath JE. (1984). *Groups: Interaction and performance.* Englewood Cliffs, NJ: Prentice-Hall.

McGrath JE. (1986). Studying groups at work: Ten critical needs. In Goodman PS (Ed.), *Designing effective work groups* (pp. 362–391). San Francisco: Jossey-Bass.

Mintzberg H. (1979). *The structuring of organizations: A synthesis of the research.* Englewood Cliffs, NJ: Prentice-Hall.

O'Reilly CA, Roberts KH. (1977). Task group structure, communication, and effectiveness. *Journal of Applied Psychology*, 62, 674–681.

Pasmore W, Francis C, Haldeman J. (1982). Sociotechnical systems: A North American reflection on empirical studies of the seventies. *Human Relations*, 35, 1179–1204.

Pearce JA, Ravlin EC. (1987). The design and activation of self-regulating work groups. *Human Relations*, 40, 751–782.

Pearson CAL. (1987). Participative goal setting as a strategy for improving performance and job satisfaction: A longitudinal evaluation with railway track maintenance gangs. *Human Relations*, 40, 473–488.

Porter LW, Lawler EE III, Hackman JR. (1987). Ways groups influence individual work effectiveness. In Steers RM, Porter LW (Eds.), *Motivation and work behavior* (4th ed., pp. 271–279). New York: McGraw-Hill.

Poza EJ, Markus ML. (1980). Success story: The team approach to work restructuring. *Organizational Dynamics*, 8(4), 3–25.

Pritchard RD, Jones S, Roth P, Stuebing K, Ekeberg, S. (1988). Effects of group feedback, goal setting, and incentives on organizational productivity [Monograph]. *Journal of Applied Psychology*, 73, 337–358.

Roberts KH, Hulin CL, Rousseau DM. (1978). *Developing an interdisciplinary science of organizations.* San Francisco: Jossey-Bass.

Salas E, Dickinson TI, Converse SA, Tannenbaum SI. (1992). Toward an understanding of team performance and training. In Swezey RW, Salas E (Eds.), *Teams: Their training and performance* (pp. 3–29). Norwood, NJ: ABLEX.

Shaw ME. (1983). Group composition. In Blumberg HH, Hare AP, Kent V, Davies M (Eds.), *Small groups and social interaction* (Vol. 1, pp. 89–96). New York: Wiley.

Shea GP, Guzzo RA. (1987). Groups as human resources. In Rowland KM, Ferris GR (Eds.), *Research in human resources and personnel management* (Vol. 5, pp. 323–356). Greenwich, CT: JAI Press.

Steiner ID. (1972). *Group process and productivity.* New York: Academic Press.

Strauss A, Corbin J. (1990). *Basics of qualitative research: Grounded theory procedures and techniques.* Newbury Park, CA: Sage.

Sundstrom E, De Meuse KP, Futrell D. (1990). Work teams: Applications and effectiveness. *American Psychologist*, 45, 120–133.
Tannenbaum SI, Beard RL, Salas E. (1992). Team building and its influence on team effectiveness: An examination of conceptual and empirical developments. In Kelley K (Ed.), *Issues, theory, and research in industrial/organizational psychology* (pp. 117–153). Amsterdam, Holland: Elsevier.
Thompson JD. (1967). *Organizations in action.* New York: McGraw-Hill.
Van de Ven AH, Ferry DL. (1980). *Measuring and assessing organizations.* New York: Wiley.
Vroom VH. (1964). *Work and motivation.* New York: Wiley.
Wainer H. (1976). Estimating coefficients in linear models: It don't make no nevermind. *Psychological Bulletin*, 83, 213–217.
Wall TD, Kemp NJ, Jackson PR, Clegg CW. (1986). Outcomes of autonomous workgroups: A long-term field experiment. *Academy of Management Journal*, 29, 281–304.
Walton RE. (1972). How to counter alienation in the plant. *Harvard Business Review*, 50(6), 70–81.
Warwick DP, Lininger CA. (1975). *The sample survey: Theory and practice.* New York: McGraw-Hill.
Webster's Seventh New Collegiate Dictionary. (1965). Springfield, MA: G.C. Merriam.
Whyte WF. (1955). *Money and motivation: An analysis of incentives in industry.* New York: Harper.
Wicker A, Kirmeyer SL, Hanson L, Alexander D. (1976). Effects of manning levels on subjective experiences, performance, and verbal interaction in groups. *Organizational Behavior and Human Performance*, 17, 251–274.
Wong CS, Campion MA. (1991). Development and test of a task level model of motivational job design. *Journal of Applied Psychology*, 76, 825–837.
Woodman RW, Sherwood JJ. (1980). The role of team development in organizational effectiveness: A critical review. *Psychological Bulletin*, 88, 166–186.
Zajonc RB. (1965). Social facilitation. *Science*, 149, 269–274.
Zander A. (1979). The psychology of group processes. In Rosenzweig MR, Porter LW (Eds.), *Annual Review of Psychology* 30, 417–451.

Appendix
Work Group Characteristics Measure

Self-management

1. The members of my team are responsible for determining the methods, procedures, and schedules with which the work gets done.
2. My team rather than my manager decides who does what tasks within the team.
3. Most work-related decisions are made by the members of my team rather than by my manager.

Participation

4. As a member of a team, I have a real say in how the team carries out its work.
5. Most members of my team get a chance to participate in decision making.
6. My team is designed to let everyone participate in decision making.

Task variety

7. Most members of my team get a chance to learn the different tasks the team performs.

8. Most everyone on my team gets a chance to do the more interesting tasks.
9. Task assignments often change from day to day to meet the work load needs of the team.

Task significance (importance)

10. The work performed by my team is important to the customers in my area.
11. My team makes an important contribution to serving the company's customers.
12. My team helps me feel that my work is important to the company.

Task identity (mission)

13. The team concept allows all the work on a given product to be completed by the same set of people.
14. My team is responsible for all aspects of a product for its area.
15. My team is responsible for its own unique area or segment of the business.

Task interdependence (interdependence)

16. I cannot accomplish my tasks without information or materials from other members of my team.
17. Other members of my team depend on me for information or materials needed to perform their tasks.
18. Within my team, jobs performed by team members are related to one another.

Goal interdependence (goals)

19. My work goals come directly from the goals of my team.
20. My work activities on any given day are determined by my team's goals for that day.
21. I do very few activities on my job that are not related to the goals of my team.

Interdependent feedback and rewards (feedback and rewards)

22. Feedback about how well I am doing my job comes primarily from information about how well the entire team is doing.
23. My performance evaluation is strongly influenced by how well my team performs.
24. Many rewards from my job (e.g., pay, promotion, etc.) are determined in large part by my contributions as a team member.

Heterogeneity (membership)

25. The members of my team vary widely in their areas of expertise.
26. The members of my team have a variety of different backgrounds and experiences.
27. The members of my team have skills and abilities that complement each other.

Flexibility (member flexibility)

28. Most members of my team know each other's jobs.
29. It is easy for the members of my team to fill in for one another.
30. My team is very flexible in terms of changes in membership.

Relative size (size)

31. The number of people in my team is too small for the work to be accomplished. (Reverse scored)

Preference for group work (team work preferences)

32. If given the choice, I would prefer to work as part of a team rather than work alone.
33. I find that working as a member of a team increases my ability to perform effectively.
34. I generally prefer to work as part of a team.

Training

35. The company provides adequate technical training for my team.
36. The company provides adequate quality and customer service training for my team.
37. The company provides adequate team skills training for my team (e.g., communication, organization, interpersonal, etc.).

Managerial support

38. Higher management in the company supports the concept of teams.
39. My manager supports the concept of teams.

Communication/cooperation between work groups

40. I frequently talk to other people in the company besides the people on my team.
41. There is little competition between my team and other teams in the company.
42. Teams in the company cooperate to get the work done.

Potency (spirit)

43. Members of my team have great confidence that the team can perform effectively.
44. My team can take on nearly any task and complete it.
45. My team has a lot of team spirit.

Social support

46. Being in my team gives me the opportunity to work in a team and provide support to other team members.
47. My team increases my opportunities for positive social interaction.
48. Members of my team help each other out at work when needed.

Workload sharing (sharing the work)

49. Everyone on my team does their fair share of the work.
50. No one in my team depends on other team members to do the work for them.
51. Nearly all the members on my team contribute equally to the work.

Communication/cooperation within the work group

52. Members of my team are very willing to share information with other team members about our work.
53. Teams enhance the communication among people working on the same product.
54. Members of my team cooperate to get the work done.

Note. Headings in parentheses are the labels in the questionnaire if they were different from the headings in Table 1. Heterogeneity items have been modified (see Discussion). Instructions: "This questionnaire consists of statements about your team, and how your team functions as a group. Please indicate the extent to which each statement describes your team." Common response scale: "(5) Strongly agree, (4) Agree, (3) Neither agree nor disagree, (2) Disagree, or (1) Strongly disagree. (Leave blank if you don't know or the statement is not applicable)."

51

Collective Mind in Organizations: Heedful Interrelating on Flight Decks

Karl E. Weick and Karlene H. Roberts

Source: *Administrative Science Quarterly* 38 (3) (1993): 357–381.

Some organizations require nearly error-free operations all the time because otherwise they are capable of experiencing catastrophes. One such organization is an aircraft carrier, which an informant in Rochlin, LaPorte, and Roberts' (1987: 78) study described as follows:

> . . . imagine that it's a busy day, and you shrink San Francisco Airport to only one short runway and one ramp and one gate. Make planes take off and land at the same time, at half the present time interval, rock the runway from side to side, and require that everyone who leaves in the morning returns that same day. Make sure the equipment is so close to the edge of the envelope that it's fragile. Then turn off the radar to avoid detection, impose strict controls on radios, fuel the aircraft in place with their engines running, put an enemy in the air, and scatter live bombs and rockets around. Now wet the whole thing down with sea water and oil, and man it with 20-year-olds, half of whom have never seen an airplane close-up. Oh and by the way, try not to kill anyone.

Even though carriers represent "a million accidents waiting to happen" (Wilson, 1986: 21), almost none of them do. Here, we examine why not. The explanation we wish to explore is that organizations concerned with reliability enact aggregate mental processes that are more fully developed than those found in organizations concerned with efficiency. By fully developed mental processes, we mean that organizations preoccupied with reliability may spend more time and effort organizing for controlled information processing (Schneider and Schiffrin, 1977), mindful attention (Langer, 1989), and heedful action (Ryle, 1949). These intensified efforts enable people to understand more of the complexity they face, which then enables them to respond with fewer errors. Reliable systems are smart systems.

Before we can test this line of reasoning we need to develop a language of organizational mind that enables us to describe collective mental processes in organizations. In developing it, we move back and forth between concepts of mind and details of reliable performance in flight operations on a modern super carrier.[1] We use flight operations to illustrate organizational mind for a number of reasons: The technology is relatively simple, the coordination among activities is explicit and visible, the socialization is continuous, agents working alone have less grasp of the entire system than they do when working together, the system is constructed of interdependent know-how, teams of people think on their feet and do the "right thing" in novel situations, and the consequences of any lapse in attention are swift and disabling. Because our efforts to understand deck operations got us thinking about the possibility that performance is mediated by collective mental processes, we use these operations to illustrate that thinking, but the processes of mind we discuss are presumed to be inherent in all organizations. What may vary across organizations is the felt need to develop these processes to more advanced levels.

The Idea of Group Mind

Discussions of collective mental processes have been rare, despite the fact that people claim to be studying "social" cognition (e.g., Schneider, 1991). The preoccupation with individual cognition has left organizational theorists ill-equipped to do much more with the so-called cognitive revolution than apply it to organizational concerns, one brain at a time. There are a few exceptions, however, and we introduce our own discussion of collective mind with a brief review of three recent attempts to engage the topic of group mind.

Wegner and his associates (Wegner, Giuliano, and Hertel, 1985; Wegner, 1987; Wegner, Erber, and Raymond, 1991) suggested that group mind may take the form of cognitive interdependence focused around memory processes. They argued that people in close relationships enact a single transactive memory system, complete with differentiated responsibility for remembering different portions of common experience. People know the locations rather than the details of common events and rely on one another to contribute missing details that cue their own retrieval. Transactive memory systems are integrated and differentiated structures in the sense that connected individuals often hold related information in different locations. When people trade lower-order, detailed, disparate information, they often discover higher-order themes, generalizations, and ideas that subsume these details. It is these integrations of disparate inputs that seem to embody the "magical transformation" that group mind theorists sought to understand (Wegner, Giuliano, and Hertel, 1985: 268). The important point Wegner contributes to our understanding of collective mental processes is that group mind is *not* indexed by within-group similarity of attitudes, understanding, or language, nor can it be understood without close attention to communications processes among group members (Wegner, Giuliano, and Hertel, 1985: 254–255). Both of these lessons will be evident in our reformulation.

Work in artificial intelligence provides the backdrop for two additional attempts to conceptualize group mind: Sandelands and Stablein's (1987) description of organizations as mental entities capable of thought and Hutchins' (1990, 1991) description of organizations as distributed information-processing systems. The relevant ideas are associated with theories of "connectionism," embodied in so-called "neural networks." Despite claims that their work is grounded in the brain's microanatomy, connectionists repeatedly refer to "neurological plausibility" (Quinlan, 1991: 41), "neuron-like units" (Churchland, 1992: 32), "brain-style processing" (Rumelhart, 1992: 69), or "neural inspiration" (Boden, 1990: 18). This qualification is warranted because the "neural" networks examined by connectionists are simply computational models that involve synchronous parallel processing among many interrelated unreliable and/or simple processing units (Quinlan, 1991: 40). The basic idea is that knowledge in very large networks of very simple processing units resides in patterns of connections, not in individuated local symbols. As Boden (1990: 14) explained, any "unit's activity is regulated by the activity of neighboring units, connected to it by inhibitory or excitatory links whose strength can vary according to design and/or learning." Thus, any one unit can represent several different concepts, and the same concept in a different context may activate a slightly different network of units.

Connectionism by itself, however, is a shaky basis on which to erect a theory of organizational mind. The framework remains grounded in a device that models a single, relatively tightly coupled actor as opposed to a loosely coupled system of multiple actors, such as an organization. Connectionists have difficulty simulating emotion and motivation (Dreyfus and Dreyfus, 1990), as well as everyday thought and reasoning (Rumelhart, 1992). In computational models there is no turnover of units akin to that found in organizations, where units are replaced or moved to other locations. And the inputs connectionists investigate are relatively simple items such as numerals, words, or phrases, with the outputs being more or less accurate renderings of these inputs (e.g., Elman, 1992). This contrasts with organizational researchers who pay more attention to complex inputs, such as traditional competitors who make overtures to cooperate, and to outputs that consist of action as well as thought.

What connectionism contributes to organizational theory is the insight that complex patterns can be encoded by patterns of activation and inhibition among simple units, if those units are richly connected. This means that relatively simple actors may be able to apprehend complex inputs if they are organized in ways that resemble neural networks. Connectionists also raise the possibility that mind is "located" in connections and the weights put on them rather than in entities. Thus, to understand mind is to be attentive to process, relating, and method, as well as to structures and content.

Sandelands and Stablein (1987: 139–141) found parallels between the organization of neurons in the brain and the organization of activities in organizations. They used this parallel to argue that connected activities encode concepts and ideas in organizations much like connected neurons encode concepts and ideas in brains. Ideas encoded in behaviors appear to interact in ways that suggest operations of intelligent processing. These parallels are consistent with the idea that organizations are minds. The important lessons from Sandelands and Stablein's analysis are that connections between behaviors, rather than people, may be the crucial "locus" for mind and that intelligence is to be found in patterns of behavior rather than in individual knowledge.

Hutchins (1990, 1991: 289) has used connectionist networks, such as the "constraint satisfaction network," to model how interpretations based on distributed cognitions are formed. These simulations are part of a larger inquiry into how teams coordinate action (Hutchins, 1990) and the extent to which distributed processing amplifies or counteracts errors that form in individual units. Hutchins' analysis suggests that systems maintain the flexible, robust action associated with mindful performance if individuals have overlapping rather than mutually exclusive task knowledge. Overlapping knowledge allows for redundant representation that enables people to take responsibility for all parts of the process to which they can make a contribution (Hutchins, 1990: 210).

The potential fit between connectionist imagery and organizational concepts can be inferred from Hutchins' (1990: 209) description of coordination by mutual constraint in naval navigation teams:

> [The] sequence of action to be taken [in group performance] need not be explicitly represented anywhere in the system. If participants know how to coordinate their activities with the technologies and people with which they interact, the global structure of the task performance will emerge from the local interactions of the members. The structure of the activities of the group is determined by a set of local computations rather than by the implementation of the sort of global plan that appears in the solo performer's procedure. In the team situation, a set of behavioral dependencies are set up. These dependencies shape the behavior pattern of the group.

The lessons we use from Hutchins' work include the importance of redundant representation, the emergence of global structure from local interactions, and behavioral dependencies as the substrate of distributed processing.

Our own attempt to describe group mind has been informed by these three sources but is based on a different set of assumptions. We pay more attention to the form of connections than to the strength of connections and more attention to mind as activity than to mind as entity. To make this shift in emphasis clear, we avoid the phrases "group mind" and "organizational mind" in favor of the phrase "collective mind." The word "collective," unlike the words "group" or "organization," refers to individuals who act as if they are a group. People who act as if they are a group interrelate their actions with more or less care, and focusing on the way this interrelating is done reveals collective mental processes that differ in their degree of development. Our focus is at once on individuals and the collective, since only individuals can contribute to a collective mind, but a collective mind is distinct from an individual mind because it inheres in the pattern of interrelated activities among many people.

We begin the discussion of collective mind by following the lead of Ryle (1949) and developing the concept of mind as a disposition to act with heed. We then follow the lead of Asch (1952) and develop the concept of collective interrelating as contributing, representing, and subordinating, and illustrate these activities with examples from carrier operations. We next combine the notions of heed and interrelating into the concept of collective mind as heedful interrelating and suggest social processes that may account for variations in heedful interrelating. Finally, we describe three examples of heedful interrelating, two from carrier operations and one from the laboratory, and present an extended example of heedless interrelating that resulted in a $38-million accident.

Mind as Disposition to Heed

"Mind" is a noun similar to nouns like faith, hope, charity, role, and culture. "Mind" is not the name of a person, place, or thing but, rather, is a dispositional term that denotes a propensity to act in a certain manner or style. As Ryle (1949: 51) said,

> The statement "the mind is its own place," as theorists might construe it, is not true, for the mind is not even a metaphorical "place." On the contrary, the chessboard, the platform, the scholar's desk, the judge's bench, the lorry-driver's seat, the studio and the football field are among its places. These are where people work and play stupidly or intelligently.

That mind is actualized in patterns of behavior that can range from stupid to intelligent can be seen in the example Ryle (1949: 33) used of a clown who trips and stumbles just as clumsy people do. What's different is that "he trips and stumbles on purpose and after much rehearsal and at the golden moment and where the children can see him and so as not to hurt himself." When a clown trips artfully, people applaud the style of the action, the fact that tripping is done with care, judgment, wit, and appreciation of the mood of the spectators. In short, the tripping is done with heed. Heed is not itself a behavior but it refers to the way behaviors such as tripping, falling, and recovering are assembled. Artful tripping is called heedful, not so much because the tripping involves action preceded by thought but because the behaviors patterned into the action of tripping suggest to the observer qualities such as "noticing, taking care, attending, applying one's mind, concentrating, putting one's heart into something, thinking what one is doing, alertness, interest, intentness, studying, and trying" (Ryle, 1949: 136). These inferences, based on the style of the action, are called "heed concepts" and support the conclusion that the behaviors were combined intelligently rather than stupidly.

The word "heed" captures an important set of qualities of mind that elude the more stark vocabulary of cognition. These nuances of heed are especially appropriate to our interest in systems preoccupied with failure-free performance. People act heedfully when they act more or less carefully, critically, consistently, purposefully, attentively, studiously, vigilantly, conscientiously, pertinaciously (Ryle, 1949: 151). Heed adverbs attach qualities of mind directly to performances, as in the description, "the airboss monitored the pilot's growing load of tasks attentively." Notice that the statement does not say that the airboss was doing two things, monitoring and also checking to be sure that the monitoring was done carefully. Instead, the statement asserts that, having been coached to monitor carefully, his present monitoring reflects this style. Mind is in the monitoring itself, not in some separate episode of theorizing about monitoring.

Heedful performance is not the same thing as habitual performance. In habitual action, each performance is a replica of its predecessor, whereas in heedful performance, each action is modified by its predecessor (Ryle, 1949: 42). In heedful performance, the agent is still learning. Furthermore, heedful performance is the outcome of training and experience that weave together thinking, feeling, and willing. Habitual performance is the outcome of drill and repetition.

When heed declines, performance is said to be heedless, careless, unmindful, thoughtless, unconcerned, indifferent. Heedless performance suggests a failure of intelligence rather than a failure of knowledge. It is a failure to see, to take note of, to be attentive to. Heedless performance is not about ignorance, cognition (Lyons, 1980: 57), and facts. It is about stupidity, competence, and know-how. Thus, mind refers to stretches of human behavior that exhibit qualities of intellect and character (Ryle, 1949: 126).

Group as Interrelated Activity

Ryle's ideas focus on individual mind. To extend his ideas to groups, we first have to specify the crucial performances in groups that could reflect a disposition to heed. To pinpoint these crucial performances, we derive four defining properties of group performance from Asch's (1952: 251–255) discussion of "mutually shared fields" and illustrate these properties with carrier examples.[2]

The first defining property of group performance is that individuals create the social forces of group life when they act as if there were such forces. As Asch (1952: 251) explained it,

> We must see group phenomena as both *the product and condition* of actions of individuals. . . . There are no forces between individuals as organisms; yet to all intents and purposes they act as if there were, and they actually create social forces. Group action achieves the kind of result that would be understandable if all participants were acting under the direction of a single organizing center. No such center exists; between individuals is a hiatus, which nevertheless, they succeed in overcoming with surprising effectiveness.

An example from carriers occurs during flight operations. The men in the tower (Air Department) monitor and give instructions to incoming and departing aircraft. Simultaneously, the men on the landing signal officers' platform do the same thing. They are backed up by the men in Air Operations who monitor and instruct aircraft at some distance from the ship. From the aviator's viewpoint, he receives integrated information about his current status and future behavior from an integrated source when, in reality, the several sources are relatively independent of one another and located in different parts of the ship.

The second defining property of group performance is that when people act as if there are social forces, they construct their actions (contribute) while envisaging a social system of joint actions (represent), and interrelate that constructed action with the system that is envisaged (subordinate). Asch (1952: 251–252) explained this as follows:

> There are group actions that are possible only when each participant has a representation that includes the actions of others and their relations. The respective actions converge relevantly, assist and supplement each other only when the joint situation is represented in each and when the representations are structurally similar. Only when these conditions are given can individuals subordinate themselves to the requirements of joint action. These representations and the actions that they initiate/bring group facts into existence and produce the phenomenal solidity of group process.

The simultaneous envisaging and interrelating that create a system occur when a pilot taxies onto the catapult for launching, is attached to it, and advances his engines to full power. Even though pilots have to rely on the catapult crew, they remain vigilant to see if representations are similar. Pilots keep asking themselves questions like, "Does it feel right?" or "Is the rhythm wrong?" The referent for the question, "Does *it* feel right?" however, is not the aircraft but the joint situation to which he has subordinated himself. If a person on the deck signals the pilot to reduce his engines from full power, he won't do so until someone stands in front of the plane, directly over the catapult, and signals for a reduction in power. Only then is the pilot reasonably certain that the joint situation has changed. He now trusts that the catapult won't be triggered suddenly and fling his underpowered aircraft into a person and then into the ocean.

The third defining property of group performance is that contributing, representing, and subordinating create a joint situation of interrelations among activities, which Asch (1952: 252) referred to as a system:

> When these conditions are given we have a social system or a process of a definite form that embraces the actions of a number of individuals. Such a system does not reside in the individuals taken separately, though each individual contributes to it; nor does it reside outside them; it is present in the interrelations between the activities of individuals.

An example from carriers is a pilot landing an aircraft on a deck. This is not a solitary act. A pilot doesn't really land; he is "recovered." And recovery is a set of interrelated activities among air traffic controllers, landing signal officers, the control tower, navigators, deck hands, the helmsman driving the ship, etc. As the recovery of a single aircraft nears completion in the form of a successful trap, nine to ten people on the landing signal officer's platform, up to 15 more people in the tower, and two to three more people on the bridge observe the recovery and can wave the aircraft off if there is a problem. While this can be understood as an example of redundancy, it can also be interpreted as activities that can be interrelated more or less adequately, depending on the care with which contributing, representing, and subordinating are done.

The fourth and final defining property of group performance suggested by Asch is that the effects produced by a pattern of interrelated activities vary as a function of the style (e.g., heedful-heedless) as well as the strength (e.g., loose-tight) with which the activities are tied together. This is suggested by the statement that, in a system of interrelated activities, individuals can work with, for, or against each other:

> The form the interrelated actions take — on a team or in an office — is a datum of precisely the same kind as any other fact. One could say that all the facts of the system can be expressed as the sum of the actions of individuals. The statement is misleading, however, if one fails to add that the individuals would not be capable of these particular actions unless they were responding to (or envisaging the possibility of) the system. Once the process described is in motion it is no longer the individual "as such" who determines its direction, nor the group acting upon the individual as an external force, but individuals working with, for, or against each other. (Asch, 1952: 252)

It is these varying forms of interrelation that embody collective mind. An example of interrelating on carriers can be seen when ordnance is loaded onto an aircraft and its safety mechanisms are removed. If there is a sudden change of mission, the live ordnance must be disarmed, removed, and replaced by other ordnance that is now activated, all of this under enormous time pressure. These interrelated activities, even though tightly coupled, can become more or less dangerous depending on how the interrelating is done.

In one incident observed, senior officers kept changing the schedule of the next day's flight events through the night, which necessitated a repeated change in ordnance up to the moment the day launches began. A petty officer changing bombs underneath an aircraft, where the pilot couldn't see him, lost a leg when the pilot moved the 36,000-pound aircraft over him. The petty officer should have tied the plane down before going underneath to change the load but failed to do so because there was insufficient time, a situation created by continual indecision at the top. Thus, the senior officers share the blame for this accident because they should have resolved their indecision in ways that were more mindful of the demands it placed on the system.

Although Asch argued that interrelated activities are the essence of groups, he said little about how these interrelations occur or how they vary over time. Instead, he treated interrelations as a variable and interrelating as a constant. If we treat interrelations as a variable and interrelating as a process, this suggests a way to conceptualize collective mind.

Heedful Interrelating as Collective Mind

The insights of Ryle and Asch can be combined into a concept of collective mind if we argue that dispositions toward heed are expressed in actions that construct interrelating. Contributing, representing, and subordinating, actions that form a distinct pattern external to any given individual, become the medium through which collective mind is manifest. Variations in heedful interrelating correspond to variations in collective mind and comprehension.

We assume, as Follett (1924: 146–153) did, that mind begins with actions, which we refer to here as contributions. The contributions of any one individual begin to actualize collective mind to the degree that heedful representation and heedful subordination define those contributions. A heedful contribution enacts collective mind as it begins to converge with, supplement, assist, and become defined in relation to the imagined requirements of joint action presumed to flow from some social activity system.

Similar conduct flows from other contributing individuals in the activity system toward others imagined to be in that system. These separate efforts vary in the heedfulness with which they interrelate, and these variations form a pattern. Since the object of these activities ("the envisaged system," to use Asch's phrase) is itself being constituted as these activities become more or less interrelated, the emergent properties of this object are not contained fully in the representation of any one person nor are they finalized at any moment in time. A single emergent property may appear in more than one representation, but seldom in all. And different properties are shared in common by different subsets of people. Asch seems to have had this distributed representation of the envisaged system in mind when he referred to "structurally similar representations." This pattern of distributed representation explains the transindividual quality of collective mind. Portions of the envisaged system are known to all, but all of it is known to none.

The collective mind is "located" in the process of interrelating just as the individual mind for Ryle was "located" in the activities of lorry driving, chess playing, or article writing. Collective mind exists potentially as a kind of capacity in an ongoing activity stream and emerges in the style with which activities are interrelated. These patterns of interrelating are as close to a physical substrate for collective mind as we are likely to find. There is nothing mystical about all this. Collective mind is manifest when individuals construct mutually shared fields. The collective mind that emerges during the interrelating of an activity system is more developed and more capable of intelligent action the more heedfully that interrelating is done.

A crude way to represent the development of a collective mind is by means of a matrix in which the rows are people and the columns are either the larger activities of contributing, representing, and subordinating, or their component behaviors (e.g., converging with, assisting, or supplementing). Initially, the cell entries can be a simple "yes" or "no." "Yes" means a person performs that action heedfully; "no" means the action is done heedlessly. The more "yeses" in the matrix, the more developed the collective mind.

We portray collective mind in terms of method rather than content, structuring rather than structure, connecting rather than connections. Interrelations are not given but are constructed and reconstructed continually by individuals (Blumer, 1969: 110) through the ongoing activities of contributing, representing, and subordinating. Although these activities are done by individuals, their referent is a socially structured field. Individual activities are shaped by this envisioned field and are meaningless apart from it. When people make efforts to interrelate, these efforts can range from heedful to heedless. The more heed reflected in a pattern of interrelations, the more developed the collective mind and the greater the capability to comprehend unexpected events that evolve rapidly in unexpected ways. When we say that a collective mind "comprehends" unexpected events, we mean that heedful interrelating connects sufficient individual know-how to meet situational demands. For organizations concerned with reliability, those demands often consist of unexpected, non-sequential interactions among small failures that are hard to see and hard to believe. These incomprehensible failures often build quickly into catastrophes (Perrow, 1984: 7–12, 22, 78, 88).

An increase in heedful interrelating can prevent or correct these failures of comprehension in at least three ways. First, longer stretches of time can be connected, as when more know-how is brought forward from the past and is elaborated into new contributions and representations that extrapolate farther into the future. Second, comprehension can be improved if more activities are connected, such as when interrelations span earlier and later stages of task sequences. And third, comprehension can be increased if more levels of experience are connected, as when newcomers who take nothing for granted interrelate more often with old-timers who think they have seen it all. Each of these three changes makes the pattern of interrelations more complex and better able to sense and regulate the complexity created by unexpected events. A system that is tied together more densely across time, activities, and experience comprehends more of what is occurring because the scope of heedful action reaches into more places. When heed is spread across more activities and more connections, there should be more understanding and fewer errors. A collective mind that becomes more comprehensive, comprehends more.

Variations in Heed

If collective mind is embodied in the interrelating of social activities, and if collective mind is developed more or less fully depending on the amount of heedfulness with which that interrelating is done, we must address the issue of what accounts for variations in heed. We suspect the answer lies in Mead's (1934: 186) insight that mind is "the individual importation of social process." We understand the phrase "social process" to mean a set of ongoing interactions in a social activity system from which participants continually extract a changing sense of self-interrelation and then re-enact that sense back into the system. This ongoing interaction process is recapitulated in individual lives and continues despite the replacement of people.

Mead stressed the reality of recapitulation, as did others. Ryle (1949: 27), for example, observed that "this trick of talking to oneself in silence is acquired neither quickly nor without effort; and it is a necessary condition to our acquiring it that we should have previously learned to talk intelligently aloud and have heard and understood other people doing so. Keeping our thoughts to ourselves is a sophisticated accomplishment." Asch (1952: 257) described the relationship between the individual and the group as the only part-whole relation in nature "that depends on recapitulation of the structure of the whole in the part." The same point is made by Morgan (1986) and Hutchins (1990: 211), using the more recent imagery of holograms: System capacities that are relevant for the functioning of the whole are built into its parts. In each of these renderings, social processes are the prior resources from which individual mind, self, and action are fashioned (Mead, 1934: 191–192). This means that collective mind precedes the individual mind and that heedful interrelating foreshadows heedful contributing.

Patterns of heedful interrelating in ongoing social processes may be internalized and recapitulated by individuals more or less adequately as they move in and out of the system. If heedful interrelating is visible, rewarded, modeled, discussed, and preserved in vivid stories, there is a good chance that newcomers will learn this style of responding, will incorporate it into their definition of who they are in the system, and will reaffirm and perhaps even augment this style as they act. To illustrate, Walsh and Ungson (1991: 60) defined organization as a "network of intersubjectively shared meanings that are sustained through the development and use of a common language and everyday social interactions." Among the shared meanings and language on carriers we heard these four assertions: (1) If it's not written down you can do it; (2) Look for clouds in every silver lining; (3) Most positions on this deck were brought in blood; and (4) Never get into something you can't get out of. Each of these guidelines, if practiced openly, represents an image of heedful interrelating that can be internalized and acted back into the system. If such guidelines are neglected, ignored, or mocked, however, interrelating still goes on, but it is done with indifference and carelessness.

Whether heedful images survive or die depends importantly on interactions among those who differ in their experience with the system. While these interactions have been the focus of preliminary discussions of communities of practice (e.g., Lave and Wenger, 1991: 98–100) involving apprentices and experts, we highlight a neglected portion of the process, namely, the effects of socialization on the insiders doing the socializing (Sutton and Louis, 1987).

When experienced insiders answer the questions of inexperienced newcomers, the insiders themselves are often resocialized. This is significant because it may remind insiders how to act heedfully and how to talk about heedful action. Newcomers are often a pretext for insiders to reconstruct what they knew but forgot. Heedful know-how becomes more salient and more differentiated when insiders see what they say to newcomers and discover that they thought more thoughts than they thought they did.

Whether collective mind gets renewed during resocialization may be determined largely by the candor and narrative skills of insiders and the attentiveness of newcomers. Candid insiders who use memorable stories to describe failures as well as successes, their doubts as well as their certainties, and what works as well as what fails, help newcomers infer dispositions of heed and carelessness. Insiders who narrate richly also often remind themselves of forgotten details when they reconstruct a previous event. And these reminders increase the substance of mind because they increase the number of examples of heed in work.

Narrative skills (Bruner, 1986; Weick and Browning, 1986; Orr, 1990) are important for collective mind because stories organize know-how, tacit knowledge, nuance, sequence, multiple causation, means-end relations, and consequences into a memorable plot. The ease with which a single story integrates diverse themes of heed in action foreshadows the capability of individuals to do the same. A coherent story of heed is mind writ small. And a repertoire of war stories, which grows larger through the memorable exercise of heed in novel settings, is mind writ large.

The quality of collective mind is heavily dependent on the way insiders interact with newcomers (e.g., Van Maanen, 1976). If insiders are taciturn, indifferent, preoccupied, available only in stylized performances, less than candid, or simply not available at all, newcomers are in danger of acting without heed because they have only banal conversations to internalize. They have learned little about heedful interdependence. When these newcomers act and try to anticipate the contributions of others, their actions will be stupid, and mistakes will happen. These mistakes may represent small failures that produce learning (Sitkin, 1992). More ominous is the possibility that these mistakes may also represent a weakening of system capacity for heedful responding. When there is a loss of particulars about how heed can be expressed in representation and subordination, reliable performance suffers. As seasoned people become more peripheral to socialization, there should be a higher incidence of serious accidents.

We have dwelt on insider participation simply because this participation is a conspicuous phenomenon that allows us to describe collective mind, but anything that changes the ongoing interaction (e.g., preoccupation with personalities rather than with the task) can also change the capability of that interaction to preserve and convey dispositions of heed. Those changes in turn should affect the quality of mind, the likelihood of comprehension, and the incidence of error.

Illustrations of Heed in Interrelating

The concepts of heed, interrelating, contributing, representing, subordinating, intelligent action, comprehension, recapitulation, and resocialization come together in the concept of collective mind as heedful interrelating. Applying the language of collective mind to four examples of complex systems, we illustrate the adequate comprehension produced by heedful interrelating and the problematic comprehension produced by heedless interrelating.

Heedful Interrelating

The first example of interrelating that is heedful involves a laboratory analogue of collective mind (Weick and Gilfillan, 1971). Three people who can neither see nor talk with one another are given target numbers between 1 and 30.

Whenever a target number is announced, each person is to contribute some number between 0 and 10 such that, when all three contributions are added together, they sum to the target number.

There are many ways to solve this problem (e.g., a target number of 13 can be achieved with a 3s strategy, 4-4-5, or a 10s strategy, 10-3-0). Once a group evolves a strategy, people are removed one at a time, and strangers, who know nothing of the strategy in use, enter. The questions are, how do old-timers interrelate with newcomers, what strategy emerges, how soon does it emerge, and how stable is it?

Austere as these operations are, they have the rudiments of a collective mind. A newcomer knows a number of things: (1) There are others in the activity system but they must be envisioned, since it is impossible to communicate with them (representation); (2) the two other people have had some experience with the system and with the game (there are imagined requirements to which one must subordinate); (3) each contribution is important and must interrelate with the others (contributions must converge, supplement, assist, and be defined in relation to one another); (4) to learn the existing system or to create a new one requires attention, careful calculations, and clear signals of intent (heedful contribution, representation, and subordination); and, (5) casual, indifferent interrelating will not be punished severely, because people are anonymous, and the rewards for participation are trivial (heedless responding is an option).

Just as the newcomers know these things, so do the old-timers. When these three people try to work out and maintain a system that hits each target on the first try, they are attempting to interrelate. They contribute, represent, and subordinate with varying amounts of heed. Their interrelating is better able to distinguish a mistake from an intentional effort to change strategy the more heedfully it was assembled. Likewise, heedful interrelating can "read" a newcomer's intentions quickly, whereas heedless interrelating cannot. These discriminations are not accomplished by single individuals but are accomplished by interrelated activities and the heedfulness with which those activities are defined in relation to one another. Heedful action at any one of these three positions can be undermined if it is not reciprocated at the other two. What is undermined, however, is a pattern of interrelations, not a person. A pattern of nonreciprocated heedfulness represents a loss of intelligence that is reflected in missed targets and slow change.

Heedful interrelating on carriers looks a lot like the pattern of interrelating seen in the common-target game. A vivid example of this similarity is Gillcrist's (1990: 287–288) account of what it feels like to land and taxi on a carrier deck at night. Having successfully trapped onto the deck, Gillcrist watched the flight director's two amber wands:

> I raised the hook handle with my right hand and simultaneously added a lot of power to get the Crusader moving forward. There was an urgency in the taxi signal movement of the wands, telling me that there must be another plane close behind me in the groove. They wanted to get my airplane completely across the foul line as quickly as possible. Taxiing at night was more carefully done than in the light of day, however. We'd had enough airplanes taxi over the side at night to learn that lesson. . . . The wands pointed to another set of wands further up the flight deck and I began to follow their direction as my F-8 was taxied all the way to the first spot on the bow. "God, how I hate this," I muttered to myself. "Do they really have to do this or are they just trying to scare me?" In spotting me in the first taxi spot on the bow, the taxi director was turning the F-8 so close to the edge of the flight deck that the cockpit actually swung in an arc over the deck's edge. All I could see was black rushing water eighty feet below. "Jesus" I said to myself, "I hope that guy knows what he is doing."

The taxi director does know what he is doing, as does the pilot, but that alone does not keep the plane from dropping off the deck. The interrelating of their know-how keeps the plane on the deck. A command from the director that is not executed by the pilot or a pilot deviation that is not corrected by the director are equally dangerous and not controllable by either party alone. The activities of taxiing and directing remain failure-free to the extent that they are interrelated heedfully.

A third example of heedful interrelating is of special interest because so much of it appears to involve the mind of one individual, in this case, the person responsible for deck operations (the bos'n). One of the people in this position who was interviewed had 23 years of experience on 16 carriers. At the time he joined this carrier's crew, it took six hours to spot 45 aircraft on the deck. He reduced that time to two hours and 45 minutes, which gave his crew more time to relax and maintain their alertness.

This person tries constantly to prevent the four worst things that can happen on a deck: It catches fire, becomes fouled, locked (nothing can move), or a plane is immobilized in the landing area. The more times a plane is moved to prevent any of these conditions, the higher the probability that it will brush against another plane ("crunch"), be damaged, and be out of service until repaired.

This bos'n, who is responsible for the smooth functioning of deck operations, gets up an hour early each day just to think about the kind of environment he will create on the deck that day, given the schedule of operations. This thinking is individual mind at work, but it also illustrates how collective mind is represented in the head of one person. The bos'n is dealing with collective mind when he represents the capabilities and weaknesses of imagined crewmembers' responses in his thinking, when he tailors sequences of activities so that improvisation and flexible response are activated as an expected part of the day's adaptive response, and when he counts on the interrelations among crewmembers themselves to "mind" the day's activities. The bos'n does not plan specific step-by-step operations but, rather, plans which crews will do the planning and deciding, when, and with what resources at hand. The system will decide the operations, and the bos'n sets up the system that will do this. The bos'n does this by attempting to recognize the strengths and weaknesses of the various crews working for him. The pieces of the system he sets up may interrelate stupidly or intelligently, in large part because they will either duplicate or undermine the heedful contributing, representing, and subordinating he anticipates.

Heedless Interrelating

When interrelating breaks down, individuals represent others in the system in less detail, contributions are shaped less by anticipated responses, and the boundaries of the envisaged system are drawn more narrowly, with the result that subordination becomes meaningless. Attention is focused on the local situation rather than the joint situation. People still may act heedfully, but not with respect to others. Interrelating becomes careless. Key people and activities are overlooked. As interrelating deteriorates and becomes more primitive, there is less comprehension of the implications of unfolding events, slower correction of errors, and more opportunities for small errors to combine and amplify. When these events are set in motion and sustained through heedless interrelating, there is a greater chance that small lapses can enlarge into disasters.

An incident that happened during a nighttime launch and recovery, which was described to us in interviews and correspondence, illustrates the steady loss of collective mind as interrelating became less heedful. This incident began to unfold during a night launch in which one-third of the planes in the mission were still on deck waiting to be launched, even though other planes were already beginning to be recovered.

Aircraft A, which was in the air and the fourth plane in line to land, had an apparent hydraulic failure, although the pilot was able to get his gear and tail hook down. This failure meant that if the plane were landed, its wings could not be folded, and it would take up twice the space normally allotted to it. This complicates the landing of all planes behind it.

While the pilot of plane A was trying to get help for his problem on a radio channel, plane B, an F-14, which was number three in order of landing, had a compound hydraulic failure, and none of his back-up hydraulic systems appeared to work, something that was unheard of. Plane C, which was fifth in line to land, then developed a control problem. Thus, the airboss was faced, first, with several A-7 aircraft that still had to be launched. This is not a trivial complication, because the only catapult available for these aircraft was the one whose blast-deflector panel extends part way into the area where planes land. Second, the airboss had a string of planes about to land that included (1) a normally operating A-7, (2) a normally operating A-7, (3) plane B with a compound hydraulic failure, (4) plane A with a hydraulic failure but gear and tail hook down, and (5) plane C with an apparent control problem.

The first plane was taken out of the landing pattern and the second was landed. Plane B, the one with the most severe problems, was told to land and then had to be waved off because the person operating the deflector panel for launches lowered the panel one second too late to allow B to land. The deflector operator had not been informed that an emergency existed. Plane B and its increasingly frightened pilot were reinserted into the landing pattern behind plane C for a second pass at the deck. Plane B then experienced an additional hydraulic failure. Plane A landed without incident, as did plane C. Plane C had corrected its control problem, but no one was informed. Thus, plane B's second pass was delayed longer than necessary because he had to wait for C to land in the mistaken belief that C still had a problem. The pilot of plane B became increasingly agitated and less and less able to help diagnose what might be wrong with his aircraft. The decision was made to send plane B to a land base, but it ran out of fuel on the way and the pilot and his RIO (radar intercept officer) had to eject. Both were rescued, but a $38-million aircraft was lost. If aircraft B had not been waved off the first time it tried to land, it would have been safely recovered. If we analyze this incident as a loss of collective mind produced by heedless interrelating, we look for two things: events that became incomprehensible, signifying a loss of mind, and increasingly heedless interrelating.

There were several events that became harder to comprehend. The failure of the hydraulic system in aircraft B was puzzling. The triggering of additional hydraulic failures was even more so. To have three of five aircraft on final approach declare emergencies was itself something that was hard to comprehend, as was the problem of how to recover three disabled planes while launching three more immediately.

Incomprehensible events made interrelating more difficult, which then made the events even harder to comprehend. The loss of heed in interrelating was spread among contributions, representations, and subordinations. The squadron representative who tried to deal with the stressed pilot in plane B was not himself a pilot (he was an RIO), and he did not scan systematically for possible sources of the problem. Instead, he simply told the pilot assorted things to try, not realizing that, in the pilot's doing so, additional systems on the plane began to fail. He didn't realize these growing complications because the pilot was both imprecise in his reports of trouble and slow to describe what happened when he tested some hypothesis proposed by the representative. And the representative did nothing to change the pilot's style of contributing.

But heedless interrelating was not confined to exchanges between pilot and representative. The RIO in plane B made no effort to calm the pilot or help him diagnose. The deflector operator was not treated as a person in the *recovery* system. Three different problems were discussed on two radio frequencies, which made it difficult to sort out which plane had which problem. No one seemed to register that the squadron representative was himself getting farther behind and making increasingly heedless contributions. The airboss in command of the tower was an F-14 pilot, but he was preoccupied with the five incoming and the three outgoing aircraft and could not be pulled completely into the activity system that was dealing with the F-14 problem. As heed began to be withdrawn from the system, activities and people became isolated, the system began to pull apart, the problems became more incomprehensible, and it became harder for individuals to interrelate with a system of activities that was rapidly losing its form. The pattern of interrelated activities lost intelligence and mind as contributions became more thoughtless and less interdependent.

Had the pattern of interrelations been more heedful, it might have detected what was subsequently said to be the most likely cause of the failures in plane B. Although the aircraft was never recovered, the navy's investigation of the incident concluded that too many demands were placed on the emergency back-up systems, and the plane became less and less flyable. Sustained heedful interrelating might well have registered that the growing number of attempted solutions had in fact created a new problem that was worse than any problem that was present to begin with.[3]

It is important to realize that our analysis, using the concepts of collective mind and heedful interrelating, implies something more than the simple advice, "be careful." People can't be careful unless they take account of others and unless others do the same. Being careful is a social rather than a solitary act. To act with care, people have to envision their contributions in the context of requirements for joint action. Furthermore, to act with care does not mean that one plans how to do this and then applies the plan to the action. Care is not cultivated apart from action. It is expressed in action and through action. Thus people can't *be* careful, they *are* careful (or careless). The care is in the action.

The preceding analysis also suggests that it is crucial to pay attention to mind, because accidents are not just issues of ignorance and cognition, they are issues of inattention and conduct as well. The examples of incomprehension mentioned above are not simply issues of fact and thinking. Facts by themselves are of no help if they cannot be communicated or heard or applied or interpreted or incorporated into activities or placed in contexts, in short, if they are not addressed mindfully. One "fact" of this incident is that plane B could have landed had it not been waved off because of the extended deflector. Furthermore, individuals within the system were not ignorant of crucial details (e.g., the pilot of plane C knew he no longer had a problem).

One interpretation of this incident is that individuals were smarter than the system, but the problem was more complex than any one individual could understand. Heedful interrelating of activities constructs a substrate that is more complex and, therefore, better able to comprehend complex events than is true for smart but isolated individuals. The F-14 may have been lost because heedful interrelating was lost. Heightened attentiveness to social process might have prevented both losses.

Discussion

We conclude from our analysis that carrier operations are a struggle for alertness and that the concept of heedful interrelating helps capture this struggle. We began with the question, How can we analyze a complex social activity system in which fluctuations in comprehension seem to be consequential? We focused on heed (understood as dispositions to act with attentiveness, alertness, and care), conduct (understood as behavior that takes into account the expectations of others), and mind (understood as integration of feeling, thinking, and willing).

We were able to talk about group mind without reification, because we grounded our ideas in individual actions and then treated those actions as the means by which a distinct higher-order pattern of interrelated activities emerged. This pattern shaped the actions that produced it, persisted despite changes in personnel, and changed despite unchanging personnel. Thus, we did not reify social entities, because we argued that they emerge from individual actions that construct interrelations. But neither did we reify individual entities, because we argued that they emerge through selective importation, interpretation, and re-enactment of the social order that they constitute.

In broadening our focus, we conceptualized mind as action that constructs mental processes rather than as mental processes that construct action. We proposed that variations in contributing, representing, and subordinating produce collective mind. Common hallmarks of mind such as alertness, attentiveness, understanding, and relating to the world were treated as coincident with and immanent in the connecting activities. To connect *is* to mind.

For the collective mind, the connections that matter are those that link distributed activities, and the ways those connections are accomplished embody much of what we have come to mean by the word "mind." The ways people connect their activities make conduct mindful. Mindless actions ignore interrelating or accomplish it haphazardly and with indifference (Bellah et al., 1991).

As a result of our analysis, we now see the importance of disentangling the development of mind from the development of a group. In Asch's description of the essence of group life, as well as in other discussions of group cognition, the development of mind is confounded with the development of the group. As a group matures and moves from inclusion through control to affection (Schutz, 1958), or as it moves from forming through storming, norming, and performing (Tuckman, 1965), both interrelating and intimacy develop jointly. If a mature group has few accidents or an immature group has many, it is difficult to see what role, if any, mind may play in this. An immature group of relative strangers with few shared norms, minimal disclosure, and formal relationships might well find it hard to cope with nonroutine events. But this has nothing to do with mind.

In our analysis we have assumed that there is something like a two-by-two matrix in which a group can be developed or undeveloped and a collective mind can be developed or undeveloped. And we assume that the combinations of developed-group–undeveloped mind and undeveloped-group–developed mind are possible. These two combinations are crucial to any proposal that collective mind is a distinct process in social life.

The combination of developed-group–undeveloped mind is found in the phenomenon of groupthink (Janis, 1982), as well as in cults (Galanter, 1989), interactions at NASA prior to the *Challenger* disaster (Starbuck and Milliken, 1988), and ethnocentric research groups (Weick, 1983). Common among these examples is subordination to a system that is envisaged carelessly, or, as Janis (1982: 174) put it, there is an overestimation of the group's power, morality, and invulnerability. Furthermore, contributions are made thoughtlessly; as Janis (1982: 175) put it, there is self-censorship of deviations, doubts, and counterarguments. And, finally, representations are careless; members maintain the false assumption that silence means consent (Janis, 1982: 175). In the presence of heedless interrelating, comprehension declines, regardless of how long the group has been together, and disasters result.

The combination of undeveloped-group–developed mind is found in ad hoc project teams, such as those that produce television specials (e.g., Peters, 1992: 189–200) or motion pictures (Faulkner and Anderson, 1987), and in temporary systems such as those that form in aircraft cockpits (Ginnett, 1990), around jazz improvisation (Eisenberg, 1990), in response to crises (Rochlin, 1989), or in high-velocity environments (Eisenhardt, 1993). The common feature shared among these diverse settings is best captured by Eisenberg (1990: 160), who characterized them as built from nondisclosive intimacy that "stresses coordination of action over alignment of cognitions, mutual respect over agreement, trust over empathy, diversity over homogeneity, loose over tight coupling, and strategic communication over unrestricted candor."

Translated into the language of heedful interrelating, what Eisenberg depicted were relationships in which shared values, openness, and disclosure, all hallmarks of a developed group, were *not* fully developed, but in which collective mind was developed. Nondisclosive intimacy is characterized by heedful contributing (e.g., loose coupling, diversity, strategic communication), heedful representing (e.g., mutual respect, coordination of action), and heedful subordinating (e.g., trust).

If heedful interrelating can occur in an undeveloped group, this changes the way we think about the well-known stages of group development. If people are observed to contribute, represent, and subordinate with heed, these actions can be interpreted as operations that construct a well-developed collective mind; however, those same actions can also be seen as the orienting, clarifying, and testing associated with the early stages of a new group just beginning to form (McGrath, 1984: 152–162). By one set of criteria, that associated with group formation, people engaged in forming are immature. By another set of criteria, that associated with collective mind, these acts of forming represent well-developed mental processes.

These opposed criteria suggest that groups may be smartest in their early stages. As they grow older, they lose mind when interrelating becomes more routine, more casual, more automatic. This line of reasoning is consistent with Gersick's (1988) demonstration that groups tend to reform halfway through their history. In our language, this midcourse reshuffling can be understood as redoing the pattern of interrelations that constitute mind, thereby renewing mind itself. If groups steadily lose mind and comprehension as they age, their capability for comprehension may show a dramatic increase halfway through their history. If that is plausible, the sudden surge in comprehension should be accompanied by a sudden decrease in the number of accidents they produce.

The Conceptualization of Topics in Organizational Theory

Our analysis of collective mind and heedful interrelating throws new light on several topics in organizational theory, including organizational types, the measurement of performance, and normal accidents.

The concept of mind may be an important tool in comparative analysis. LaPorte and Consolini (1991) argued that high-reliability organizations such as aircraft carriers differ in many ways from organizations usually portrayed in organizational theory as (for convenience) high-efficiency organizations. Typical efficiency organizations practice incremental decision making, their errors do not have a lethal edge, they use simple low-hazard technologies, they are governed by single rather than multilayered authority systems, they are more often in the private than the public sector, they are not preoccupied with perfection, their operations are carried on at one level of intensity, they experience few nasty surprises, and they can rely on computation or judgment as decision strategies (Thompson and Tuden, 1959) but seldom need to employ both at the same time. LaPorte and Consolini (1991: 19) concluded that existing organizational theory is inadequate to understand systems in which the "consequences and costs associated with major failures in some technical operations are greater than the value of the lessons learned from them."

Our analysis suggests that most of these differences can be subsumed under the generalization that high-efficiency organizations have simpler minds than do high-reliability organizations. If dispositions toward individual and collective heed were increased in most organizations in conjunction with increases in task-related interdependence and flexibility in the sequencing of tasks, then we would expect these organizations to act more like high-reliability systems. Changes of precisely this kind seem to be inherent in recent interventions to improve total quality management (e.g., U.S. General Accounting Office, 1991).

Our point is simply that confounded in many comparisons among organizations that differ on conspicuous grounds, such as structure and technology, are less conspicuous but potentially more powerful differences in the capability for collective mind. A smart system does the right thing regardless of its structure and regardless of whether the environment is stable or turbulent. We suspect that organic systems, because of their capacity to reconfigure themselves temporarily into more mechanistic structures, have more fully developed minds than do mechanistic systems.

We also suspect that newer organizational forms, such as networks (Powell, 1990), self-designing systems (Hedberg, Nystrom, and Starbuck, 1976), cognitive oligopolies (Porac, Thomas, and Baden-Fuller, 1989: 413), and interpretation systems (Daft and Weick, 1984) have more capacity for mind than do M forms, U forms, and matrix forms. But all of these conjectures, which flow from the idea of collective mind, require that we pay as much attention to social processes and microdynamics as we now pay to the statics of structure, strategy, and demographics.

The concept of mind also suggests a view of performance that complements concepts such as activities (Homans, 1950), the active task (Dornbusch and

Scott, 1975), task structure (Hackman, 1990: 10), group task design (Hackman, 1987), and production functions (McGrath, 1990). It adds to all of these a concern with the style or manner of performance. Not only can performance be high or low, productive or unproductive, or adequate or inadequate, it can also be heedful or heedless. Heedful performance might or might not be judged productive, depending on how productivity is defined.

Most important, the concept of mind allows us to talk about careful versus careless performance, not just performance that is productive or unproductive. This shift makes it easier to talk about performance in systems in which the next careless error may be the last trial. The language of care is more suited to systems concerned with reliability than is the language of efficiency.

Much of the interest in organizations that are vulnerable to catastrophic accidents can be traced to Perrow's (1981) initial analysis of Three Mile Island, followed by his expansion of this analysis into other industries (Perrow, 1984). In the expanded analysis, Perrow suggested that technologies that are both tightly coupled and interactively complex are the most dangerous, because small events can escalate rapidly into a catastrophe. Nuclear aircraft carriers such as those we have studied are especially prone to normal accidents (see Perrow, 1984: 97) because they comprise not one but several tightly coupled, interactively complex technologies. These include jet aircraft, nuclear weapons carried on aircraft, nuclear weapons stored on board the ship, and nuclear reactors used to power the ship. Furthermore, the marine navigation system and the air traffic control system on a ship are tightly coupled technologies, although they are slightly less complex than the nuclear technologies.

Despite their high potential for normal accidents, carriers are relatively safe. Our analysis suggests that one of the reasons carriers are safe is because of, not in spite of, tight coupling. Our analysis raises the possibility that technological tight coupling is dangerous in the presence of interactive complexity, unless it is mediated by a mutually shared field that is well developed. This mutually shared field, built from heedful interrelating, is itself tightly coupled, but this tight coupling is social rather than technical. We suspect that normal accidents represent a breakdown of social processes and comprehension rather than a failure of technology. Inadequate comprehension can be traced to flawed mind rather than flawed equipment.

The Conceptualization of Practice

The mindset for practice implicit in the preceding analysis has little room for heroic, autonomous individuals. A well-developed organization mind, capable of reliable performance is thoroughly social. It is built of ongoing interrelating and dense interrelations. Thus, interpersonal skills are not a luxury in high-reliability systems. They are a necessity. These skills enable people to represent and subordinate themselves to communities of practice. As people move toward individualism and fewer interconnections, organization mind is simplified and soon becomes indistinguishable from individual mind. With this change comes heightened vulnerability to accidents. Cockpit crews that function as individuals rather than teams show this rapid breakdown in ability to understand what is happening (Orlady and Foushee, 1987). Sustained success in coping with emergency conditions seems to occur when the activities of the crew are more fully interrelated and when members' contributions, representations, and subordination create a pattern of joint action. The chronic fear in high-reliability systems that events will prove to be incomprehensible (Perrow, 1984) may be a realistic fear only when social skills are underdeveloped. With more development of social skills goes more development of organization mind and heightened understanding of environments.

A different way to state the point that mind is dependent on social skills is to argue that it is easier for systems to lose mind than to gain it. A culture that encourages individualism, survival of the fittest, macho heroics, and can-do reactions will often neglect heedful practice of representation and subordination. Without representation and subordination, comprehension reverts to one brain at a time. No matter how visionary or smart or forward-looking or aggressive that one brain may be, it is no match for conditions of interactive complexity. Cooperation is imperative for the development of mind.

Reliable performance may require a well-developed collective mind in the form of a complex, attentive system tied together by trust. That prescription sounds simple enough. Nevertheless, conventional understanding seems to favor a different configuration: a simple, automatic system tied together by suspicion and redundancy. The latter scenario makes sense in a world in which individuals can comprehend what is going on. But when individual comprehension proves inadequate, one of the few remaining sources of comprehension is social entities. Variation in the development of these entities may spell the difference between prosperity and disaster.

Notes

1. Unless otherwise cited, aircraft carrier examples are drawn from field observation notes of air operations and interviews aboard Nimitz class carriers made by the second author and others over a five-year period. Researchers spent from four days to three weeks aboard the carriers at any one time. They usually made observations from different vantage points during the evolutions of various events. Observations were entered into computer systems and later compared across observers and across organizational members for clarity of meaning. Examples are also drawn from quarterly workshop discussions with senior officers from those carriers over the two years. The primary observational research methodology was to triangulate observations made by three faculty researchers, as suggested by Glaser and Strauss (1967) and Eisenhardt (1989). The methodology is more fully discussed in Roberts, Stout, and Halpern (1993). Paper-and-pencil data were also collected and are discussed elsewhere (Roberts, Rousseau, and LaPorte, 1993).

2. We could just as easily have used Blumer's (1969: 78–79) discussion of "the mutual alignment of action."

3. There is a limit to heedfulness, given the number and skill of participants, and on this night this ship was at that limit. The system was overloaded, and the situation was one that managers of high-technology weapons systems worry about all the time. They call it OBE (overcome by events). Given perhaps only minor differences in the situation, the outcomes might have been different. In this situation, for example, had the carrier air group commander

come to the tower (which he often does), he would have added yet another set of eyes and ears, with their attendant skills. Perhaps he could have monitored one aspect of the situation while the boss and mini boss took charge of others, and the situation would have been a more heedful one. Had the squadron representative in the tower been a pilot, he might have searched through his own repertoire of things that can go wrong and helped the F-14's pilot calm down and solve his problem, increasing the heedfulness of the situation.

References

Asch, Solomon E. 1952 *Social Psychology*. Englewood Cliffs, NJ: Prentice-Hall.

Bellah, Robert N., Richard Madsen, William M. Sullivan, Ann Swidler, and Steven M. Tipton 1991 *The Good Society*. New York: Knopf.

Blumer, Herbert 1969 *Symbolic Interaction*. Berkeley: University of California Press.

Boden, Margaret A. 1990 "Introduction." In Margaret A. Boden (ed.), *The Philosophy of Artificial Intelligence*: 1–21. New York: Oxford University Press.

Bruner, Jerome 1986 *Actual Minds. Possible Worlds*. Cambridge, MA: Harvard University Press.

Churchland, Paul M. 1992 "A deeper unity: Some Feyerabendian themes in neurocomputational form." In Steven Davis (ed.), *Connectionism: Theory and Practice*: 30–50. New York: Oxford University Press.

Daft, Richard, and Karl E. Weick 1984 "Toward a model of organizations as interpretation systems." *Academy of Management Review*, 9: 284–295.

Dornbusch, Sandford M., and W. Richard Scott 1975 *Evaluation and the Exercise of Authority*. San Francisco: Jossey-Bass.

Dreyfus, Hubert L., and Stuart E. Dreyfus 1990 "Making a mind versus modeling the brain: Artificial intelligence back at a branch point." In Margaret A. Boden (ed.), *The Philosophy of Artificial Intelligence*: 309–333. New York: Oxford University Press.

Eisenberg, Eric 1990 "Jamming: Transcendence through organizing." *Communication Research*, 17: 139–164.

Eisenhardt, Kathleen M. 1989 "Building theories from case study research." *Academy of Management Review*, 14: 532–550.

Eisenhardt, Kathleen M. 1993 "High reliability organizations meet high velocity environments: Common dilemmas in nuclear power plants, aircraft carriers, and microcomputer firms." In Karlene Roberts (ed.), *New Challenges to Understanding Organizations*: 117–135. New York: Macmillan.

Elman, Jeffrey L. 1992 "Grammatical structure and distributed representations." In Steven Davis (ed.), *Connectionism: Theory and Practice*: 138–178. New York: Oxford University Press.

Faulkner, Robert R., and A. B. Anderson 1987 "Short-term projects and emergent careers: Evidence from Hollywood." *American Journal of Sociology*, 92: 879–909.

Follett, Mary Parker 1924 *Creative Experience*. New York: Longmans, Green.

Galanter, Marc 1989 *Cults*. New York: Oxford University Press.

Gersick, Connie G. 1988 "Time and transition in work teams: Toward a new model of group development." *Academy of Management Journal*, 31: 9–41.

Gillcrist, P. T. 1990 *Feet Wet: Reflections of a Carrier*. Novato, CA: Presidio Press.

Ginnett, Robert C. 1990 "Airline cockpit crew." In J. Richard Hackman (ed.), *Groups That Work (and Those That Don't)*: 427–448. San Francisco: Jossey-Bass.

Glaser, Barney, and Anselm L. Strauss 1967 *The Discovery of Grounded Theory: Strategies for Qualitative Research*. Chicago: Aldine.

Hackman, J. Richard 1987 "The design of work teams." In Jay Lorsch (ed.), *Handbook of Organizational Behavior*: 315–342. Englewood Cliffs, NJ: Prentice-Hall.

Hackman, J. Richard (ed.) 1990 *Groups That Work (and Those That Don't)*. San Francisco: Jossey-Bass.

Hedberg, Bo L. T., Paul C. Nystrom, and William H. Starbuck 1976 "Camping on seesaws: Prescriptions for a self-designing organization." *Administrative Science Quarterly*, 21: 41–65.

Homans, George C. 1950 *The Human Group*. New York: Harcourt.
Hutchins, Edwin 1990 "The technology of team navigation." In Jolene Galegher, Robert E. Kraut, and Carmen Egido (eds.), *Intellectual Teamwork*: 191–220. Hillsdale, NJ: Erlbaum. 1991 "The social organization of distributed cognition." In Lauren B. Resnick, John M. Levine, and Stephanie D. Teasley (eds.), *Perspectives on Socially Shared Cognition*: 283–307. Washington, DC: American Psychological Association.
Janis, Irving 1982 *Groupthink*, 2d.ed. Boston: Houghton-Mifflin.
Langer, Eleanor J. 1989 "Minding matters: The consequences of mindlessness-mindfulness." In Leonard Berkowitz (ed.), *Advances in Experimental Social Psychology*, 22: 137–173. New York: Academic Press.
LaPorte, Todd R., and Paula M. Consolini 1991 "Working in practice but not in theory: Theoretical challenges of high-reliability organizations." *Journal of Public Administration Research and Theory*, 1: 19–47.
Lave, Jean, and Etienne Wenger 1991 *Situated Learning: Legitimate Peripheral Participation*. New York: Cambridge University Press.
Lyons, William 1980 *Gilbert Ryle: An Introduction to His Philosophy*. Atlantic Highlands, NJ: Humanities Press.
McGrath, Joseph E. 1984 *Groups: Interaction and Performance*. Englewood Cliffs, NJ: Prentice-Hall.
McGrath, Joseph E. 1990 "Time matters in groups." In Jolene Galegher, Robert E. Kraut, and Carmen Egido (eds.), *Intellectual Teamwork*: 23–61. Hillsdale, NJ: Erlbaum.
Mead, George Herbert 1934 *Mind, Self, and Society*. Chicago: University of Chicago Press.
Morgan, Gareth 1986 *Images of Organization*. Beverly Hills, CA: Sage.
Orlady, Harry W., and H. Clayton Foushee 1987 *Cockpit Resource Management Training*. Springfield, VA: National Technical Information Service (N87-22634).
Orr, Julian E. 1990 "Sharing knowledge, celebrating identity: Community memory in a service culture." In David Middleton and Derek Edwards (eds.), *Collective Remembering*: 169–189. Newbury Park, CA: Sage.
Perrow, Charles 1981 "The President's Commission and the normal accident." In D. Sills, C. Wolf, and V. Shelanski (eds.), *The Accident at Three Mile Island: The Human Dimensions*: 173–184. Boulder, CO: Westview Press.
Perrow, Charles 1984 *Normal Accidents*. New York: Basic Books.
Peters, Tom 1992 *Liberation Management*. New York: Knopf.
Porac, Joseph F., Howard Thomas, and Charles Baden-Fuller 1989 "Competitive groups as cognitive communities: The case of Scottish knitwear manufacturers." *Journal of Management Studies*, 26: 397–416.
Powell, Walter W. 1990 "Neither market nor hierarchy: Network forms of organization." In Barry M. Staw and Larry L. Cummings (eds.), *Research in Organizational Behavior*, 12: 295–336. Greenwich, CT: JAI Press.
Quinlan, Phillip 1991 *Connectionism and Psychology*. Chicago: University of Chicago Press.
Roberts, Karlene H., Denise M. Rousseau, and Todd R. LaPorte 1993 "The culture of high reliability: Quantitative and qualitative assessment aboard nuclear powered aircraft carriers." *Journal of High Technology Management Research* (in press).
Roberts, Karlene H., Susan Stout, and Jennifer J. Halpern 1993 "Decision dynamics in two high reliability military organizations." *Management Science* (in press).
Rochlin, Gene I. 1989 "Organizational self-design is a crisis-avoidance strategy: U.S. naval flight operations as a case study." *Industrial Crisis Quarterly*, 3: 159–176.
Rochlin, Gene I., Todd R. LaPorte and Karlene H. Roberts 1987 "The self-designing high-reliability organization: Aircraft carrier flight operations at sea." *Naval War College Review*, 40(4): 76–90
Rumelhart, David E. 1992 "Towards a microstructural account of human reasoning." In Steven Davis (ed.), *Connectionism: Theory and Practice*: 69–83. New York: Oxford University Press.
Ryle, Gilbert 1949 *The Concept of Mind*. Chicago: University of Chicago Press.
Sandelands, Lloyd E., and Ralph E. Stablein 1987 "The concept of organization mind." In Samuel Bacharach and Nancy DiTomaso (eds.), *Research in the Sociology of Organizations*, 5: 135–161. Greenwich, CT: JAI Press.

Schneider, David J. 1991 "Social cognition." In Lyman W. Porter and Mark R. Rosenzweig (eds.), *Annual Review of Psychology*, 42: 527–561. Palo Alto, CA: Annual Reviews.
Schneider, W., and R. M. Shiffrin 1977 "Controlled and automatic human information processing: I. Detection, search and attention." *Psychological Review*, 84: 1–66.
Schutz, William C. 1958 *FIRO: A Three-Dimensional Theory of Interpersonal Behavior*. New York: Holt, Rinehart, and Winston.
Sitkin, Sim 1992 "Learning through failure: The strategy of small losses." In Barry Staw and Larry Cummings (eds.), *Research in Organizational Behavior*, 14: 231–266. Greenwich, CT: JAI Press.
Starbuck, William H., and Francis J. Milliken 1988 "Challenger: Fine-tuning the odds until something breaks." *Journal of Management Studies*, 25: 319–340.
Sutton, Robert I., and Meryl R. Louis 1987 "How selecting and socializing newcomers influences insiders." *Human Resource Management*, 26: 347–361.
Thompson, James D., and Arthur Tuden 1959 "Strategies, structures, and processes of organizational decision." In James D. Thompson (ed.), *Comparative Studies in Organization*: 195–216. Pittsburgh: University of Pittsburgh Press.
Tuckman, Bruce W. 1965 "Developmental sequence in small groups." *Psychological Bulletin*, 63: 384–399.
U.S. General Accounting Office 1991 *Management Practices: U.S. Companies Improve Performance through Quality Efforts*. Document GAO/NSIAD-91-190. Washington, DC: U.S. Government Printing Office.
Van Maanen, John 1976 "Breaking in: Socialization to work." In Robert Dubin (ed.), *Handbook of Work, Organization and Society*: 67–130. Chicago: Rand McNally.
Walsh, James P., and Gerardo R. Ungson 1991 "Organizational memory." *Academy of Management Review*, 16: 57–91.
Wegner, Daniel M. 1987 "Transactive memory: A contemporary analysis of the group mind." In Brian Mullen and George R. Goethals (eds.), *Theories of Group Behavior*: 185–208. New York: Springer-Verlag.
Wegner, Daniel M., Ralph Erber, and Paula Raymond 1991 "Transactive memory in close relationships." *Journal of Personality and Social Psychology*, 61:923–929.
Wegner, Daniel M., Toni Giuliano, and Paula T. Hertel 1985 "Cognitive interdependence in close relationships." In William J. Ickes (ed.), *Compatible and Incompatible Relationships*: 253–276. New York: Springer-Verlag.
Weick, Karl E. 1983 "Contradictions in a community of scholars: The cohesion-accuracy tradeoff." *Review of Higher Education*, 6(4): 253–267.
Weick, Karl E., and Larry Browning 1986 "Arguments and narration in organizational communication." *Journal of Management*, 12: 243–259.
Weick, Karl E., and David P. Gilfillan 1971 "Fate of arbitrary traditions in a laboratory microculture." *Journal of Personality and Social Psychology*, 17: 179–191.
Wilson, G. C. 1986 *Supercarrier*. New York: Macmillan.

52

Why Differences Make a Difference: A Field Study of Diversity, Conflict, and Performance in Workgroups

Karen A. Jehn, Gregory B. Northcraft and Margaret A. Neale

Source: *Administrative Science Quarterly* 44 (4) (1999): 741–763.

In response to changing economic conditions, organizations recently have embraced new structural forms designed to reduce costs while simultaneously maximizing flexibility and responsiveness to customer demands (e.g., Boyett and Conn, 1991; Byrne, 1993; Donnellon, 1996). The resulting flatter, more decentralized organizational forms tend to be built around groups and depend on rich synchronous communication provided by teams and task forces to a much greater extent than more traditional hierarchical and centralized organizations (Nohria, 1991). In addition, groups have become important vehicles for identifying high-quality solutions to emerging organizational problems (Dumaine, 1991).

While groups have become central to organizations, they present their own intrinsic problems of coordination, motivation, and conflict management (Gladstein, 1984; Jehn, 1995). In large part, the use of groups as fundamental building blocks of organizational structure and strategy seems to be premised on the assumption that groups can gather together the diversity of information, backgrounds, and values necessary to make things happen (Jackson, 1992), to produce effective organizational action. If groups are to provide forums for sharing information across functional and cultural boundaries (Lipnack and Stamps, 1993), however, the diverse views and backgrounds members bring with them to the group must be successfully managed. Moreover, the work-force is becoming increasingly diverse on a number of dimensions (e.g., age, gender, ethnicity). Although differences among members of workgroups are the norm, Byrne's (1971) similarity-attraction theory suggests that people prefer similarity in their interactions. Likewise, theories of selection (Chatman, 1991) and socialization (Van Maanen and Schein, 1979) promote similarity in values and demographics as the basis for maintaining effective work environments. Recently, however, diversity theorists (Jackson, 1992; Williams and O'Reilly, 1998), group researchers (Lipnack and Stamps, 1993; Gruenfeld, 1995; Gruenfeld et al., 1996), and creativity theorists (Amabile, 1994; Oldham and Cummings, 1998) have been singing the praises of diversity in workgroups. But empirical research on the effects of diversity has produced mixed results.

In some studies, diverse groups have been shown to outperform homogenous groups (Hoffman and Maier, 1961; Hoffman, 1978; Nemeth, 1986; Jackson,

1992). In contrast, other studies have demonstrated that homogenous groups avoid the process loss associated with poor communication patterns and excessive conflict that often plague diverse groups (Steiner, 1972; O'Reilly and Flatt, 1989; Ancona and Caldwell, 1992). These inconsistent results should not be all that surprising. No theory suggests that a workgroup's diversity on outward personal characteristics such as race and gender should have benefits except to the extent that diversity creates other diversity in the workgroup, such as diversity of information or perspective. For instance, social category diversity may not always reflect other types of diversity (Tsui and O'Reilly, 1989) — age does not necessarily reflect values or even work experience. Even when workgroups do possess that "other" diversity (e.g., information or perspective), performance benefits should be expected only to the extent that workgroup members successfully manage the difficulties of interacting effectively with dissimilar others (e.g., Tsui and O'Reilly, 1989).

In light of these concerns, it is also not surprising that Williams and O'Reilly's (1998) review of forty years of diversity research concluded that there are no consistent main effects of diversity on organizational performance. They proposed that a more complex framework and a more complex conceptualization of the nature of diversity are needed to study the impact of diversity. Specifically, they called for the incorporation of contextual aspects (e.g., task and organizational characteristics), types of diversity (informational and demographic), and intervening variables (e.g., communication and conflict). Our study addresses these concerns by examining the effects of three specific types of diversity (informational diversity, social category diversity, and value diversity), a key intervening process (conflict), and two contextual moderators of these effects (task interdependence and task type) on workgroup outcomes. We thus provide a more detailed model of the process by which various types of workgroup diversity affect performance than past theorizing. For example, differences in gender may not affect member satisfaction if all members express similar values, and information diversity may have little effect on performance when tasks are highly routine.

Our research builds on prior research investigating various aspects of contextual and intervening variables to articulate a more comprehensive understanding of the relationship between diversity and performance. Pelled (1996a, 1996b), for example, suggested that a workgroup's social category diversity (group differences in social category membership) enhances its performance and that task conflict — disagreement about task issues — mediates the effects of social category diversity. In contrast to task conflict, however, relationship conflicts, which are often caused by social category diversity, can negatively influence group outcomes (Jehn, 1995). Thus, while personality conflicts may interfere with task performance, conflict about the best way to perform the task may lead to insights that increase task performance. We investigate three types of conflict to determine how different types of diversity influence workgroup outcomes.

Effects of Diversity in Workgroups

Diversity and Conflict

Researchers have devoted considerable attention to how workgroups can generate knowledge and insights beyond the reach of their individual members (e.g., Murray, 1983; Doise and Mugny, 1984; Perret-Clermont, Ferret, and Bell, 1991; Garton, 1992). This research on emergent knowledge in groups suggests that social interaction among diverse perspectives can lead to the emergence of new insights through conceptual restructuring within the groups (e.g., Levine and Resnick, 1993). The creation of knowledge and the discovery of insight by groups appears to depend on the presence of diverse viewpoints and perspectives about the task (Damon, 1991; Levine and Resnick, 1993; Nonaka and Takeuchi, 1995). We explore three categories of diversity discussed in past research on groups: informational diversity, social category diversity, and value diversity. These three types of diversity are not always distinct in practice. For example, two individuals from different races (social category diversity) may (though not necessarily) have experienced different educational cultures (informational diversity) and may consequently espouse different values (value diversity). Each of these different kinds of diversity implies different challenges and opportunities for workgroups, and consequently, each should differentially influence workgroup outcomes.

Informational Diversity

Informational diversity refers to differences in knowledge bases and perspectives that members bring to the group. Such differences are likely to arise as a function of differences among group members in education, experience, and expertise. These differences in educational background, training, and work experience increase the likelihood that diverse perspectives and opinions exist in a workgroup (Stasser, 1992). Recent research has demonstrated that differences in educational background lead to an increase in task-related debates in work teams (Jehn, Chadwick, and Thatcher, 1997). Task-related debates can be about either the content or the process of the task. Task content is about what to do (e.g., a new marketing campaign), in contrast to task process, which is about how to do it (e.g., delegation of responsibilities). Following Jehn (1995, 1997), we refer to disagreements about task content as task conflict and disagreements about task process as process conflict. We expect that informational diversity will increase the potential for task conflict:

> Hypothesis 1a (H1a): Informational diversity will increase task conflict in workgroups.

Workgroups often fail to realize the potential benefits of informational diversity and task conflict for two reasons. First, when groups form naturally in organizations, the most common bases for group formation are similarity (e.g.,

Newcomb, 1960; Ancona and Caldwell, 1992), proximity (e.g., Festinger, Schachter, and Back, 1950), and familiarity (e.g., Tenbrunsel et al., 1994; Mannix, Goins, and Carroll, 1996). These natural group formation processes typically overselect members from the same social networks. Because the knowledge, experiences, and perspectives of group members from the same social networks may be more redundant than diversified (Granovetter, 1973), naturally formed groups are likely to lack diversity, undermining their potential for learning, insight, and problem-solving effectiveness (Jackson, 1992).

Organizations often counter the tendency of groups to form based on shared social networks (i.e., similarity, proximity, familiarity) by creating cross-functional teams, or teams with members of different functional training, to enhance the informational diversity available in the group (Northcraft et al., 1995). Even when group membership is specifically managed to enhance informational diversity, however, the potential of this diversity often is not realized (Steiner, 1972; Hackman, 1990). Dougherty (1992), for example, found that cross-functional new product teams had difficulty getting their products to market, and Ancona and Caldwell (1992) found managers' ratings of innovativeness to be lower when teams were functionally diverse than when they were homogeneous. Similarly, O'Reilly and Flatt (1989) found that top management teams with homogeneous patterns of organizational tenure were more creative than teams whose tenure patterns were more diverse.

Groups with diverse members often prove ineffective at capitalizing on the potential benefits of their informational diversity (Stasser and Titus, 1985, 1987). Managers have expressed frustration with the time and resource demands of functionally diverse teams, while team members have bemoaned the difficulty of motivating their members to work together effectively (Dumaine, 1994). Even in groups demonstrating performance benefits from membership diversity, group members report finding the experience frustrating and dissatisfying (e.g,. Baron, 1990; Amason and Schweiger, 1994).

The second reason groups often fail to realize the benefit of informational diversity is that what makes a group informationally diverse may also prevent the group from realizing the benefits of its informational diversity. Disagreements in workgroups could be disagreements about task content (task conflict), but they could also be disagreements about how to do the task or how to delegate resources, reflecting process conflict (Jehn, 1997). For example, a group member with an engineering background will probably want to proceed differently (in terms of how to identify potential courses of action and choose among them) than a group member with a marketing or accounting background. Therefore, process conflict — disagreements about delegation of duties and resources — are often distinct from task content conflicts — potentially productive disagreements about the task or problem at hand, such as the interpretation of market analysis. Recent research has demonstrated that groups with members of diverse educational majors experience more difficulty defining how to proceed than groups in which members have similar educational backgrounds (Jehn, Chadwick, and Thatcher, 1997). This gives rise to a second hypothesis:

Hypothesis 1b (H1b): Informational diversity will increase process conflict in workgroups.

Social Category Diversity

While informational diversity is clearly an important resource for organizations, social category diversity is most often what people are referring to when talking about diversity (McGrath, Berdahl, and Arrow, 1996). Social category diversity refers to explicit differences among group members in social category membership, such as race, gender, and ethnicity (Jackson, 1992; Pelled, 1996a). Explicit social category membership characteristics provide a particularly salient basis by which individuals can categorize themselves and others. Social category diversity is likely to influence group interactions by virtue of social identity effects (e.g., Tajfel and Turner, 1986).

According to social identity theory, group members establish positive social identity and confirm affiliation by showing favoritism to members of their own social category (e.g., Billig and Tajfel, 1973), an effect, via discrimination and self-segregation, that disrupts group interaction. Social category membership provides naturally occurring lines along which conflicts can be drawn; categorizing individuals into different groups can provoke hostility or animosity within the work-group. This intragroup hostility can surface as relationship conflict — conflict over workgroup members' personal preferences or disagreements about interpersonal interactions, typically about nonwork issues such as gossip, social events, or religious preferences (Jehn, 1995, 1997). This leads to another hypothesis:

Hypothesis 2 (H2): Social category diversity will increase relationship conflict in workgroups.

Value Diversity

Value diversity occurs when members of a workgroup differ in terms of what they think the group's real task, goal, target, or mission should be. In many cases, these differences can lead to task conflict — disagreements about task content such as disagreements about appropriate advertisements (Jehn, 1994). They also could lead to process conflicts — disagreements about delegation and resource allocation. For instance, group members who value effectiveness (e.g., quality) are likely to have disagreements about duty and resource allocation with group members who value efficiency (e.g., units produced). In addition, similarity in group members' goals and values enhances interpersonal relations within the group (Hackman, 1990). This similarity of values will likely decrease relationship conflict among members (Jehn, 1994). This leads to a third hypothesis:

Hypothesis 3 (H3): Value diversity will increase task conflict, process conflict, and relationship conflict in workgroups.

Diversity and Performance

Research addressing the determinants of group performance in organizations suggests that success often hinges on the ability of the workgroup to embrace, experience, and manage (rather than avoid) disagreements that arise (Tjosvold, 1991; Gruenfeld et al., 1996). Considerable evidence points to the detrimental effects of unmanaged conflicts (e.g., Pruitt and Rubin, 1986; Bettenhausen, 1991; Jehn, 1997). Schwenk and Valacich (1994) found that evaluating and critiquing — engaging conflicts about the task — yielded better decisions in workgroups than when members avoided conflicts or smoothed over their disagreements. Similarly, Putnam (1994) showed that explicit task disagreements helped group members better identify issues, and Baron (1991) showed that disagreements within groups encouraged group members to develop new ideas and approaches.

Mischel and Northcraft (1997) noted that a workgroup's success depends not only on its ability to do the task but also on the group's ability to manage its own interactions effectively, including communicating, cooperating, and coordinating its collective efforts. Similarly, Nonaka and Takeuchi (1995), in their discussion of the organizational conditions that facilitate group performance in knowledge-creating companies, suggested that informational diversity can offer little benefit to a workgroup whose members cannot work together effectively to capitalize on it. They suggested that total diversity among workgroup members is not desirable; rather, some similarity in perspective among group members is necessary to ensure enough common ground to facilitate successful group interaction. Given the aforementioned negative effects of value and social category diversity (i.e., increased relationship conflict), similarity is likely to be most effective in the areas of value and social category diversity. In effect, low value diversity and low social category diversity create conditions for a workgroup to take advantage of its informational diversity, which should be reflected in work-group performance:

> Hypothesis 4 (H4): The effects of informational diversity on work-group performance will be moderated by value diversity and social category diversity within the group; informational diversity is more likely to increase workgroup performance when value diversity and social category diversity in the group are low than when they are high.

Performance is not the only outcome of interest to organizational workgroups. Also at stake are the morale and commitment of the workers, which have long-term implications for group performance as well as for costs associated with absenteeism and turnover. Individuals do not enjoy being immersed in interpersonal conflict (Walton and Dutton, 1969; Peterson, 1983; Ross, 1989), and such conflict makes individuals less likely to remain (Pervin and Rubin, 1967; Emmons, Diener, and Larsen, 1986; Chatman, 1991). Significantly, it is not necessarily differences resulting from informational diversity in how to solve the problem or make the decision that creates the ill-will and bad feelings leading to physical or psychological withdrawal; rather, it typically comes from the relationship conflict often caused by social category diversity and value diversity:

> Hypothesis 5 (H5): High value diversity and social category diversity will decrease worker morale.

Moderators of Diversity Effects

The effects of workgroup diversity on workgroup performance are likely to be affected by structural aspects of the task (e.g., Brehmer, 1976; Van de Ven and Ferry, 1980). Evidence suggests that when a task is simple and well understood, group members can rely on standard operating procedures. Under these circumstances, debates about task strategy are unnecessary and likely to prove disruptive and counterproductive (Barnard, 1938; Gladstein, 1984; Jehn, 1995). This is consistent with Jehn's (1997) finding that process conflict interferes with effective performance of simple, routine tasks. When a task is complex and not well understood, however, discussing and debating competing perspectives and approaches is essential for group members to identify appropriate task strategies and to increase the accuracy of members' assessments of the situation (e.g., Fiol, 1994; Amason and Schweiger, 1994; Putnam, 1994; Jehn, 1995). Such complex tasks require problem solving, have a high degree of uncertainty, and have few set procedures (Van de Ven, Delbecq, and Koenig, 1976), while routine tasks have a low level of variability, are repetitive (Hall, 1972), and are generally familiar and done the same way each time (Thompson, 1967). The constructive discussions and debates needed to accomplish complex tasks depend on the availability of informational diversity:

> Hypothesis 6 (H6): Informational diversity is more likely to increase workgroup performance when tasks are complex rather than routine.

Prior research also suggests that task interdependence can influence diversity effects in workgroups. Task interdependence is the extent to which group members rely on one another to complete their jobs (Van de Ven, Delbecq, and Koenig, 1976). When tasks are interdependent, the demand for smooth interaction among group members (communication, cooperation, and coordination of effort) is heightened (Thibaut and Kelley, 1959; Salancik and Pfeffer, 1977; Saavedra et al., 1993). The disruptive effect of value diversity and social category diversity will be exacerbated when tasks are interdependent:

> Hypothesis 7 (H7): The moderating effects of value diversity and social category diversity on the relationship between informational diversity and workgroup performance will be stronger when tasks are interdependent rather than independent.

The inhibiting effect of value and social diversity on the positive relationship between informational diversity and performance (H4) will be increased when members must interact closely to perform a task. Similarly, because task interdependence heightens the disruptive roles of value diversity and social category diversity on group interaction, task interdependence also should strengthen the negative effects of value diversity and social category diversity on worker morale:

> Hypothesis 8 (H8): Value diversity and social category diversity will be more likely to decrease morale when tasks are interdependent than when they are independent.

Mediators of Diversity Effects

Finally, because we have hypothesized that informational, value, and social category diversity give rise to conflict in workgroups and that conflict in turn has been linked to work-group performance (e.g., Jehn, 1995), we also hypothesize that the effects of workgroup diversity will be mediated by the types of conflict in the workgroup they give rise to, based on the previous discussions. Relationship and process conflict have been negatively linked to performance and morale, while task conflict has been shown to have positive effects on performance (Jehn, 1995, 1997; Amason, 1996). Therefore, we propose the following hypotheses:

> Hypothesis 9a (H9a): Task conflict will mediate the effects of informational diversity on workgroup performance.
>
> Hypothesis 9b (H9b): Process conflict will mediate the effects of informational diversity on workgroup performance.
>
> Hypothesis 9c (H9c): Process conflict will mediate the effects of value diversity on worker morale.
>
> Hypothesis 9d (H9d): Relationship conflict will mediate the effects of value diversity and social category diversity on worker morale.

The hypotheses were tested in a field study of organizational groups.

Methods

Research Site and Sample

The sample consisted of 545 employees in one of the top three firms in the household goods moving industry. The sample (as reported in Jehn, 1995) was taken from the international headquarters for this firm, which houses all functional areas: divisions include marketing and sales, accounting, information systems, domestic and international operations, etc. The featured diversity constructs and measures are unique to this study.

This firm had formally designated work units (teams). A work unit is defined in the organization as a group in which all personnel report directly to the same supervisor and interact to complete unit tasks. We verified the organization's delineation of work units by examining departmental reports and organizational charts, which indicated that members were batched together to perform tasks and were seen by others as a group. The organization's delineation of work units was quite accurate and corresponded with the supervisors' and employees' view of who their fellow group members were.

Work units completed all functions within the organization, from sorting and delivering mail to making corporate strategy. The work units included sales units selling services to corporations moving their employees to other domestic and international locations, data entry and coding units that process this information, and groups that oversee the governmental regulations on state and national cross-border transit. This organization provides a fitting arena in which to test our hypotheses, since it has well-delineated work units that vary on a wide range of demographic variables and our other variables of interest (e.g., conflict, task type, interdependence) yet were relatively similar in size ($x = 6.21$, s.d. = .47).

Survey Procedure

We distributed a survey to all employees in the firm. Although the survey was voluntary, the chief executive officer requested that all employees participate in the confidential study, supervisors and employees were told in advance that we would be there to administer the survey, and employees were given company time to complete it. The response rate of the survey (89 percent, 485 employees) was quite high and included 92 complete work units. Later, we followed up with employees who were absent or off-site (e.g., sales teams) when we administered the survey. The high response rate allowed us to include in the analysis only units with a 100 percent response rate. Thirteen units that did not achieve full response were dropped from the study.

The survey consisted of 85 self-report, Likert-style questions, randomly ordered. We used personnel records to verify the demographic information collected by the survey and, at the same time, collected archival data, such as performance appraisals and departmental output reports. Sixty supervisors, managers, and vice presidents received and returned a packet of materials to evaluate their work unit(s). Information collected in this packet included organizational charts, group and individual effectiveness ratings, and departmental output reports.

Measures

Diversity

Perceived *value diversity* among group members was measured by six 5-point Likert scales anchored by 1 = "Strongly disagree" and 5 = "Strongly agree."

Members were asked if the values of all group members were similar, if the work unit as a whole had similar work values, if the work unit as a whole had similar goals, whether members had strongly held beliefs about what was important within the work unit, whether members had similar goals, and if all members agreed on what was important to the group. The coefficient alpha for this scale was .85. Items were reverse-coded so that higher scores reflected higher diversity.

Following past research (e.g., McGrath, Berdahl, and Arrow, 1996; Jehn, Chadwick, and Thatcher, 1997), *informational diversity* measures assessed heterogeneity of education (i.e., major), functional area in the firm (e.g., marketing, mailroom, operations), and position in the firm (i.e., hourly employee or management). *Social category diversity* measures assessed heterogeneity of sex and age. The firm's executives declined to provide data on the ethnic background or nationality of the employees.

As is typical in the treatment of categorical variables, we used the entropy-based index (Teachman, 1980; Ancona and Caldwell, 1992) to form an aggregate measure of the informational and social category diversity within workgroups:

$$\text{Diversity} = \Sigma - P_i(\ln P_i),$$

where P_i represents the proportion of the work unit that has each diversity characteristic. If a demographic characteristic is not represented in the team, the value assigned is zero. Thus, the diversity index represents the sum of the products of each characteristic's proportion in the work unit's makeup and the natural log of its proportion. The higher the diversity index, the greater the distribution of characteristics within the work unit. If the work unit is composed of six individuals, one female and five male, their diversity index is .4506; if all six members are female, the diversity index is 0.00; and if three members are female and three are male, the diversity index is .6931. Likewise, a group with three engineers and three accountants would have a diversity index of .6931, and if all members are engineers, the diversity index is 0.00.

Intragroup Conflict

We used the items of the intragroup conflict scale developed by Jehn (1995) to measure the amount and type of perceived relationship and task conflict in the work units. The 12 items on the presence of conflict were rated on a 5-point Likert scale anchored by 1 = "None" and 5 = "A lot." Four items measured *relationship conflict* ("How much friction is there among members in your work unit?" "How much are personality conflicts evident in your work unit?" "How much tension is there among members of your work unit?" and "How much emotional conflict is there among members in your work unit?"). Examples of the five items measuring *task conflict* include the following: "How frequently are there conflicts about ideas in your work unit?" and "How often do people in your work unit disagree about opinions?" The coefficient alphas for relationship and task conflict were .90 and .88, respectively.

Three items measuring *process conflict* were taken from Shah and Jehn (1993): "How often do members of your work unit disagree about who should do what?" "How frequently do members of your work unit disagree about the way to complete a group task?" and "How much conflict is there about delegation of tasks within your work unit?" The coefficient alpha for process conflict was .78.

High correlations among the conflict variables led us to conduct a number of analyses to examine the discriminant validities of the conflict variables, using Howell's (1987) approach. While it is important to discriminate between these measures in our analyses, it is not unreasonable to expect that the different types of conflict may overlap. For example, conflicts originating in personal relationships have been shown to spill over into disagreements about how to do the task (Jehn, 1997). The test of discriminant validity computes the upper limit for the confidence interval of the observed correlations and assesses whether this limit is smaller than the maximum possible correlation between the scores as computed from their reliability coefficients. All of the conflict construct pairs meet the discriminant validity test at $p < .0013$. In addition, in conducting a factor analysis with oblique rotation, we found results similar to Shah and Jehn (1993), Amason (1996), and others (Amason and Sapienza, 1997; Janssen, Van De Vliert, and Veenstra, 1998) who used the intragroup conflict scale (Jehn, 1995) and found that relationship, task, and process conflict items load separately (see Simons and Peterson, 1999, for a review of these studies and the intercorrelations between the types of conflict).

Task Moderator Variables

To measure *task interdependence*, we used Van de Ven, Delbecq, and Koenig's (1976) workflow interdependence scale, which provides diagrams depicting the workflow within a unit to measure interdependence. Group members indicated on a 5-point Likert scale the degree to which the level of interdependence in their work unit was similar to the diagram. The average standard deviation among members within units was quite low (s.d. = .34), indicating that members viewed their level of task interdependence similarly. We also included Likert-style questions on task interdependence: "Within my work unit, people have one-person jobs: that is, people can complete most of the jobs on their own, with no help from others" (reverse coded); "Often, all the work unit members meet together to discuss how each task, case, or claim should be performed or treated in order to do the work in this unit." The Cronbach alpha of these items (including the diagram Likerts) was .78. In addition, we verified the reported interdependence with respondents' supervisors and via observation.

Task type was measured using an adaptation and combination of Perrow's (1970) index of routinization and Van de Ven, Delbecq, and Koenig's (1976) dimension of task variety. Examples of items from the 12-item, agree-disagree 5-point scale are "I encounter a lot of variety in my normal working day" (reverse coded), "The methods I follow in my work are about the same for dealing with all types of work, regardless of the activity," "My job is very routine," and "I feel like I am doing the same thing over and over again," and are similar to Jehn's (1995) routineness adaptation of the same scales. The coefficient alpha for this scale was .94, with high scores reflecting routineness. We verified the scores for the reported team task type scores with supervisors' reports and observation of the task units at work.

Worker Morale

We used three different measures of *worker morale*: satisfaction, intent to remain, and commitment. Individual satisfaction with the group was measured by a 5-point Likert question ("How satisfied are you working in this work unit?") anchored by 1 = "Not at all" and 5 = "Very" and the Kunin Faces Scale (1955). Members responded to the Kunin Faces Scale by circling the face that indicated how happy they were working in their groups. The coefficient alpha for the two satisfaction items was .85. Members also reported on their intent to remain in the group by responding to Kraut's (1975) measure of tenure intentions: "How long do you expect to stay in this work unit?" "If you have your own way, will you be working in this same work unit three years from now?" and "Do you want to change work units?" The Cronbach alpha of the three-item scale was .96.

We rated the commitment of group members by the degree to which members agreed or disagreed on a 5-point Likert scale with the following items: "I talk up this work unit to my friends as a great group to work in," "I am very committed to my work unit," "I am proud to tell others that I am part of this work unit," and "I feel a sense of ownership for this work unit rather than being just an employee." This adaptation of O'Reilly and Chatman's (1986) commitment questionnaire had a coefficient alpha of .85.

Workgroup Performance

Perceived group performance was measured as members' responses to the following questions on a 5-point Likert scale: "How well do you think your work unit performs?" and "How effective is your work unit?" The coefficient alpha was .93. Actual group performance was assessed by departmental records (computerized production records and error reports) provided and standardized by the firm, and efficiency was assessed by supervisors' ratings of the groups. This firm has developed well-established outcome measures that are comparable across work units and that are updated biannually. Its Quality Assurance department is specifically designed to assess the productivity of work units. For example, to assess the performance of one top management team, Quality Assurance designed a 360-degree feedback system that included ratings of members' performance by one another, users of their work (e.g., subordinates), and their vice presidents. We put this outcome into a standardized form to compare with other work units. To measure the performance of more routine task groups, such as one data entry group, the Quality Assurance team measured the number of data fields entered in a specified time period and deducted for data errors, along with measuring other unit tasks. Once again, we put the outcome into a standardized form for comparison with other units. This firm is considered a leader by others in the industry for the performance measures developed by its Quality Assurance department.

Workgroup efficiency was assessed by supervisors' ratings of two items measured on 7-point Likert scales, "How effective is this group at getting things done quickly?" and "How efficient is this work unit?" (1 = "Not at all Effective" to 7 = "Very Effective"). The Cronbach alpha for this two-item measure was .88.

Results

Table 1 provides the means, standard deviations, and correlations for all variables in the model. Our three types of diversity are all statistically independent of each other.

Diversity and Conflict

Table 2 provides the regression analyses that tested H1 through H3. Supporting H1a, informational diversity was positively related to task conflict in workgroups. H1b, predicting that informational diversity would increase process conflict, was not supported. Support also was found for H2: social category diversity increased relationship conflict in workgroups. As predicted by H3, value diversity was positively and significantly related to all three types of conflict. Social category diversity and value diversity explained 21.9 percent of the variance in relationship conflict within the groups. Informational and value diversity explained 13.9 percent of the variance in task conflict; value diversity alone explained 10.3 percent of the variance in process conflict within workgroups.

Impact of Diversity on Performance and Worker Morale

We conducted regression analyses to test our hypotheses predicting the effects of workgroup diversity on worker morale and performance (H4 through H8). As shown in tables 3 and 4, below, the hypothesized relationships explain between 6.6 percent (workgroup efficiency) and 37.8 percent (commitment to workgroup) of workgroup performance and worker morale. Utilizing a procedure for cross-level analysis (Rousseau, 1985), we averaged individual responses on each of the independent and moderator variables for each work unit to create a group-level measure for the analysis of group-level dependent variables only (i.e., workgroup performance). We identified workgroups from a listing of who reports to whom, which was verified by the unit members. The average intragroup interrater agreement for each variable aggregated for the group performance equations was between .75 and .87. In addition, we calculated the ε^2, which indicates whether any two people in the same group are more similar than two people who are members of different groups (Florin et

Table 1: Means, Standard Deviations, and Intercorrelations*

Variable	1	2	3	4	5	6	7	8	9	10	11	12	13
1. Value diversity	–												
2. Informational diversity	.05												
3. Social category diversity	.09	.08											
4. Relationship conflict	.16	.04	.07										
5. Process conflict	.13	−.07	.03	.63									
6. Task conflict	.33	.09	.06	.55	.55								
7. Interdependence	−.00	.19	.10	.05	.07	.06							
8. Task type	.29	−.24	−.25	.18	.07	−.02	−.15						
9. Commitment	−.19	.16	.09	−.41	−.30	−.31	.15	.16					
10. Satisfaction	−.17	.08	.14	−.50	−.39	−.41	.10	−.23	.28				
11. Intent to remain	−.18	.08	.08	−.42	−.29	−.29	.07	−.22	.29	.36			
12. Perceived performance	−.08	.02	.14	−.12	−.18	−.09	.05	−.09	.33	.26	.09		
13. Actual group performance	−.13	.20	−.07	−.29	−.36	−.29	−.01	−.01	.45	.46	.24	.29	
14. Group efficiency	−.16	−.03	−.03	.02	.06	.09	.06	−.10	.02	−.05	.12	.06	−.11
Mean	.99	.19	.35	2.22	1.87	2.64	2.22	3.01	3.44	4.44	3.37	4.22	3.97
S.d.	.33	.27	.28	1.21	1.02	1.02	0.74	0.59	0.99	1.20	1.45	0.57	0.73

*Correlations above .09 are significant at the .05 level.

Table 2: Regression Analyses Predicting Conflict ($N=518$)

	Relationship conflict	Task conflict	Process conflict
Informational diversity (H1)	.02	.09*	−.06
Social category diversity (H2)	.08*	.05	.03
Value diversity (H3)	.15**	.37**	.13**
Adjusted R^2	.219	.139	.103
F	32.06***'	16.21***	8.34**

*$p<.05$; **$p<.01$; ***$p<.001$.

Table 3: Hierarchical Regression Analyses Predicting Workgroup Performance

	Perceived performance ($N=508$)	Actual group performance ($N=87$)	Group efficiency ($N=90$)
Step 1: Main effects			
Informational diversity (ID)	.05	.30***	−.05
Social category diversity (SC)	.16***	−.07	−.03
Value diversity (V)	−.10*	−.12*	−.17**
Task type (T)	−.10*	−.05	−.06
R^2	.309	.127	.072
F	14.57***	4.70***	1.87*
Step 2: Interactions			
H4 ID × V	−.06	−.34*	−.75**
H4 ID × SC	.18	.25	−.44*
H6 ID × T	−.35*	−.74***	−.53*
Change in R^2	.007	.021	.059
F change	.601	1.66*	2.33**
R^2	.316	.148	.131
Adjusted R^2	.284	.108	.066
F	10.13***	3.78***	2.90***

*$p<.05$; **$p<.01$; ***$p<.001$.

al., 1990). Our results, averaging .54, exceeded Georgopoulos's (1986) minimum criteria of .20, indicating that it was appropriate to aggregate the variables into group-level variables for the analysis of workgroup performance.

Workgroup Performance

Table 3 presents the hierarchical regression analyses conducted to test the hypotheses about informational diversity and workgroup performance. Step 1 of the hierarchical regression includes the main effects of informational diversity, value diversity, social category diversity, and task type; step 2 includes the three hypothesized interactions (informational diversity × value diversity; informational diversity × social category diversity; and informational diversity × task type).

Informational diversity was positively related to actual work-group performance. In support of H4, value diversity moderated the effect of informational

Table 4: Hierarchical Regression Analyses Predicting Worker Morale

	Satisfaction ($N=491$)	Intent to remain ($N=412$)	Commitment ($N=488$)
Step 1: Main effects			
Informational diversity (ID)	.09	.06	.10*
H5 Social category diversity (SC)	.14**	.12**	.16***
H5 Value diversity (V)	−.11*	−.19***	−.19**
Interdependence (I)	.04	.03	.10*
R^2	.318	.167	.391
F	16.09***	6.76***	22.70**
Step 2: Interactions			
H8 V × I	−.19	−.22	−.10
H8 SC × I	.36*	−.13	.24*
Change in R^2	.009	.007	.002
F change	1.005	.900	.442
R^2	.327	.174	.393
Adjusted R^2	.313	.138	.378
F	13.22**	4.98***	17.02**

*$p<.05$; **$p<.01$; ***$p<.001$.

diversity on actual performance and efficiency; informational diversity was more beneficial when there were low levels of value diversity than when there were high levels. In further support of H4, informational diversity was more positively related to efficiency when social category diversity was low. In support of H6, the interaction between informational diversity and task type was significant for all three measures of workgroup performance — perceived, actual, and efficiency; informational diversity was more likely to increase performance and efficiency when tasks were complex.

Hypothesis 7 predicted that the moderating effects of value diversity and social category diversity on the relationship between informational diversity and workgroup performance would be strongest when tasks are highly interdependent. Given the binormal distribution of interdependent and independent task groups, to get a clear picture of the three-way interactions, we dichotomized the groups into those that were highly interdependent ($N=57$) and those that were highly independent ($N=35$). In partial support of H7, the interaction between informational diversity and value diversity was more strongly related to performance when groups were interdependent ($B=-.35$, $p<.01$) than when members were independent ($B=.09$, n.s.), but interdependence did not similarly moderate the effects of the social category diversity and informational diversity interaction on workgroup performance.

Worker Morale

H5 predicted that high value diversity and social category diversity would decrease the morale of work-group members. As shown in table 4, more value diversity in the workgroup decreased satisfaction, intent to remain, and commitment of group members. In contrast, a higher level of social category diversity increased satisfaction, intent to remain, and commitment, opposite to what we had hypothesized. Hypothesis 8 predicted that value diversity and social category diversity would be more likely to affect morale when task interdependence is high. As shown in table 4, members in interdependent groups were more satisfied and felt more committed when high levels of social category diversity were present; however, these interactions did not significantly add to the variance explained by the main effects. The interaction between value diversity and task interdependence was not significant.

Mediators of Diversity Effects

Hypothesis 9a predicted that task conflict would mediate the effects of informational diversity on workgroup performance. Using the procedure suggested by Baron and Kenny (1986), we found that the significant effect of informational diversity on actual group performance ($B=.30$, $p<.001$) became nonsignificant ($B=.07$, n.s.) when task conflict was controlled for. Thus; the mediating role of task conflict between informational diversity and actual group performance was confirmed. The results did not confirm hypothesis 9b, that process conflict would mediate the effects of informational diversity on workgroup performance. Results did confirm hypothesis 9c, which predicted that process conflict would mediate the effects of value diversity on worker morale. Value diversity was significantly related to the following dependent variables: satisfaction ($B=-.11$, $p<.05$), intent to remain ($B=-.19$, $p<.001$), commitment ($B=-.19$, $p<.01$), perceived performance ($B=-.10$, $p<.05$), actual group performance ($B=-.12$, $p<.05$), and group efficiency ($B=-.17$, $p<.01$). The effect of value diversity became nonsignificant when process conflict was included in the regression analyses on satisfaction ($B=-.04$), intent to remain ($B=-.05$), commitment ($B=-.05$), perceptual performance ($B=.04$), and actual group performance ($B=.03$), meaning that value diversity accounts for the variation in these outcome variables through process conflict. Process conflict thus has a mediating role in the relationship of value diversity to satisfaction, intent to remain, commitment, perceptual performance, and actual group performance.

H9d predicted that relationship conflict would mediate the effects of value diversity and social category diversity on worker morale. Relationship conflict (mediator) was regressed on value diversity and social category diversity (independent variables) and found to be significant (Bs$=.15$, $.08$, $p<.01$, $.05$, respectively). Second, value diversity was significantly related to the following dependent variables: satisfaction ($B=-.11$, $p<.05$), intent to remain ($B=-.19$, $p<.001$), commitment ($B=-.19$, $p<.01$), perceptual performance ($B=-.10$, $p<.05$), actual group performance ($B=-.12$, $p<.05$), and group efficiency ($B=-.17$, $p<.01$). Social category diversity was significantly related to the following dependent variables: satisfaction ($B=.14$, $p<.01$), intent to remain ($B=.12$, $p<.01$), commitment ($B=.16$, $p<.001$), and perceptual performance ($B=.16$, $p<.001$). Value diversity's effect became nonsignificant when relationship conflict was included in the regression analyses for satisfaction ($B = .01$), intent to remain ($B = -.02$), perceptual performance ($B = .04$), and actual group performance ($B = .02$), meaning that value diversity accounts for the variation in these outcome variables through relationship conflict. Thus, for satisfaction and intent to remain, relationship conflict mediates between value diversity and worker morale.

Social category diversity's effect also became nonsignificant when relationship conflict was included in the regression analyses for satisfaction ($B = -.01$), intent to remain ($B = .05$), commitment ($B = .06$), and perceptual performance ($B = .01$). Thus, the mediating role of relationship conflict between social category diversity and worker morale was confirmed for satisfaction, intent to remain, and commitment.

Discussion

The purpose of this study was to explore the differential impact of three group compositional factors (social category diversity, value diversity, and informational diversity) and two moderating variables (task type and task interdependence) on workgroup performance. With few exceptions (see Gruenfeld et al., 1996; Jehn, Chadwick, and Thatcher, 1997), past research has lumped social category diversity and informational and value diversity under the general heading of diversity in attempting to understand the impact of diversity on workgroup performance. In addition, even when distinctions have been made about types of diversity, previous research has typically been limited to studying one type of diversity but not others (e.g., O'Reilly and Flatt, 1989; Ancona and Caldwell, 1992). As a result, it is not surprising that a review of this literature produces different results across studies that purport to study the same thing — diversity and its impact on performance.

The present study was successful in distinguishing among three types of diversity and their impact on workgroup performance. While previous research has demonstrated the influence of conflict on workgroup outcomes (Jehn, 1995, 1997; Amason, 1996), the study described here takes the additional step of exploring how different types of diversity evoke conflict. The results show that different forms of diversity exacerbate different forms of conflict (within different task configurations), which in turn affects perceived performance, actual performance, satisfaction, intent to remain, and commitment. How these different types of diversity ultimately influence performance, both perceived and actual, is no simple story.

Before we review that story, however, we wish to acknowledge the limitations of this study. Because we did not have access to the ethnic diversity of the participants in this study, our social category diversity results are based on the measurement of only two factors: age and gender. In addition, the study is cross-sectional, so that no causal inferences can be drawn. Finally, some of the variable measures are self-reported, and we cannot rule out the possibility of response-response bias in some of our analyses. To minimize this possiblity, however, we also included measures from archival data and multiple sources. For example, while different types of conflict, value diversity, the moderators, and many of the affective dependent variables (e.g., measures of morale, intent to remain, etc.) were self-reported, the two other diversity measures (social category and informational diversity), actual group performance, and workgroup efficiency were based on archival data or supervisory ratings. In addition, besides conducting multiple tests to assess the discriminant validity of the three types of conflict, future research should examine the transformation of one type of conflict into another. For instance, arguments about who is capable of doing what can often lead to relationship conflicts, and vice versa.

While most of our hypotheses received support in the predicted direction, we did have an unexpected finding. The finding that social category diversity resulted in increased relationship conflict, even though group members reported increased morale, runs counter to both conventional wisdom and past research. For example, Jehn (1995, 1997), among others (Pruitt and Rubin, 1986; Bettenhausen, 1991; Schwenk and Valacich, 1994), illustrated how relationship conflict is associated with a general reduction in worker morale. One explanation for this finding may be the particular variables that compose the social category diversity variable in this study. It may be that at least for age, diversity on this factor reflects lower levels of intragroup competition, as workers are more likely to be competing with similar (in age) others for various valued organizational resources. But this inconsistency, coupled with the cross-sectional nature of the data, suggests at least one plausible, alternative explanation. It may be that high performance leads to high morale and low task conflict rather than, in our interpretation, that low task conflict leads to high morale and high performance.

We explored this alternative explanation by positing that a third variable, group performance, may be affecting the relationship between social category diversity and worker morale such that it would overwhelm the negative effects of relationship conflict on morale. In fact, follow-up analyses demonstrated that performance mediated the impact of social category diversity on morale. While, as noted above, social category diversity was significantly related to satisfaction, intent to remain, perceived performance, and commitment, these effects became nonsignificant when group performance was included in the regression analyses. Thus, the mediating role of performance between social category diversity and worker morale was confirmed for satisfaction, intent to remain, perceived performance, and commitment. Diverse groups performed better and perhaps, therefore, were more pleased with the group in which they were working, independent of its level of social category diversity.

From this study we can identify the types of diversity that are associated with various types of performance. For a team to be effective, members should have high information diversity and low value diversity. For a team to be efficient, members should have low value diversity. For a team to have high morale (higher satisfaction, intent to remain, and commitment) or to perceive itself as effective, it should be composed of participants with low value diversity. What these consistent findings suggest is the value, for most measures of group performance, of low value diversity among members. Moreover, it may also be that value diversity, which is often not immediately discernible, becomes more important as a predictor of group performance over time, while age and gender diversity, characteristics that are readily apparent, become less relevant over time. The importance of low value diversity on workgroup performance over time is also supported by results of a recent field study of research and development teams (Owens and Neale, 1999).

The most arresting aspect of this study may be the window it provides into our understanding of the importance of value diversity to both workgroup performance and worker morale. Thus, it seems that certain types of similarity are dramatically more important than others, despite the assumption that people generally strive for similarity among those with whom they interact (Byrne, 1971). It is the diversity associated with values, and not social category, that causes the biggest problems in and has the greatest potential for enhancing both workgroup performance and morale.

This study suggests, like Williams and O'Reilly (1998), that the impact of diversity goes well beyond simple main effects. Task interdependence and task type moderate the relationships between diversity and various measures of performance. Informational diversity is more likely to lead to improved performance when tasks are nonroutine. Again, social category diversity unexpectedly led to greater satisfaction and commitment when task interdependence was high than when it was low. It is more difficult here to explain away this finding by deferring to the performance-morale path. It may actually be that social category diversity results in higher morale in interdependent tasks. Being able to work together successfully, even when the group is diverse with respect to age and gender composition, may result in greater morale because the group has overcome a serious challenge to its effectiveness. Further, these groups may have discovered that the social category differences were not good signals of value diversity. This interpretation received support from earlier research. For example, several studies in the 1960s (Byrne and Wong, 1962; Stein, Hardyck, and Smith, 1965) found that whites preferred blacks with attitudes similar to their own over whites with opposing attitudes, but this effect of value similarity on racial attitudes apparently has been ignored in recent years, as researchers have used similarity in attributes such as race or gender as surrogates for value similarity. Further, it appears from the work of Owens and Neale (1999) that groups are aware of some of the impact of different types of diversity on performance and can select team members for their contributions along multiple diversity dimensions that enhance group performance. Taken as a whole, these results provide direction for creating and managing diverse teams to enhance performance. Our results suggest that diversity itself is not enough to ensure innovation; the nature of the team's diversity is critical. For group members to be willing to engage in the difficult and conflictful processes that may lead to innovative performance, it seems that group members must have similar values.

Our results also shed light on the difficulty of studying social category diversity. One problem associated with attempting to make predictions about the effects of social category diversity on workgroup performance is that social category diversity may represent informational diversity, value diversity, both, or neither. Since social category diversity is not necessarily associated with either informational or value diversity, it poses prediction problems for researchers and signaling problems for group members. What does being the only woman in an otherwise all-male group mean about the unique perspectives that an individual brings to the group? If the task of the group is to define a strategic direction for the organization, and all group members have backgrounds in finance, it is not likely that the gender of one member will make a significant difference in the information that an individual brings to the group. If the group must select product features for a new model of automobile, however, the experience of being a woman may bring a different orientation to the discussion, even if that woman is an engineer, just like everyone else in the group. Finally, are all categorical variables equally influential? Are there ebbs and flows of influence of these categorical variables as teams age and evolve? While clearly important questions, we must await future research for the answers.

Unlike demographic characteristics, the characteristics of value and informational diversity are not easily discernible from a quick physical inspection of fellow group members the way that social category characteristics often are. Thus, because of their ease of observation, demographic characteristics, in particular, are more likely to be incorporated into the heuristic information processing of group members as they develop mechanisms to manage group processes and complete assigned tasks. Just as past researchers may have relied on social category diversity as a surrogate for informational and value diversity, social category similarity may lead group members to overlook important sources of informational and value diversity or to assume similarity where it does not exist.

Our findings — specifically, distinguishing among different types of diversity and their differential effects — may help reconcile some of the inconsistencies in past research. If the type of diversity measured is informational diversity, group performance may be enhanced by diversity. If the type of diversity measured is social category diversity, the most positive effects will likely be on worker morale (satisfaction, intent to remain, commitment, and perceived performance). In contrast, groups that have greater diversity as measured in terms of values may suffer significant performance decrements (being less effective and efficient as well as having poorer perceived performance) and diminished worker morale (decreased satisfaction, commitment, and intent to remain in the group). While the story told in previous research, even with its contradictory findings and inconsistent empirical support, may have been easier to tell — heterogeneity leads to better workgroup performance and homogeneity leads to easier workgroup process — the more complex representation of these relationships as provided by this paper should enhance our understanding of the ways to create, intervene in, and manage high-performance groups and teams.

References

Amabile, Teresa M. 1994 "The atmosphere of pure work: Creativity in research and development." In William R. Shadish and Steve Fuller et al. (eds.), *The Social Psychology of Science*: 316–328. New York: Guilford Press.

Amason, Allen C. 1996 "Distinguishing the effects of functional and dysfunctional conflict on strategic decision making: Resolving a paradox for top management teams." *Academy of Management Journal*, 39: 123–148.

Amason, Allen C., and Harry J. Sapienza 1997 "The effects of top management team size and interaction norms on cognitive and affective conflict." *Journal of Management*, 23: 495–516.

Amason, Allen C., and David M. Schweiger 1994 "Resolving the paradox of conflict, strategic decision making and organizational performance." *International Journal of Conflict Management*, 5: 239–253.

Ancona, Deborah, and David Caldwell 1992 "Demography and design: Predictors of new product team performance." *Organization Science*, 3: 321–341.

Barnard, Chester I. 1938 *The Functions of the Executive*. Cambridge, MA: Harvard University Press.

Baron, R. A. 1990 "Countering the effects of destructive criticism: The relative efficacy of four interventions." *Journal of Applied Psychology*, 75: 235–245.

Baron, R. A. 1991 "Positive effects of conflict: A cognitive perspective." *Employees Responsibilities and Rights Journal*, 4: 25–36.

Baron, R. M., and D. A. Kenny 1986 "The moderator-mediator variable distinction in social psychological research: Conceptual, strategic, and statistical considerations." *Journal of Personality and Social Psychology*, 51: 1173–1182.

Bettenhausen, Kenneth L. 1991 "Five years of group research: What we have learned and what needs to be addressed." *Journal of Management*, 17: 345–381.

Billig, Michael, and Henri Tajfel 1973 "Social categorization and similarity in intergroup behavior." *European Journal of Social Psychology*, 3 (1): 27–52.

Boyett, J. H., and H. P. Conn 1991 *Workplace 2000: The Revolution Reshaping American Business*. New York: Dutton.

Brehmer, Berndt 1976 "Social judgement theory and the analysis of interpersonal conflict." *Psychological Bulletin*, 83 (6): 985–1003.

Byrne, D. 1971 *The Attraction Paradigm*. New York: Academic Press.

Byrne, D., and T. J. Wong 1962 "Racial prejudice, interpersonal attraction, and assumed dissimilarity of attitudes." *Journal of Abnormal Psychology*, 65: 246–253.

Byrne, John A. 1993 "The horizontal corporation." *Business Week*, December 20: 76–81.

Chatman, Jennifer A. 1991 "Matching people and organizations: Selection and socialization in public accounting firms," *Administrative Science Quarterly*, 36: 459–484.

Damon, William 1991 "Problems of direction in socially shared cognition." In Lauren B. Resnick, John M. Levine, and S. D. Teasley (eds.), *Perspectives on Socially-Shared Cognition*: 384–397. Washington, DC: American Psychological Association.

Doise, W., and G. Mugny 1984 *The Social Development of the Intellect*. Oxford: Pergamon Press.

Donnellon, Anne 1996 *Team Talk: The Power of Language in Team Dynamics*. Cambridge, MA: Harvard Business School Press.

Dougherty, Deborah 1992 "Interpretive barriers to successful product innovation in large firms." *Organization Science*, 3: 179–202.

Dumaine, Brian 1991 "The bureaucracy busters." *Fortune*, June 17: 36–50.

Dumaine, Brian 1994 "The trouble with teams." *Fortune*, September 5: 52–59.

Emmons, Robert A., Ed Diener, and Randy J. Larsen 1986 "Choice and avoidance of everyday situations among men and women in managerial, professional, and technical positions." *Journal of Applied Psychology*, 75: 539–546.

Festinger, Leon, S. Schacter, and K. Back 1950 *Social Pressure in Informal Groups*. New York: Harper and Brothers.

Fiol, C. Marlene 1994 "Consensus, diversity, and learning in organizations." *Organization Science*, 5: 403–420.

Florin, Paul, Gary A. Giamartino, David A. Kenny, and Abraham Wandersman 1990 "Levels of analysis and effects: Clarifying group influence and climate by separating individual and group effects." *Journal of Applied Social Psychology*, 20: 881–900.

Garton, A. F. 1992 *Social Interaction and the Development of Language and Cognition*. Hillsdale, NJ: Erlbaum.

Georgopoulos, Basil Spyros 1986 *Organizational Structure, Problem Solving, and Effectiveness*. San Francisco: Jossey-Bass.

Gladstein, Deborah L. 1984 "A model of task group effectiveness." *Administrative Science Quarterly*, 29: 499–517.

Granovetter, Mark 1973 *Getting a Job: A Study of Contacts and Careers*. Cambridge, MA: Harvard University Press.

Gruenfeld, Deborah H. 1995 "Status, ideology, and integrative complexity on the U.S. Supreme Court: Rethinking the politics of political decision making." *Journal of Personality and Social Psychology*, 68: 5–20.

Gruenfeld, Deborah H., Elizabeth A. Mannix, Katherine Y. Williams, and Margaret A. Neale 1996 "Group composition and decision making: How member familiarity and information distribution affect process and performance." *Organizational Behavior and Human Decision Processes*, 67: 1–15.

Hackman, J. Richard 1990 *Groups that Work (and Those that Don't)*. San Francisco: Jossey-Bass.

Hall, R. H. 1972 *Organizations, Structure, and Process*. Englewood Cliffs, NJ: Prentice-Hall.

Hoffman, L. R. 1978 "The group problem-solving process." In L. Berkowitz (ed.), *Group Processes*: 101–114. New York: Academic Press.

Hoffman, L. R., and N. R. F. Maier 1961 "Quality and acceptance of problem solutions by members of homogeneous and heterogeneous groups." *Journal of Abnormal and Social Psychology*, 62: 401–407.

Howell, R. D. 1987 "Covariance structure modeling and measurement issues: A note on interrelations among a channel entitity's power sources." *Journal of Marketing Research*, 24: 119–126.

Jackson, Susan 1992 "Team composition in organizations." In S. Worchel, W. Wood, and J. Simpson (eds.), *Group Process and Productivity*: 1–12. London: Sage.

Janssen, Onne, Evert Van De Vliert, and Christian Veenstra 1998 "How task and person conflict shape the role of positive interdependence in management teams." *Journal of Management*, 25: 117–142.

Jehn, Karen A. 1994 "Enhancing effectiveness: An investigation of advantages and disadvantages of value-based intragroup conflict." *International Journal of Conflict Management*, 5: 223–238.

Jehn, Karen A. 1995 "A multimethod examination of the benefits and detriments of intragroup conflict." *Administrative Science Quarterly*, 40: 256–282.

Jehn, Karen A. 1997 "A qualitative analysis of conflict types and dimensions in organizational groups." Administrative Science Quarterly, 42: 530–557.

Jehn, Karen A., Clint Chadwick, and Sherry Thatcher 1997 "To agree or not to agree: Diversity, conflict, and group outcomes." *International Journal of Conflict Management*, 8: 287–306.

Kraut, Allen I. 1975 "Predicting turnover of employees for measured job attitudes." *Organizational Behavior and Human Decision Processes*, 13: 233–243.

Kunin, Theodore 1955 "The construction of a new type of attitude measure." *Personnel Psychology*, 8: 65–77.

Levine, John M., and L. B. Resnick 1993 "Social foundations of cognitions." *Annual Review of Psychology*, 41: 585–612. Palo Alto, CA: Annual Reviews.

Lipnack, J., and J. Stamps 1993 *The Teamnet Factor: Bringing the Power of Boundary Crossing in the Hearts of Your Business*. Essex Junction, VT: Oliver Wright.

Mannix, Elizabeth A., Sheila Goins, and S. Carroll 1996 "Starting at the beginning: Linking team performance to process and performance." Working paper, School of Business, Columbia University.

McGrath, Joseph E., Jennifer L. Berdahl, and Holly Arrow 1996 "No one has it but all groups do: Diversity as a collective, complex, and dynamic property of groups." In Susan E. Jackson and Marian N. Ruderman (eds.), *Diversity in Work Teams: Research Paradigms for a Changing World*: 42–66. Washington, DC: APA Publications.

Mischel, Leann Joan, and Gregory B. Northcraft 1997 "I think we can, I think we can . . .: The role of self-efficacy beliefs in group and team effectiveness." In Barry Markovsky and

Michael J. Lovaglia (eds.), *Advances in Group Processes*, 14: 177–197. Greenwich, CT: JAI Press.

Murray, F. B. 1983 "Learning and development through social interaction and conflict: A challenge to social learning theory." In L. Liben (ed.), *Piaget and the Foundation of Knowledge*: 231–247. Hillsdale, NJ: Erlbaum.

Nemeth, Charlan 1986 "Differential contributions of majority and minority influence." *Psychological Review*, 93: 23–32.

Newcomb, T. M. 1960 *The Acquaintance Process*. New York: Holt, Rinehart, and Winston.

Nohria, Nitin 1991 "Garcia-Pont, Carlos Global strategic linkages and industry structure." *Strategic Management Journal*, 12: 105–124.

Nonaka, Ikujiro, and H. Takeuchi 1995 *The Knowledge-Creating Company*. New York: Oxford University Press.

Northcraft, Gregory B., Jeffrey T. Polzer, Margaret A. Neale, and Roderick Kramer 1995 "Productivity in cross-functional teams: Diversity, social identity, and performance." In Susan E. Jackson and Marian N. Ruderman (eds.), *Diversity in Work Teams: Research Paradigms for a Changing World*: 69–96. Washington, DC: APA Publications.

Oldham, Greg, and Anne Cummings 1998 "Creativity in the organizational context." *Productivity*, 39 (2): 187–194.

O'Reilly, Charles A.,and Jennifer Chatman 1986 "Organizational commitment and psychological attachment: The effects of compliance, identification, and internalization of prosocial behavior." *Journal of Applied Psychology*, 71: 492–499.

O'Reilly, Charles A., and Sylvia Flatt 1989 "Executive team demography: Organizational innovation and firm performance." Working paper, School of Business, University of California at Berkeley.

Owens, David A., and Margaret A. Neale 1999 "The dubious benefit of group heterogeneity in highly uncertain situations: Too much of a good thing?" Working paper, School of Business, Vanderbilt University.

Pelled, Lisa 1996a "Demographic diversity, conflict, and work group outcomes: An intervening process theory." *Organization Science*, 17: 615–631.

Pelled, Lisa 1996b "Relational demography and perceptions of group conflict and performance: A field investigation." *International Journal of Conflict Resolution*, 7: 230–246.

Perret-Clermont, A., Jean-Francois Perret, and Nancy Bell 1991 "The social construction of meaning and cognitive activity in elementary school children." In Lauren B. Resnick, John M. Levine, and S. D. Teasley (eds.), *Perspectives on Socially-Shared Cognition*: 41–62. Washington, DC: American Psychological Association.

Perrow, Charles 1970 *Organizational Analysis: A Sociological View*. Belmont, CA: Brooks/Cole.

Pervin, Lawrence A., and Donald B. Rubin 1967 "Student dissatisfaction with college and the college drop-out: A transactional approach." *Journal of Social Psychology*, 72: 285–295.

Peterson, Donald R. 1983 "Conflict." In Harold H. Kelley et al. (eds.), *Close Relationships*: 360–396. New York: W.H. Freeman.

Pruitt, Dean G., and Jeffrey Z. Rubin 1986 *Social Conflict: Escalation, Stalemate, and Settlement*. New York: Random House.

Putnam, Linda 1994 "Productive conflict: Negotiation as implicit coordination." *International Journal of Conflict Management*, 5: 285–299.

Ross, Raymond 1989 "Conflict." In R. Ross and J. Ross (eds.), *Small Groups in Organizational Settings*: 139–178. Englewood Cliffs, NJ: Prentice-Hall.

Rousseau, Denise 1985 "Issues of level in organizational research." In L. L. Cummings and Barry M. Staw (eds.), *Research in Organizational Behavior*, 7: 1–37. Greenwich, CT: JAI Press.

Saavedra, Richard, P. Christopher Earley, Linn Van Dyne, and Cynthia Lee 1993 "Complex interdependence in task-performing groups." *Journal of Applied Psychology*, 71: 61–72.

Salancik, Gerald R., and Jeffrey Pfeffer 1977 "A social information processing approach to job attitudes and task design." *Administrative Science Quarterly*, 23: 224–253.

Schwenk, Charles, and Joseph S. Valacich 1994 "Effects of devil's advocacy and dialectical inquiry on individuals versus groups." *Organizational Behavior and Human Decision Processes*, 59: 210–222.

Shah, P. P., and Karen A. Jehn 1993 "Do friends peform better than acquaintances: The interaction of friendship, conflict, and task." *Group Decision and Negotiation*, 2: 149–166.

Simons, T., and R. Peterson 1999 "Task conflict and relationship conflict in top management teams: The pivotal role of intragroup trust." *Journal of Applied Psychology* (in press).

Stasser, Garold 1992 "Information salience and the discovery of hidden profiles by decision-making groups: A 'thought experiment.'" *Organizational Behavior and Human Decision Processes*, 52: 156–181.

Stasser, Garold, and William Titus 1985 "Pooling of unshared information in group decision making: Biased information sampling during discussion." *Journal of Personality and Social Psychology*, 48: 1467–1478.

Stasser, Garold, and William Titus 1987 "Effects of information load and percentage of shared information on the dissemination of unshared information during group discussion." *Journal of Personality and Social Psychology*, 53: 81–93.

Stein, D. D., J. E. Hardyck, and M. B. Smith 1965 "Race and belief: An open and shut case." *Journal of Personality and Social Psychology*, 1: 21–289.

Steiner, Ivan D. 1972 *Group Process and Productivity*. San Diego: Academic Press.

Tajfel, Henri, and J. C. Turner 1986 "The social identity theory of intergroup behavior." In S. Worchel and W. G. Austin (eds.), *Psychology of Intergroup Relations*: 33–47. Chicago: Nelson-Hall.

Teachman, J. D. 1980 "Analysis of population diversity." *Sociological Methods and Research*, 8: 341–362.

Tenbrunsel, Ann E., Kim A. Wade-Benzoni, Joseph Moag, and Max H. Bazerman 1994 "The costs of friendship: The effects of strong ties on matching efficiency, individual effectiveness and barriers to learning." Unpublished manuscript, Kellogg Graduate School of Management (DRRC), Northwestern University.

Thibaut, John W., and Hal H. Kelley 1959 *The Social Psychology of Groups*. New York: Wiley.

Thompson, James D. 1967 *Organizations in Action*. New York: McGraw-Hill.

Tjosvold, Dean 1991 *The Conflict Positive Organization*. Reading, MA: Addison-Wesley.

Tsui, Anne S., and Charles A. O'Reilly 1989 "Beyond simple demographic effects: The importance of relational demography in superior-subordinate dyads." *Academy of Management Journal*, 32: 402–423.

Van de Ven, Andrew H., Andre Delbecq, and R. Koenig 1976 "Determinants of coordination modes within organizations." *American Sociological Review*, 41: 322–338.

Van de Ven, Andrew H., and Diane Ferry 1980 *Measuring and Assessing Organizations*. New York: Wiley.

Van Maanen, John, and Edgar Schein 1979 "Toward a theory of organizational socialization." In L. L. Cummings and B. M. Staw (eds.), *Research in Organizational Behavior*, 1: 209–264. Greenwich, CT: JAI Press,

Walton, Richard E., and John M. Dutton 1969 "The management of interdepartment conflict: A model and review." *Administrative Science Quarterly*, 14: 73–84.

Williams, Katherine Y., and Charles A. O'Reilly 1998 "Demography and diversity in organizations." In Barry M. Staw and Robert M. Sutton (eds.), *Research in Organizational Behavior*, 20: 77–140. Stamford, CT: JAI Press.

53

Exploring the Black Box: An Analysis of Work Group Diversity, Conflict, and Performance

Lisa Hope Pelled, Kathleen M. Eisenhardt and Katherine R. Xin

Source: *Administrative Science Quarterly* 44 (1) (1999): 1–28.

If we cannot now end our differences, at least we can help make the world safe for diversity.
— John F. Kennedy

In the past decade, demographic diversity has become one of the foremost topics of interest to managers and management scholars. The term demographic diversity refers to the degree to which a unit (e.g., a work group or organization) is heterogeneous with respect to demographic attributes. Attributes classified as demographic generally include "immutable characteristics such as age, gender, and ethnicity; attributes that describe individuals' relationships with organizations, such as organizational tenure or functional area; and attributes that identify individuals' positions within society, such as marital status" (Lawrence, 1997: 11). The heightened concern with demographic diversity (hereafter referred to simply as diversity) stems not only from the growing presence of women and minorities in the work force (Buhler, 1997) but also from modern organizational strategies that require more interaction among employees of different functional backgrounds (e.g., Dean and Snell, 1991). One of the most significant bodies of research to arise from this trend is a stream of field studies linking group composition to cognitive task performance — i.e., performance on tasks that involve generating plans or creative ideas, solving problems, or making decisions. The impact of diversity on cognitive task performance has been examined in studies of top management teams (e.g., Bantel and Jackson, 1989; Murray, 1989; Eisenhardt and Schoonhoven, 1990) and lower-level work groups (e.g., Kent and McGrath, 1969; Murnighan and Conlon, 1991).

Despite this spotlight on diversity in work groups, there is more to be done. Investigations of diversity and work group performance have largely been what Lawrence (1997) referred to as "black box" studies, which do not measure intervening process variables. Further, the effects on performance are still unclear. Some studies (e.g., Bantel and Jackson, 1989) have linked diversity to favorable performance on cognitive tasks, and some (e.g., Murnighan and Conlon, 1991) have linked it to unfavorable performance on such tasks. Others (e.g., Watson, Kumar, and Michaelson, 1993) have shown that group diversity both enhances and diminishes cognitive task performance. To capture fully the complex relationship between work group diversity and performance, we need more sophisticated theories and empirical work incorporating intervening variables and multiple types of diversity. The objective of the current investigation is to begin to meet these needs, offering an intervening process theory — one that attempts to untangle the complicated set of relationships among five types of diversity and performance — and providing a test of that theory.

Two prior studies that have empirically assessed whether process variables intervene between group diversity and performance are particularly important to our efforts. Ancona and Caldwell (1992) examined the intervening role of internal task process (i.e., the setting of goals and priorities) and external communication. Later, Smith et al.'s (1994) top management team study looked at three potential intervening variables (social integration, informality of communication, and communication frequency). The authors of both studies discovered that the process variables they measured did not fully explain the observed effects of diversity on performance, and both then suggested that the mediating effect of conflict should be assessed in future research. Hence, in the model we propose and test here, conflict plays an intervening role.

The model proposes that work group diversity indirectly affects cognitive task performance through two kinds of conflict: intragroup task conflict and intragroup emotional conflict. Task conflict is a condition in which group members disagree about task issues, including goals, key decision areas, procedures, and the appropriate choice for action, and emotional conflict is a condition in which group members have interpersonal clashes characterized by anger, frustration, and other negative feelings (Jehn, 1994; Eisenhardt, Kahwajy, and Bourgeois, 1997a).[1] We suggest that job-related types of diversity largely drive task conflict. In contrast, emotional conflict is shaped by a complex web of diversity types that increase emotional conflict based on stereotyping and decrease emotional conflict based on social comparison. Task routineness and group longevity moderate these diversity-conflict relationships. Each type of diversity indirectly affects performance via its relationship with conflict: task conflict tends to enhance performance, while emotional conflict tends to diminish performance. Thus, we offer a model that postulates that the black box between diversity and performance contains a more elaborate set of relationships than previously thought.

Theoretical Background and Hypotheses

Link between Diversity and Task Conflict

When the members of a work group have different demographic backgrounds, they may have dissimilar belief structures (Wiersema and Bantel, 1992), i.e., priorities, assumptions about future events, and understandings of alternatives (Hambrick and Mason, 1984: 195), based on previous training and experiences. As Eisenhardt, Kahwajy, and Bourgeois (1997b: 48) recently noted, executives "who have grown up in sales and marketing typically see opportunities and issues from vantage points that differ from those who have primarily engineering experience." Such distinct perspectives may stem, in part, from resource allocation and reward disparities (Donnellon, 1993), which encourage contrasting views of what is important. Due to their respective belief structures, group members with different demographic backgrounds may have divergent preferences and interpretations of tasks (Dearborn and Simon, 1958; Walsh, 1988; Waller, Huber, and Glick, 1995). These divergences are likely to manifest themselves as intragroup task conflict. As diversity within a work group increases, such task conflict is likely to increase. Increased diversity generally means there is a greater probability that individual exchanges will be with dissimilar others. Members are more likely to hear views that diverge from their own, so intragroup task conflict may become more pronounced.

While any type of diversity may trigger task conflict, some are more likely to do so than others, based on the relevance of their corresponding belief structures. People hold multiple belief structures about a variety of information domains (Walsh, 1988); those belief structures most relevant to the information processing task at hand tend to influence interpretation of that task (Wickens, 1989; Waller, Huber, and Glick, 1995). Thus, demographic attributes corresponding to highly relevant belief structures should be especially influential in the perception of work group tasks. The job-relatedness of a demographic attribute is the degree to which that attribute captures experiences and skills germane to cognitive tasks at work (Pelled, 1996; Pelled, Cummings, and Kizilos, 1999). If work group members differ with respect to a demographic attribute that is low in job-relatedness, then their divergent experiences and knowledge may not pertain to the work they do, and opposing task perceptions may not emerge in the group. If work group members differ with respect to a highly job-related demographic attribute, however, then their divergent experiences and knowledge are apt to be pertinent to the task, and incongruent task perceptions are likely to emerge. Diversity with respect to highly job-related attributes is therefore apt to have a stronger relationship with task conflict than is diversity with respect to less job-related attributes.

Functional background and tenure are highly job-related attributes. Both are defined by one's workplace experiences, specifically, whether one is exposed to a particular functional area and how much time one has worked for a company. Also, cognitive tasks in organizations typically demand the experience and knowledge obtained through exposure to functional areas and organizational tenure. Ancona and Caldwell (1992) noted that for tasks such as those of product development teams, functional background and company tenure are likely to be particularly important because they determine one's technical skills, information, expertise, and one's perspective on an organization's history. Others (Sessa and Jackson, 1995; Milliken and Martins, 1996; Pelled, 1996) have similarly argued that functional background and tenure are especially pertinent to work group tasks.

Age, gender, and race, in contrast, are low in job-relatedness. In a recent editorial, one scholar even went as far as to argue, "there is no such thing as a woman's approach to mathematics or an African American approach to physics" (Heriot, 1996: M5). A less extreme and perhaps more realistic assertion is that of Zenger and Lawrence (1989: 357): "Although age similarity may produce similarity in general attitudes about work . . ., such attitudinal similarity is unlikely to have much direct bearing on conversations about technical work." The same logic applies to race and gender. For example, the attribute *race* tends to capture a broad collection of experiences, such as traditions followed, treatment received from teachers, and clubs joined. Work experiences may only be a fraction of the total set of experiences it captures (Pelled, Ledford, and Mohrman, 1998). Sessa and Jackson (1995: 137) have observed that race, gender, and age "form the context of more general social relationships" and, compared with tenure and department membership, are less directly associated with team objectives. Functional background and tenure, then, are apt to have a stronger impact on perceptions of work group tasks than are race, gender, and age:

> Hypothesis 1 (H1): Functional background and tenure diversity will have stronger positive associations with intragroup task conflict than will diversity in age, gender, and race.

Link between Diversity and Emotional Conflict

While task conflict is largely shaped by the job-relatedness of diversity, emotional conflict is shaped by a more complex set of forces. One key factor is categorization, the subconscious tendency of individuals to sort each other into social categories, often on the basis of demographic attributes (Tajfel et al., 1971; Tajfel, 1972, 1982). Because there is an abundance of information about people and things in our environment, categorization is a useful way to simplify and "make our perceived world more predictable and controllable" (Zimbardo and Leippe, 1991: 236). Once categorization takes place, people strive for self-esteem by developing positive opinions of their own category and negative opinions of other categories (Turner, 1975; Tajfel, 1978). They perceive members of their own social category as superior and engage in stereotyping, distancing, and disparaging members of other categories (Tajfel, 1982). Members of other social categories, in turn, resent such stereotyping and disparaging treatment, and hostile interactions erupt in the group (Reardon, 1995). These hostile interactions constitute emotional conflict, clashes characterized by anger, resentment, and other negative feelings. As diversity within a work group increases, individuals generally will have more exchanges with those in different social categories. People in different social categories will be directly confronted with each other's negative stereotypes and self-serving biases, and emotional conflict may become more pronounced.

Although any kind of diversity may provoke categorization — and thus emotional conflict — in this manner, some kinds have a greater tendency to do so than others, depending on the permeability of their defining demographic attributes. The permeability of an attribute is the degree to which that attribute can be altered, moving a person from one social category to another social category. Diversity based on relatively impermeable attributes is particularly likely to yield intercategory clashes. When attributes are not easily penetrated, it is difficult for employees to "stand in the shoes" of those in another social category. Consequently, employees feel especially polarized — and are therefore especially likely to stereotype — members of another category (Nelson, 1989; Kramer, 1991).

Race, gender, age, and tenure are not easily permeated. A person cannot change his or her race or gender and must wait a period of time for noteworthy increases in age and tenure. Also, a person can never regress to a younger age or a lesser amount of tenure in the same company. Functional background, in contrast, is more permeable. Employees often can transfer from one functional area to another if they simply want exposure to different areas or if the conditions in another area are better, and many companies "rotate employees in and out of both technical and business-oriented positions to help them round out their skills" (Ryan, 1991: 76). Hence, employees may find it easier to identify with those of a different functional background than with those of a different race, gender, age, or tenure. Thus, categorization theory leads to the following prediction:

> Hypothesis 2a (H2a): Diversity in race, gender, age, and tenure will have stronger positive associations with intragroup emotional conflict than will diversity in functional background.

The previous hypothesis, derived from categorization theory, suggests that diversity tends to increase emotional conflict, but it is also possible that diversity diminishes emotional conflict. Festinger's (1954) social comparison theory suggests this competing prediction.[2] According to Festinger, humans have an innate tendency to evaluate themselves and their qualities — e.g., their opinions, abilities, and progress. Festinger contended that when objective information is unavailable, people prefer to compare themselves with similar others. This preference exists because comparison with a similar other is more meaningful and informative than comparison with someone who is very different. Festinger further asserted that, in Western culture, people feel a pressure to improve their abilities or other qualities continually, and, as a result, they strive to be slightly better than the targets of their comparisons. This process, according to Festinger, leads to competition among similar others.

In his original formulation of social comparison theory, Festinger used the term similarity to refer to similarity on the ability or other quality being evaluated. Over the years, however, researchers have found that people also compare themselves with those who are similar on other dimensions, such as attractiveness or gender (e.g., Miller, 1982; Major and Forcey, 1985). Based on social comparison theory, then, we can expect demographic similarity to be associated with rivalry or professional competition in work groups. Feelings of jealousy and hostility may accompany competitive, rivalrous interactions (Tjosvold, 1991). Hence, demographic similarity (group homogeneity) may ultimately be linked to intragroup emotional conflict. As a work group becomes more homogeneous (i.e., as diversity decreases), individuals generally have more exchanges with similar others. Employees may be confronted with more social comparisons that precipitate rivalrous clashes, and emotional conflict in the group may become more pronounced.

While any kind of similarity may trigger social comparison — and thus emotional conflict — in the work context, some kinds of similarity are stronger triggers than others, depending on the career-relatedness of their defining demographic attributes. The career-relatedness of an attribute is the degree to which that attribute is considered in formal and informal assessments of career progress. In a work context, people are especially attuned to career achievements, rather than outside accomplishments. Hence, when employees engage in social comparison at work, they primarily look at those demographic attributes tied to career progress evaluations. Similarity with respect to highly career-related attributes is therefore particularly likely to yield the jealous rivalry that characterizes emotional conflict.

Age, tenure, and functional background are highly career-related. There are powerful age norms encouraging employees to expect that career progress comes over time, with age (Lawrence, 1988). Hence, workers may view people who are similar in age — people at the same stage in life — as yardsticks with which to measure their own career progress, and they may be concerned about falling behind (e.g., not being as successful or powerful as) those persons. For example, a group member who is 25 years old is apt to be more concerned about a 27-year-old "shining" in the group than about a 50-year-old shining. By the same token, there are implicit career timetables and expectations for know-how associated with tenure, so employees are inclined to look to others of the same tenure to see who has achieved greater recognition, acquired more expertise, or made more career progress in other ways. Additionally, since formal evaluations of individual employees typically compare employees in the same functional area or department (Kirkpatrick, 1986), group members may be especially inclined to focus on persons from the same functional area when making social comparisons.

Race and gender, in contrast, are less highlighted in informal and formal assessments of career progress. For example, a female employee will not necessarily expect the same degree of influence and recognition as another female employee simply because both are women; each may bring a different degree of experience to a group. Moreover, comparisons based on gender and race similarity are, in general, not an explicit part of formal evaluations. Thus, consideration of social comparison processes leads to the following expectation:

> Hypothesis 2b (H2b): Diversity in age, tenure, and functional background will have stronger negative associations with intragroup emotional conflict than will diversity in race and gender.

Link between Task and Emotional Conflict

Researchers have previously suggested that task and emotional conflict may influence each other. As Ross (1989: 140) observed, it is possible for task-related disagreements "to generate emotionally harsh language, which can be taken personally. We then have both task and psychological conflicts occurring at the same time." Group members may feel strongly that their views on a particular issue are correct, and they may show impatience or intolerance when others express different views. Moreover, members whose ideas are disputed may feel that others in the group do not respect their judgment. Tjosvold (1991) observed that group members sometimes assume their competence is being challenged when their ideas are criticized. Conversely, emotional conflict may sometimes lead to task conflict. Individuals who feel frustrated or angry with other members of their group may have a propensity to dispute the ideas of those other members, for angry people are generally less compliant and agreeable than those who are cheerful (Milberg and Clark, 1988). Also, there may be a negative halo effect, such that when one feels irritated by or hostile toward another person, one is more inclined to find fault with that person's ideas. In their multiple case study of top management teams, Eisenhardt and Bourgeois (1988) concluded that executives who were engaged in political infighting tended to have distorted perceptions of each other's ideas. Even though such conflict is affective in origin, it may evolve into substantive debate:

> Hypothesis 3 (H3): There will be a positive association between task conflict and emotional conflict in work groups.

Moderators of Diversity-Conflict Linkages

Jackson (1992: 155) observed that an important but as-yet-unanswered question is, "Does the nature of the task moderate the impact of group composition?" In addressing this query, a key task feature that warrants attention is task routineness, the extent to which a task has low information processing requirements, set procedures, and stability (Van de Ven, Delbecq, and Koenig, 1976; Gladstein, 1984; Jehn, 1995). This feature is especially relevant because it determines the richness of information required for a group's task, that is, whether a group needs to draw on different knowledge bases.

There are two possible effects that task routineness can have on the positive association between diversity and task conflict. One possibility is that task routineness diminishes the association between diversity and task conflict. When tasks are well-defined and straightforward, group members have little need to exchange opinions or challenge each other. Hence, in groups with routine tasks, even if members have diverse backgrounds, there is only minimal room for task conflict based on those backgrounds. In groups with nonroutine tasks, there is more room for task conflict, so group members with diverse backgrounds are more likely to exchange opposing opinions and preferences derived from their backgrounds. The tendency for diversity to trigger task conflict may therefore be heightened by task nonroutineness, or diminished by task routineness:

> Hypothesis 4a (H4a): Task routineness will reduce the positive associations between diversity variables and task conflict in work groups.

An alternative possibility is that task routineness will have the opposite effect. Early theories of optimal arousal (Hebb, 1955; Fiske and Maddi, 1961; Berlyne, 1967) suggested that people have a preferred level of arousal, a preference for stimulation that is neither too low nor too high. Drawing on these theories, Zuckerman (1979) postulated that people engage in behaviors that decrease stimulus input when their optimal level of arousal is exceeded and increase stimulus input when they are underaroused. Empirical evidence, including results of sensory deprivation research (e.g., Vernon and McGill, 1960) and Zuckerman's own studies (1979, 1984), is consistent with this notion. An implication of Zuckerman's theory, referred to as sensation-seeking theory, is that people who are understimulated will seek experiences and interactions that offer them greater arousal. Since group members performing routine tasks may experience suboptimal levels of stimulation, they may seek opportunities to debate about their tasks to make their work more exciting. More specifically, they may elicit opposing task perspectives from people with different backgrounds, hoping to engage in cognitively stimulating discourse. In contrast, group members performing nonroutine tasks may be sufficiently aroused by the group's task and may be less motivated to draw out additional task conflict for the sake of excitement:

> Hypothesis 4b (H4b): Task routineness will enhance the positive associations between diversity variables and task conflict in work groups.

Sensation-seeking theory is not likely to apply in the case of emotional conflict. When tasks are routine, group members are unlikely to seek emotional conflicts with people of different backgrounds, for researchers have found that people typically do not seek unpleasurable arousal when understimulated (Zuckerman, 1979; Gallagher, Diener, and Larsen, 1989). Still, task routineness may influence the relationship between diversity and emotional conflict. People performing complex tasks may be more anxious and, consequently, rely more heavily on cognitive mechanisms for simplifying information processing (Staw, Sandelands, and Dutton, 1981). In contrast, when tasks are routine, people have less need for such cognitive mechanisms (e.g., categorization). Thus, the tendency for diversity to trigger categorization and, ultimately, emotional conflict will be weaker when tasks are routine.

Research on displaced aggression also suggests that the association between diversity and emotional conflict will be weaker when tasks are routine. Studies have shown that frustrating work conditions lead to more interpersonal aggression among employees (Storms and Spector, 1987; Chen and Spector, 1992). Because routine tasks tend to be less frustrating than complex tasks, members of groups with routine tasks may have less frustration to vent and, consequently, less inclination to blame or "pick on" people of different backgrounds, compared with members of groups with complex tasks:

> Hypothesis 5 (H5): Task routineness will reduce the positive associations between diversity variables and emotional conflict in work groups.

In addition to task routineness, a second moderator likely to operate on diversity-conflict relationships is group longevity, the length of time group members have spent working together (Katz, 1982). After a period of time, group members may become familiar with the different perspectives in a diverse group. If a group member has an idea, he or she may be able to anticipate other members' criticisms and, consequently, either frame the idea to make it more acceptable or avoid expressing the idea altogether. Alternatively, through informational social influence, group members may begin to share each other's perspectives, arriving at a common understanding of the group's tasks (Katz, 1982). In this manner, group longevity may diminish any tendency for diversity to trigger task conflict.

Group longevity may also weaken any positive associations between work group diversity and emotional conflict, for social categories based on demographic attributes may eventually become blurred. The boundaries of the "in" category and the "out" category may change, so that individuals who were once considered outsiders become insiders (Kramer, 1991). Over time, as members grow accustomed to being in the same work group, the perceived "in" category is apt to become the whole work group, while other work groups are perceived as "out" categories. Since people have a limited focus of attention (Kahneman, 1973; Fiske and Taylor, 1991), their enhanced focus on the group as a category is likely to diminish their focus on demographic categories. Group longevity may therefore moderate any tendency of diversity to yield emotional conflict in work groups:

> Hypothesis 6 (H6): Group longevity will diminish the positive associations between diversity variables and conflict in work groups.

Task and Emotional Conflict as Mediators: Their Links to Performance

The task and emotional conflict triggered by a group's diversity may, in turn, affect the cognitive task performance of the group, although the mediating roles of the two types of conflict are apt to differ. The task conflict that diversity yields is likely to enhance group performance on cognitive tasks. Exposure to opposing points of view encourages group members to gather new data, delve into issues more deeply, and develop a more complete understanding of problems and alternative solutions (Tjosvold, 1986). Also, the constructive criticism associated with task conflict can facilitate vigilant problem solving, an approach that Janis (1989) recommended for making important decisions. If group members fail to criticize each other's ideas because they are too concerned about maintaining unanimity, they may overlook important details, succumbing to "groupthink" (Janis, 1982).

Previous case study and empirical findings support the notion that task conflict enhances cognitive task performance. Hoffman and Maier (1961) found in a lab study that groups with conflicting opinions produced better solutions to standardized sets of problems. Later, using a sample of student groups, Jehn (1994) showed that task conflict was positively associated with group performance on a class project. In a multiple case study, Eisenhardt, Kahwajy, and Bourgeois (1997b) found that top management teams in high-performing firms had higher task conflict than teams in low-performing firms. Amason's (1996) large-scale study of top management teams was consistent with this pattern, revealing that task conflict was positively associated with decision quality. Thus,

> Hypothesis 7 (H7): Task conflict will have a positive association with the cognitive task performance of work groups.

In contrast to task conflict, the emotional conflict yielded by diversity is likely to impair the cognitive task performance of work groups. First, since anxiety (an emotion that characterizes emotional conflict) often leads to cognitive interference (Sarason, 1984), group members may not take relevant information into consideration when solving problems. Second, the hostility that characterizes affective conflict may make group members reluctant to share or listen to each other's potentially useful ideas or information. Third, when there is emotional conflict, group members are likely to consume time and energy making — or defending themselves against — personal attacks; as a result, they may have little remaining time and energy to devote to critical task-related matters (Evan, 1965; Jehn, 1994).

The results of prior research are consistent with the notion that emotional conflict impairs cognitive task performance. Evan's (1965) study of research and development teams suggested that interpersonal attacks diminished team productivity. Also, Jehn's (1994) above-mentioned study of student groups showed that intragroup emotional conflict was negatively associated with group performance on a class project. Similarly, Amason's (1996) study of top management teams revealed that emotional conflict, which he called affective conflict, was negatively associated with decision quality. Eisenhardt, Kahwajy, and Bourgeois (1997b), too, found that emotional conflict impaired team process and firm performance. Thus,

> Hypothesis 8 (H8): Emotional conflict will have a negative association with the cognitive task performance of work groups.

The above eight hypotheses constitute an intervening process theory of work group diversity, conflict, and performance. Below we describe the field study conducted to test our hypotheses.

Methods

Sample

Participants in this study included the members of 45 teams from the electronics divisions of three major corporations. The teams were involved in monitoring and modifying work processes with the objective of improving those processes, and often they were also involved in the design of new products. All teams were assembled to complete lengthy but time-limited projects, and many of the teams were cross-functional, including research and development (R&D) and manufacturing representatives within their functional mixtures. They also were engaged in cognitive tasks that, according to team members' assessments, ranged from moderate to high complexity.

We obtained team performance ratings from 41 out of 45 team managers and received completed questionnaires from members of all 45 teams. Although the number of teams is modest by some standards, it compares well with other field studies and, as a field study, yields valuable insights not attainable in the laboratory. A total of 443 team-member questionnaires were distributed, and 317 were returned. On average, 73 percent of the members of a team returned completed questionnaires. The teams in our sample had an average size of approximately 10 members (s.d. = 3.2). The average team had a mean age of 38.5 years (s.d. = 5.5) and a mean tenure of 10.6 years (s.d. = 4.9). Also, the average percentage of nonwhites on a team was 19 percent (s.d. = 21 percent); the average percentage of women on a team was 25 percent (s.d. = 18 percent); and the average R&D representation on a team was 44 percent (s.d. = 33 percent).

Measures

Diversity

This study included two types of work group diversity indices, one for numeric demographic data and another for categorical demographic data. The team-member questionnaire was the source of demographic data used to compute these indices. Following an approach recommended by Allison (1978) for numeric variables, we used the coefficient of variation (standard deviation divided by the mean) to measure age diversity and tenure diversity. Thus, to assess age diversity within teams, we divided each team's standard deviation of age by the team's mean age. Similarly, we assessed company tenure diversity within teams by dividing each team's standard deviation of tenure by the team's mean tenure.

To measure team diversity with respect to categorical variables (functional background, gender, and race), we used an index recommended by Teachman (1980):

$$H = -\sum_{I=1}^{I} P_i(\ln P_i).$$

The index takes into account how work group members are distributed among the possible categories of a variable. The total number of categories of a variable equals I, and P_i is the fraction of team members falling into category I. For example, the gender variable has two possible categories (I = 2): 1 corresponds to a woman and 2 to a man. If a given team of ten members has three women and seven men, then P_1 equals .3, P_2 equals .7, and H equals .61. If a team of ten members has one woman and nine men, then P_1 equals .1, P_2 equals .9, and H equals .32. As Ancona and Caldwell (1992: 328) noted, "The only exception occurs when [a category] is not represented." In such a case, one cannot set P_i equal to zero, for the natural logarithm of zero does not exist; thus, one would only use the P_i values for the other categories to compute H.

Conflict

The task conflict scale ($\alpha = .78$) comprised four questionnaire items based on Jehn's (1994) measure of task conflict, and the emotional conflict scale ($\alpha = .83$) comprised four questionnaire items adapted from Jehn's (1994) emotional conflict scale. Each item measured conflict on a 5-point Likert scale. Group-level indices of task and emotional conflict were formed by averaging individual-level indices of task and emotional conflict. Aggregation to the group level was justified by an eta square of .33 for task conflict ($F = 3.03$, $p < .001$) and an eta square of .32 for emotional conflict ($F = 2.92$, $p < .001$). In general, an eta square greater than .20 indicates that any two people within the same group are more similar in their responses than two people who are members of different groups (Georgopoulos, 1986; Florin et al., 1990).

Moderators

The task routineness measure ($\alpha = .62$) comprised three items drawn from Gladstein's (1984) task complexity measure, reverse-scored on a 5-point Likert scale. Aggregation to the group level was justified by an eta square of .31 ($F = 2.72$, $p < .001$). To ensure the discriminant validity of the task conflict, emotional conflict, and task routineness items, we conducted a confirmatory factor analysis (CFA). Table 1 shows the results of this analysis. The CFA, which used the maximum likelihood method, produced a chi square of 46.90 with 41 degrees of freedom (i.e., a chi square of 1.14 per degree of freedom). The goodness-of-fit index is .98, and the root-mean-square residual is .035. These figures indicate that there is a good fit between the data and the theoretical factor structure (Wheaton, Alwing, and Summers, 1977).

Our group longevity measure, following previous studies (Katz, 1982; Smith et al., 1994), was the average length of time the members of a team had belonged to that team. A team with a higher average has a longer history of working together.

Performance

Using 5-point Likert scales, each team's manager rated the team on two dimensions from Ancona and Caldwell's (1992) measure of manager-rated team performance: efficiency of team operations and number of innovations or new ideas introduced by the team. The group performance measure comprised these items ($\alpha = .61$). Principal components analysis revealed that the items constituted

Table 1: Results of Confirmatory Factor Analysis of Task Routineness and Intragroup Conflict Items

Item	Factor 1 Task conflict	Factor 2 Relationship conflict	Factor 3 Task routineness
1. To what extent are there differences of opinion in your team?	.68		
2. How often do the members of your team disagree about how things should be done?	.83		
3. How often do the members of your team disagree about which procedure should be used to do your work?	.76		
4. To what extent are the arguments in your team task-related?	.57		
5. How much are personality clashes evident in your team?		.80	
6. How much tension is there among the members of your team?		.83	
7. How often do people get angry while working in your team?		.62	
8. How much jealousy or rivalry is there among the members of your team?		.70	
9. The technology, required skills, and information needed by the team are constantly changing. [reverse-scored]			.67
10. During a normal work week, exceptions frequently arise that require substantially different methods or procedures for the team. [reverse-scored]			.79
11. Frequent interaction between team members is needed to do our work effectively. [reverse-scored]			.47

a single factor accounting for 74 percent of the variance in responses. Each item in the scale had a factor loading of .86.

When forming each multi-item measure in this study (i.e., task conflict, emotional conflict, task routineness, and performance), we combined items by using factor scores as weights and computed a weighted sum of scale items, as described by Pedhazur and Schmelkin (1991: 125). Using factor scores in this manner creates a more accurate measure than simply computing a mean, which assigns equal weights to items. This procedure was feasible because respondents answered all of the items in a scale.

Controls

Group size was a control variable in our study because the literature on groups has noted that size is a key variable influencing group dynamics and performance (Brewer and Kramer, 1986) and because larger teams have more potential for heterogeneity (Bantel and Jackson, 1989; Jackson et al., 1991). We also controlled for site differences in our regressions. In particular, Site B had a distinct culture. It was more conservative and bureaucratic than the other two sites, offered less training in team skills, and was heavily connected to the declining defense industry. We controlled for group longevity and task routineness (the moderators described above) because previous research has found that the average tenure of group members and the nature of group tasks often influence group interactions and performance (e.g., Katz, 1982; Weingart, 1992). Also, when testing for moderating effects, it is necessary to control for the main effect of the moderating variable (Baron and Kenny, 1986).

Data Analysis

We tested the hypotheses using seemingly unrelated regression (SURE) and ordinary least squares regression. When evaluating the significance of most predicted effects (all except those predicted in competing hypotheses), we used one-tailed tests, which are suitable for directional hypotheses (Erickson and Nosanchuk, 1977; Wonnacott and Wonnacott, 1984). LISREL was not an alternative for this study because it was inappropriate for our sample size (Breckler, 1990).

Hypotheses 1 through 6 (with the exception of hypothesis 3) were tested with SURE analyses. SURE is a statistical technique that solves a set of regression equations simultaneously and allows for error covariances among the equations (Zellner, 1962; Parker and Dolich, 1986; Ghosh, 1991). It is appropriate to use the technique in this study because the predictors in the equations with emotional conflict as a dependent variable have considerable overlap with the predictors in the equations with task conflict as a dependent variable. Also, we expected that task and emotional conflict would be significantly correlated. Each SURE analysis involved two equations, one that had task conflict as the dependent variable and another that had emotional conflict as the dependent variable.

Our first SURE analysis only included control variables as predictors. In the second SURE analysis, we included both diversity and control variables as predictors. In the third SURE analysis, we added interaction terms, having mean centered the interaction term variables to reduce potential multicollinearity effects. We used the likelihood ratio test (Kennedy, 1979; Ghosh, 1991) to determine the goodness of fit of the models. For the likelihood ratio test, the criterion is as follows:

$$- 2[\ln(\text{likelihood function}_{\text{constrained model}}) - \ln(\text{likelihood function}_{\text{unconstrained model}})].$$

The criterion has a χ^2 distribution with degrees of freedom equal to the number of constraints.

To test H3, which suggested that task and emotional conflict would be positively associated, we examined the zero-order correlation between the two conflict measures. We also performed an OLS regression (not shown but available from the authors), using the approach that Smith et al. (1994) used in their study of demography, group process, and performance when assessing the effect of one intervening process variable (informality of communication) on another intervening process variable (social integration). They treated social integration as an "intervening dependent variable," while "all team demography and control variables, plus informality of communication were treated as independent variables" (1994: 429). Here, we assessed the effect of emotional conflict on task conflict by regressing task conflict on the diversity variables, controls, and emotional conflict. Similarly, we assessed the effect of task conflict on emotional conflict by regressing emotional conflict on the diversity variables, controls, and task conflict.

When testing H7 and H8, which posited that task and emotional conflict would act as mediators between diversity and group performance, we followed a standard procedure used to test for mediating effects, "first, regressing the mediator on the independent variable; second, regressing the dependent variable on the independent variable; and third, regressing the dependent variable on both the independent variable and on the mediator . . ." (Baron and Kenny, 1986: 1177). The independent variables were the diversity variables in this study, and the proposed mediating variables were task and emotional conflict. The dependent variable was group performance. After running the regressions, we examined the results to see if a mediating effect was present. Pure mediation calls for the following conditions (Baron and Kenny, 1986): First, the independent (diversity) variables should affect the mediator (task or emotional conflict) in the first equation. Second, the independent variables should affect the dependent variable (performance) in the second equation. Third, the mediator (task or emotional conflict) should affect the dependent variable in the third equation, and the effect of the independent variables on the dependent variable should be less in the third equation than in the second equation.

Results

Table 2 shows the means, standard deviations, and correlations among all predictor, outcome, and control variables.

We performed several checks on the correlational properties of the data before testing our hypotheses. First, we reviewed the correlations among the independent variables shown in table 2. The median correlation magnitude (absolute value) was .19, and the correlation with the greatest magnitude was .48. As noted by Tsui et al. (1995: 1531), "There is no definitive criterion for the level of correlation that constitutes a serious multicollinearity problem. The general rule of thumb is that it should not exceed .75." Similarly, Kennedy (1979) indicated that correlations of .8 or higher are problematic. As a second check, we examined the variance inflation factor (VIF) of each independent variable. The largest VIF in our regressions was less than 3, a sign that multicollinearity was not a problem (Guo, Chumlea, and Cockram, 1996).

As described earlier, H1, 2a, and 2b were tested with SURE analysis involving two equations. Using the likelihood-ratio test, we determined that the goodness of fit of this two-equation model was $\chi^2 = 24.81$ (d.f. $= 10$; $p < .01$). Table 3

Table 2: Means, Standard Deviations, and Intercorrelations among Study Variables*

	Mean	S.D.	1	2	3	4	5	6	7	8	9	10	11
1. Group size	9.84	3.17	—										
2. Task routineness	5.31	.94	−.10	—									
3. Group longevity	.89	.71	.43	−.03	—								
4. Site B	—	—	−.04	.22	.08	—							
5. Emotional conflict	4.58	1.04	.35	−.43	.03	.09	—						
6. Task conflict	6.93	1.32	.28	−.23	.01	.32	.48	—					
7. Race diversity	.45	.42	−.18	.09	−.17	−.46	−.04	−.24	—				
8. Gender diversity	.48	.20	−.20	.26	−.28	.31	−.10	.03	.25	—			
9. Age diversity	.24	.08	−.14	.30	−.14	−.06	−.45	−.27	.19	.18	—		
10. Company tenure diversity	.72	.20	.18	−.06	.37	−.25	.27	−.01	−.08	−.35	−.18	—	
11. Functional background diversity	.64	.37	−.07	.22	−.24	−.07	−.21	.14	.26	.19	.33	−.16	—
12. Group performance	6.26	1.30	.16	.00	.40	−.30	−.07	.05	−.02	−.19	.09	.36	−.13

*Because the emotional conflict, task conflict, task routineness, and group performance measures were each formed by using factor scores to compute weighted sums of scale items, their means may exceed 5, even though their items had only 5 response anchors. All correlations above .19 are significant at $p < .10$, one-tailed tests. All correlations above .25 are significant at $p < .05$, one-tailed tests.

Table 3: Seemingly Unrelated Regression (SURE) Results: Equations with Task Conflict as Dependent Variable ($N=45$)†

	Model			
Independent variables	1	2	3	4
Controls				
Group size	.14**	.12**	.11**	.08
Site B	1.25***	1.34**	1.41***	1.28**
Task routineness	−.39**	−.40**	−.42**	−.52***
Group longevity	−.33	−.30	−.18	−.19
Predictors				
Race diversity (RD)		−.02	.15	.11
Gender diversity (GD)		−.04	.17	.50
Age diversity (AD)		−3.64	−3.46	−4.17*
Company tenure diversity (TD)		.64	.58	.61
Functional background diversity (FD)		.108**	.98**	1.01**
Interactions				
FD × Task routineness			.74*	
FD × Group longevity				−.153**
R-squared	.29	.39	.44	.42

*$p<.10$; **$p<.05$; ***$p<.01$.
†Entries are SURE coefficients. One-tailed tests were used for effects predicted in directional hypotheses.

presents the SURE equations with task conflict as the dependent variable. Model 2 tested H1, that functional background and tenure diversity would have positive associations with task conflict in work groups. This hypothesis was supported for functional background diversity, which had a significant positive relationship with task conflict (beta = 1.08, $p<.05$), while, consistent with the same hypothesis, diversity in race, gender, and age had nonsignificant associations with task conflict. Unexpectedly, the relationship between tenure diversity and task conflict was also nonsignificant. This suggests that functional background diversity is the key demographic driver of task conflict.

Table 4 presents the SURE equations with emotional conflict as the dependent variable. Model 2 tested H2a and 2b, predictions about diversity and emotional conflict based on categorization theory and social comparison theory, respectively. Consistent with H2a, both tenure diversity (beta = 1.76, $p<.01$) and race diversity (beta = .81, $p<.05$) had significant positive associations with emotional conflict, while functional background diversity did not. Consistent with H2b, age diversity had a significant negative association with emotional conflict (beta = −3.43, $p<.05$). Surprisingly, gender diversity was not significantly related to emotional conflict in either direction.

As a sensitivity analysis, we replaced each of our diversity measures for categorical variables (race, functional background, and gender) with an alternative measure. Instead of Teachman's index, we used Blau's (1977) heterogeneity index:

Table 4: Seemingly Unrelated Regression (SURE) Results: Equations with Emotional Conflict as Dependent Variable ($N = 45$)†

	Model			
Independent variables	1	2	3	4
Controls				
Group size	.12**	.13***	.12***	.13***
Site B	.49*	1.04***	1.20***	1.06***
Task routineness	−.47***	−.41***	−.48***	−.43***
Group longevity	−.22	−.49**	−.49***	−.52**
Predictors				
Race diversity (RD)		.81**	.89***	.76**
Gender diversity (GD)		−.27	−.19	−.36
Age diversity (AD)		−3.43**	−2.63*	−3.52**
Company tenure diversity (TD)		1.76***	1.74***	1.64***
Functional background diversity (FD)		−.23	−.38	−.12
Interactions				
RD × Task routineness			−1.12**	
TD × Task routineness			−.59**	
RD × Group longevity				−.70**
TD × Group longevity				−1.44*
R-squared	.32	.54	.60	.59

*$p < .10$; **$p < .05$; ***$p < .01$.
†Entries are SURE coefficients. One-tailed tests were used for effects predicted in directional hypotheses.

$$H = 1 - \sum_{i=1}^{I} (P_i)^2,$$

where P_i is the proportion of group members in category i, and I is the number of possible categories. For example, if a given team of ten members has three women and seven men, then P_1 equals .3, P_2 equals .7, and H equals .42. With this alternative measure, we still obtained the pattern of results described above.

Since the effect of gender diversity was nonsignificant, we conducted several additional sensitivity analyses to test the robustness of this result. We replaced our gender diversity measure with the simpler measure that South et al. (1982) used in their study of intergender relations: the proportion of women in a work group. This measure, like Teachman's, revealed no significant effect of gender diversity. Next, we used a dummy variable to indicate whether or not a group was skewed, with women constituting less than 20 percent of its membership or men constituting less than 20 percent of its membership (Kanter, 1977). We looked at the effects of skewness in favor of men (i.e., larger proportion of men), skewness in favor of women, and skewness in general, and still, we found no significant effects of gender composition on conflict in work groups.

Our initial test of H3, which predicted a positive association between task conflict and emotional conflict, was the zero-order correlation between the two variables. The two types of conflict were positively correlated ($r = .48$, $p < .01$), consistent with H3. Several OLS regressions (not shown but available from the authors) corroborated this finding. A regression of task conflict on diversity variables, controls, and emotional conflict showed that emotional conflict was a significant predictor of task conflict (beta $=.35$, $p<.05$). A regression of emotional conflict on diversity variables, controls, and task conflict showed that task conflict was a significant predictor of emotional conflict (beta $=.26$, $p<.05$).

Like H1 and 2, H4a, 4b, and 5 were tested with a SURE analysis involving two equations. Using the likelihood ratio test, we determined that the goodness of fit of this two-equation model was $\chi^2=10.36$ (d.f.$=3$; $p<.05$). Hypothesis 4a proposed that task routineness would reduce the positive relationships between diversity and task conflict, while competing hypothesis 4b suggested that task routineness would enhance the positive relationships between diversity and task conflict. In table 3 (above), model 3 presents the equation that tested these predictions.

Consistent with H4b, the interaction of task routineness and functional background diversity had a significant positive association with task conflict (beta $=.74$, $p<.10$), suggesting that functional background differences were more likely to trigger task conflict when tasks were routine than when tasks were nonroutine. A partial derivative analysis revealed that the effect of functional background diversity on task conflict was monotonic over the range of task routineness observed in our sample.[3] A graphical display showed that the effect was stronger for higher levels of task routineness.

H5 proposed that task routineness would diminish the positive relationships between diversity and emotional conflict. In table 4, model 3 presents the equation that tested this hypothesis. Consistent with H5, the interaction of task routineness and race diversity had a significant negative association with emotional conflict (beta $=-1.12$, $p<.05$), and the interaction of task routineness and tenure diversity had a significant negative association with emotional conflict (beta $=-.59$, $p<.05$). Partial derivative analyses revealed that the effects of race diversity and tenure diversity were monotonic over the range of task routineness observed in our sample. Graphical displays showed that these effects were weaker for higher levels of task routineness.

H6 proposed that group longevity would diminish the positive relationships between diversity and conflict. This hypothesis was tested with a SURE analysis involving two equations, one for task conflict and one for emotional conflict. The goodness of fit of this two-equation model was $\chi^2=9.486$ (d.f.$=3$; $p<.05$). As revealed in model 4 of tables 3 and 4, all moderating effects of longevity supported H6. The interaction of group longevity and functional background diversity had a significant negative association with task conflict (beta $=-1.53$, $p<.05$). The interaction of group longevity and race diversity had a significant negative association with emotional conflict (beta $=-.70$, $p<.05$), as did the interaction of group longevity and tenure diversity (beta $=-1.44$, $p<.10$). Partial derivative analyses revealed that the effects of the diversity variables were nonmonotonic over the range of longevity in our sample. Graphical displays showed that longevity had to reach a certain threshold to diminish the diversity-conflict relationships. For functional background diversity, this threshold was .66 years. For race diversity, the threshold was 1.09 years, and for tenure diversity, the threshold was 1.14 years.

Table 5: Regression of Group Performance on Conflict, Diversity, and Control Variables ($N=41$)†

Independent variables	Model 1	2	3	4	5
Controls					
Group size	.02	.03	−.07	.00	−.05
Task routineness	.16	.17	.22	.19	.19
Group longevity	.44**	.35**	.40**	.37*	.37*
Site B	−.39**	−.40**		−.42*	−.48**
Diversity variables					
Race diversity		−.12	−.11	−.14	−.09
Age diversity		.13	.20	.14	.18
Company tenure diversity		.14	.11	.12	.14
Functional background diversity		−.10	−.19	−.10	−.21
Conflict variables					
Task conflict			.30**		.32**
Emotional conflict				.06	−.09
R-squared	.29	.34	.40	.34	.40
F	3.70**	2.06*	2.27**	1.79	2.00*

*$p<.10$; **$p<.05$; ***$p<.01$.
†Entries are standardized regression coefficients. One-tailed tests were used for effects predicted in directional hypotheses.

The above-mentioned regressions of conflict against diversity variables constituted the first portion of our tests of mediating effects. Table 5 displays the remaining portions required by the Baron and Kenny (1986) method. Model 2 of table 5 shows the regression of our dependent variable (performance) against the independent variables (diversity measures) and controls. Model 3 shows the regression of our dependent variable against one process variable (task conflict), the independent variables, and controls. Model 4 shows the regression of our dependent variable against another process variable (emotional conflict), the independent variables, and controls. Finally, model 5 shows the regression of our dependent variable against both process variables (task and emotional conflict), the independent variables, and controls. The results do not strictly satisfy all requirements for a mediating effect, for we did not find a significant relationship between diversity and performance in model 2, although we did find evidence (model 3) that task conflict is positively associated with cognitive task performance (beta$=.30$, $p<.05$), as H7 predicted. This positive relationship remained when diversity variable predictors were removed from the equation.

Thus far, an implicit assumption in our hypotheses and analyses has been that the effects of diversity variables on conflict are independent of one another. It is conceivable, however, that this is not the case. Interactions among different types of diversity may also shape work unit dynamics (Alexander et al., 1995). Therefore, we performed additional, exploratory analyses to examine how the joint effects of diversity variables influenced conflict in these data. Given the exploratory nature of this analysis, we used two-tailed tests. Since there were five diversity variables, the number of possible combinations of two diversity variables was 10. We ran 10 SURE analyses to assess the effects of these 10 diversity variable interaction pairs on each of two conflict variables. Out of 20 possible interaction effects, four were significant. Gender diversity and age diversity had a positive interaction effect on emotional conflict (beta = 17.40, $p < .05$), as did gender diversity and functional background diversity (beta = 4.00, $p < .05$). Age diversity and tenure diversity had a negative interaction effect on emotional conflict (beta = −15.39, $p < .05$), and the interaction between race diversity and functional background diversity had a negative effect on task conflict (beta = −2.50, $p < .05$).

Discussion

Researchers of work group demography and top management team composition have often relied on the argument that diversity increases conflict, which, in turn; influences group performance. This study assesses the validity of that argument and reveals that the black box between diversity and performance is complex. Our results suggest that different types of diversity have distinctive effects. The diversity variables that drive task conflict differ from those that drive emotional conflict, and task conflict, in turn, tends to have more favorable performance consequences than emotional conflict. Additionally, our findings indicate that diversity can both increase and decrease conflict. The results also suggest that the combination of diversity types present and several contextual moderators influence the strength of the relationship between a particular diversity variable and conflict. In short, our findings point to multiple interrelated factors that must be considered when determining how work group composition will shape conflict and, ultimately, performance.

Given that a preponderance of group research has been confined to laboratory settings, the current study is noteworthy in its use of field data, including team members' questionnaires as well as managers' ratings of team performance, but the data are not without limitations. In particular, they are cross-sectional, a feature that renders causal interpretations difficult. Also, the sample size of 45 teams, while substantial for a field investigation, limits the power of statistical tests. In addition, several of our predictor variables (e.g., age diversity) had only modest variation across groups, so their effects may be underestimated. Nonetheless, this study begins to develop a multifaceted model of demographics, process variables, and outcomes, and the use of workplace data helps ensure that our findings have external validity. The findings themselves reveal some telling relationships among diversity, conflict, and performance.

Diversity and Conflict

The findings suggest that task conflict is a relatively straightforward phenomenon driven by functional background differences, a highly job-related type of diversity. Apparently, because functional background is so related to work, people are particularly likely to draw on belief structures based on functional background when addressing workplace issues; hence, functional background differences are the key source of task conflict in work groups. This result substantiates managers' use of cross-functional teams to create difference of opinion.

While task conflict is a relatively simple phenomenon driven by functional background differences, our findings suggest that emotional conflict is more complicated. On one hand, emotional conflict is increased by dissimilarity in race and tenure. It appears that, because race and tenure attributes are relatively impermeable, people find it difficult to identify with (and easy to stereotype) those of a different race or tenure. Race and tenure differences therefore tend to encourage heated interactions in work groups. Given this tendency, managers may want to pay particular attention to group process in multi-race and mixed-tenure settings. On the other hand, emotional conflict is increased by similarity in age. Any tendency for age differences to trigger emotional conflict appears to be overshadowed by the tendency for age similarity to trigger social comparison and, ultimately, emotional conflict. Age is a career-related attribute, so employees tend to measure their own career progress by looking at that of coworkers in their age cohort (Lawrence, 1988). When age similarity in a group increases, these career progress comparisons, which prompt jealous rivalry, often increase. As Hambrick (1994: 202) noted, "if group members are extremely similar, they face the prospect of head-on rivalries that could drive them apart. . . . This might particularly occur if several group members were the same age and vying to be the next group leader. . . ." Overall, these findings suggest that diverse groups face countervailing forces, such that some forms of diversity increase conflict and other forms do the reverse. Hence, managers must be prepared to meet challenges presented by heterogeneity as well as homogeneity in their work groups.

The lack of a gender diversity effect in this study is intriguing, given that other studies have found important effects of gender heterogeneity on work group outcomes, including reduced performance on cognitive tasks (Kent and McGrath, 1969; Murnighan and Conlon, 1991), reduced cross-gender support (South et al., 1982), and increased within-gender support (Ely, 1994). One possible interpretation is that gender composition, by itself, is unrelated to conflict. Alternatively, gender diversity may trigger categorization and social comparison processes that cancel each other's effects. Yet another possibility is that the absence of gender diversity effects on conflict stems from the bicategorical nature of gender. If a demographic attribute has only two possible categories, then an increase in group diversity means that the distribution in the group becomes more balanced. When this occurs, members of the category previously holding the numerical majority have less opportunity to interact with similar others; consequently, they may face more conflicts. Members of the category previously in the numerical minority, however, have a greater opportunity to interact with similar others; consequently, they may face fewer conflicts. Hence, the net level of the conflict in the work group may change very little. Future research could profitably examine gender dynamics further.

Our exploratory analysis of diversity variable interactions yielded four significant findings. Some of these lend credence to recent arguments by Alexander et al. (1995). For example, we found that age diversity and tenure diversity, which were negatively correlated, had a negative interaction effect on emotional conflict. This finding is consistent with Alexander et al.'s (1995) assertion that when two types of diversity are negatively correlated, categorization may become more difficult, and intercategory tensions may be less likely. Conversely, the positive interaction effect of gender diversity and age diversity supports Alexander et al.'s claim that categorization is easier — and intercategory tensions more likely — when two types of diversity are positively correlated. The positive interaction effect of gender diversity and functional background diversity also supports this claim.

Another exploratory finding, the negative interaction effect of race diversity and functional background diversity on task conflict, may reflect subtle racism, a form of prejudice that is more indirect and difficult to detect than old-fashioned direct racial prejudice (Dovidio and Gaertner, 1991). Due to societal norms discouraging the blatant expression of racial prejudice, employees may have reverted to a less obvious form of discrimination. That is, they may have ignored the opinions of persons from different functional areas when those persons were of a different race as well. Any interpretations of the observed diversity variable interactions must be treated with caution, however, given the exploratory nature of these analyses. Nonetheless, they do represent an avenue for future research.

Moderator Effects

Our work also extends traditional main-effects research on diversity by examining conditional effects. We found that task routineness and diversity interacted to influence conflict, although the direction of this interaction effect was not the same for both types of conflict. While task routineness reduced the positive associations between diversity and emotional conflict, it enhanced the positive associations between diversity and task conflict. The effect on emotional conflict supports our reasoning that, because routine tasks create less frustration than complex tasks, people in groups performing routine tasks are less likely to displace frustration onto dissimilar others. The effect on task conflict supports the notion that group members performing routine tasks seek task debates with dissimilar others to make their work more interesting. Consistent with previous sensation-seeking research, our study suggests that when a task is understimulating, people seek arousing experiences (e.g., task conflict based on background differences) but not unpleasant arousing experiences (e.g., emotional conflict based on background differences).

The moderating effects of group longevity were consistent with our hypotheses as well. The positive associations between diversity variables and both types of conflict were weaker in groups with more longevity. Apparently, after working together for a period of time, group members of different backgrounds either develop a shared understanding of tasks or learn to anticipate and deflect opposition to their ideas. Additionally, during that period, the boundaries of social categories may become blurred, so that individuals who were once considered out-group members become in-group members.

Conflict and Performance

Another significant finding in our sample of work groups was that task conflict had a positive association with cognitive task performance. Such conflict evidently fosters a deeper understanding of task issues and an exchange of information that facilitates problem solving, decision making, and the generation of ideas. Although we found that functional background diversity was related to task conflict and task conflict was related to performance (both with and without controlling for diversity variables), a third sign was needed to verify a mediating effect: a positive relationship between diversity and performance, without task conflict included in the equation. We did not see that third sign, perhaps because we were unable to use a full structural equation model. A LISREL analysis with a larger sample may ultimately confirm mediation.

Surprisingly, we found no evidence that emotional conflict impaired performance. In a recent study, Jehn (1995) found no relationship between emotional conflict and performance, despite having found one before (Jehn, 1994). She explained that "while relationship troubles cause great dissatisfaction, the conflicts may not influence work as much as expected, because the members involved in the conflicts choose to avoid working with those with whom they experience [emotional] conflict" (Jehn, 1995: 276). In the present study, subjects may also have found ways to cope with those with whom they had emotional conflicts.

Although there are clear theoretical and empirical distinctions between task and emotional conflict, our results show that each type of conflict tends to accompany the other. Task conflicts may be taken personally by group members and generate emotional conflict, or emotional conflict may prompt group members to criticize each other's ideas, thereby fostering task conflict.

Taken together, our findings suggest that diversity variables can influence conflict and yet, with the exception of functional background diversity, do not necessarily have much bearing on work group performance. That is, while race, tenure, and age diversity influenced emotional conflict, they lacked substantial ties to performance. Groups (at least these groups) were apparently able to manage their negative effects. At the same time, except in the case of functional background diversity, groups did not achieve sizable gains from background differences.

Directions for Future Research

This investigation opens a number of avenues for related research. One avenue is to continue exploring the role of process variables in the relationship between work group diversity and performance. Although we have measured task and emotional conflict in this study, it would also be helpful to measure some of the other process variables, such as social comparison, categorization, and sensation seeking, that were unmeasured components of our theory. Because these processes provide explanations for competing predictions, it would be useful to investigate them further. Additionally, a next logical step for researchers is to conduct a study that assesses the linkages among conflict and the process variables that other demography studies have examined — e.g., external communication (Ancona and Caldwell, 1992), social integration (O'Reilly, Caldwell, and Barnett, 1989), and informal communication (Smith et al., 1994). Another opportunity is to explore further the dynamics of gender diversity in work groups. The effects of such diversity may be difficult to capture in a deductive empirical study such as this one. Case research may be more revealing, allowing a more microscopic view of member exchanges in groups with moderate, low, and high levels of gender diversity. Finally, it would be useful to assess the robustness of the diversity variable interaction effects found here. Further study would be useful in probing the generalizability of those findings. Also, investigations with different dependent variables may determine whether diversity variable interactions influence processes other than conflict. Subtle racism, which one of our diversity variable interactions may reflect, is a particularly provocative topic for future research.

Conclusion

Demography scholars (e.g., Wagner, Pfeffer, and O'Reilly, 1984; Ancona and Caldwell, 1992) have frequently implied that conflict plays a central role in the relationship between diversity and work group outcomes. The present study has built on this notion, examining a complex set of linkages among work group diversity, conflict, and performance. The findings suggest that different types of diversity have distinct relationships with task and emotional conflict because of the specific properties of each type of diversity, because of interactions among the diversity types that are present, and because of the group's longevity and task routineness. Further, the findings suggest that task conflict has more favorable performance consequences than does emotional conflict. Overall, these results offer researchers a clearer view of the black box between work group diversity and performance.

This study also sheds light on patterns that practitioners can expect in diverse work groups. In particular, managers and members of cross-functional teams can take comfort in knowing that task conflict is likely in those teams and that such conflict may enhance performance. There is also a basis for expecting age variation to diminish emotional conflict. At the same time, race and tenure diversity may increase emotional conflict, especially in new groups with nonroutine tasks. Anticipating such a possibility may be critical if organizations hope to manage employees' background differences successfully.

Notes

1. These two types of conflict have been given a variety of labels, such as substantive and affective conflict (Guetzkow and Gyr, 1954; Pelled, 1996), cognitive and affective conflict

(Amason, 1996), substantive conflict and interpersonal conflict (Eisenhardt, Kahwajy, and Bourgeois, 1997a), and task and emotional conflict (Jehn, 1994). Although they have used different labels, these studies have offered similar definitions for the two dimensions, essentially describing the same constructs.

2. Personal communication from Blake Ashforth, 1997.

3. To interpret each significant interaction further, we took an additional analytical step not shown in this paper (but available from the authors). Specifically, we examined the functional form of the interactions. This procedure, which Schoonhoven (1981) described in detail, is appropriate for interactions involving two continuous variables and avoids the information loss associated with median split procedures. First, we took a partial derivative to determine mathematically whether the moderated relationship (i.e., the relationship between the diversity variable and conflict, moderated by either task routineness or group longevity) was monotonic or nonmonotonic. We then plotted the partial derivative over the range of the mean-centered moderating variable (either task routineness or group longevity); this plot illustrated how the relationship between diversity and conflict changed over the range of the moderator's values (i.e., the values of task routineness or group longevity).

References

Alexander, Jeffrey, Beverly Nuchols, Joan Bloom, and Shoou-Yih Lee 1995 "Organizational demography and turnover: An examination of multiform and nonlinear heterogeneity." *Human Relations*, 48: 1455–1480.

Allison, Paul D. 1978 "Measures of inequality." *American Sociological Review*, 43: 865–880.

Amason, Allen C.' 1996 "Distinguishing the effects of functional and dysfunctional conflict on strategic decision making: Resolving a paradox for top management teams." *Academy of Management Journal*, 39: 123–148.

Ancona, Deborah G., and David F. Caldwell 1992 "Demography and design: Predictors of new product team performance." *Organization Science*, 3: 321–341.

Bantel, Karen A., and Susan E. Jackson 1989 "Top management and innovations in banking: Does the composition of the top team make a difference?" *Strategic Management Journal*, 10: 107–124.

Baron, Reuben M., and David A. Kenny 1986 "The moderator-mediator variable distinction in social psychological research: Conceptual, strategic, and statistical considerations." *Journal of Personality and Social Psychology*, 51: 1173–1182.

Berlyne, Daniel E. 1967 "Arousal and reinforcement." In D. Levine (ed.), *Nebraska Symposium on Motivation*, 1967: 1–110. Lincoln, NE: University of Nebraska Press.

Blau, Peter M. 1977 *Inequality and Heterogeneity*. New York: Free Press.

Breckler, Steven J. 1990 "Applications of covariance structure modeling in psychology: Cause for concern?" *Psychological Bulletin*, 107: 260–273.

Brewer, Marilynn B., and Roderick M. Kramer 1986 "Choice behavior in social dilemmas: Effects of social identity, group size, and decision framing." *Journal of Personality and Social Psychology*, 50: 543–549.

Buhler, Patricia 1997 "Scanning the environment: Environmental trends affecting the workplace." *Supervision*, 58: 24–26.

Chen, Peter Y., and Paul E. Spector 1992 "Relationships of work stressors with aggression, withdrawal, theft, and substance use: An exploratory study." *Journal of Occupational and Organizational Psychology*, 65: 177–184.

Dean, James W., Jr., and Scott A. Snell 1991 "Integrated manufacturing and job design: Moderating effects of organizational inertia." *Academy of Management Journal*, 34: 776–804.

Dearborn, DeWitt C., and Herbert A. Simon 1958 "Selective perception: A note on the department identifications of executives." *Sociometry*, 21: 140–144.

Donnellon, Anne 1993 "Crossfunctional teams in product development: Accommodating the structure to the process." *Journal of Product Innovation Management*, 10: 377–392.

Dovidio, John F., and Samuel L. Gaertner 1991 "Changes in the expression and assessment of racial prejudice." In Harry J. Knopke, Robert J. Norrell, and Ronald W. Rogers (eds.), *Opening Doors: Perspectives on Race Relations in Contemporary America*: 119–225. Tuscaloosa, AL: University of Alabama Press.

Eisenhardt, Kathleen M., and L. J. Bourgeois 1988 "Politics of strategic decision making in high-velocity environments." *Academy of Management Journal*, 32: 543–576.

Eisenhardt, Kathleen M., Jean L. Kahwajy, and L.J. Bourgeois 1997a "How management teams can have a good fight." *Harvard Business Review*, 75 (4): 77–85.

Eisenhardt, Kathleen M., Jean L. Kahwajy, and L.J. Bourgeois 1997b "Conflict and strategic choice: How top management teams disagree." *California Management Review*, 39 (2): 42–62.

Eisenhardt, Kathleen M., and Claudia B. Schoonhoven 1990 "Organizational growth: Linking founding team strategy, environment, and growth among U.S. semiconductor ventures, 1978–1988." *Administrative Science Quarterly*, 35: 504–529.

Ely, Robin J. 1994 "The effects of organizational demographics and social identity on relationships among professional women." *Administrative Science Quarterly*, 39: 203–238.

Erickson, Bonnie H., and T. A. Nosanchuk 1977 *Understanding Data*. Toronto: McGraw-Hill Ryerson.

Evan, William 1965 "Conflict and performance in R&D organizations." *Industrial Management Review*, 7: 37–46.

Festinger, Leon 1954 "A theory of social comparison processes." *Human Relations*, 7: 117–140.

Fiske, Donald W., and Salvatore R. Maddi 1961 *Functions of Varied Experience*. Homewood, IL: Dorsey.

Fiske, Susan T., and Shelley E. Taylor 1991 *Social Cognition*. New York: McGraw-Hill.

Florin, Paul, Gary A. Giamartino, David A. Kenny, and Abraham Wandersman 1990 "Levels of analysis and effects: Clarifying group influence and climate by separating individual and group effects." *Journal of Applied Social Psychology*, 20: 881–900.

Gallagher, Dennis, Ed Diener, and Randy J. Larsen 1989 "Individual differences in affect intensity: A moderator of the relation between emotion and behavior." Unpublished manuscript, Department of Psychology, University of Illinois, Urbana-Champaign.

Georgopoulos, Basil S. 1986 *Organizational Structure, Problem Solving, and Effectiveness*. San Francisco: Jossey-Bass.

Ghosh, Sukesh K. 1991 *Econometrics: Theory and Applications*. Englewood Cliffs, NJ: Prentice-Hall.

Gladstein, Deborah L. 1984 "Groups in context: A model of task group effectiveness." *Administrative Science Quarterly*, 29: 499–517.

Guetzkow, Harold, and John Gyr 1954 "An analysis of conflict in decision-making groups." *Human Relations*, 7: 367–381.

Guo, Shumei S., William C. Chumlea, and David B. Cockram 1996 "Use of statistical methods to estimate body composition." *American Journal of Clinical Nutrition*, 64: 428S–435S.

Hambrick, Donald C. 1994 "Top management groups: A conceptual integration and reconsideration of the 'team' label." In B. M. Staw and L. L. Cummings (eds.), *Research in Organizational Behavior*, 16: 171–213. Greenwich, CT: JAI Press.

Hambrick, Donald C., and Phyllis A. Mason 1984 "Upper echelons: The organization as a reflection of its top managers." *Academy of Management Review*, 9: 193–206.

Hebb, D. O. 1955 "Drives and the CNS (conceptual nervous system)." *Psychological Review*, 62: 243–254.

Heriot, Gail 1996 "Diversity is more than skin deep." *Los Angeles Times*, July 14: M5.

Hoffman, L. Richard, and Norman R. F. Maier 1961 "Quality and acceptance of problem solutions by members of homogeneous and heterogeneous groups." *Journal of Abnormal and Social Psychology*, 62: 401–407.

Jackson, Susan E. 1992 "Team composition in organizational settings: Issues in managing an increasingly diverse workforce." In Stephen Worchel, Wendy Wood, and Jeffrey A. Simpson (eds.), *Group Process and Productivity*: 138–173. Newbury Park, CA: Sage.

Jackson, Susan E., Joan F. Brett, Valerie I. Sessa, Dawn M. Cooper, Johan A. Julin, and Karl Peyronnin 1991 "Some differences make a difference: Individual dissimilarity and group heterogeneity as correlates of recruitment, promotions, and turnover." *Journal of Applied Psychology*, 76: 675–689.

Janis, Irving L. 1982 *Victims of Groupthink*. Boston: Houghton Mifflin.

Janis, Irving L. 1989 *Crucial Decisions: Leadership in Policymaking and Crisis Management*. New York: Free Press.

Jehn, Karen A. 1994 "Enhancing effectiveness: An investigation of advantages and disadvantages of value-based intragroup conflict." *International Journal of Conflict Management*, 5: 223–238.

Jehn, Karen A. 1995 "A multimethod examination of the benefits and detriments of intragroup conflict." *Administrative Science Quarterly*, 40: 256–282.

Kahneman, Daniel 1973 *Attention and Effort*. Englewood Cliffs, NJ: Prentice-Hall.

Kanter, Rosabeth M. 1977 *Men and Women of the Corporation*. New York: Basic Books.

Katz, Ralph 1982 "The effects of group longevity on project communication and performance." *Administrative Science Quarterly*, 27: 81–104.

Kennedy, Peter 1979 *A Guide to Econometrics*. Cambridge, MA: MIT Press.

Kent, R. N., and N. E. McGrath 1969 "Task and group characteristics as factors influencing group performance." *Journal of Experimental Social Psychology*, 5: 429–440.

Kirkpatrick, David L. 1986 "Performance appraisal: When two jobs are too many." *Training: The Magazine of Human Resource Development*, 23 (3): 65–68.

Kramer, Roderick M. 1991 "Intergroup relations and organizational dilemmas: The role of categorization processes." In L. L. Cummings and B. M. Staw (eds.), *Research in Organizational Behavior*, 13: 191–228. Greenwich, CT: JAI Press.

Lawrence, Barbara S. 1988 "New wrinkles in the theory of age: Demography, norms, and performance ratings." *Academy of Management Journal*, 31: 309–337.

Lawrence, Barbara S. 1997 "The black box of organizational demography." *Organization Science*, 8: 1–22.

Major, B., and B. Forcey 1985 "Social comparisons and pay evaluations: Preferences for same-sex and same-job wage comparisons." *Journal of Experimental Social Psychology*, 21: 393–405.

Milberg, Sandra, and Margaret S. Clark 1988 "Moods and compliance." *British Journal of Social Psychology*, 27: 79–90.

Miller, C. T. 1982 "The role of performance-related similarity in social comparison of abilities: A test of the related attributes hypothesis." *Journal of Personality and Social Psychology*, 46: 1222–1228.

Milliken, Frances J., and Luis L. Martins 1996 "Searching for common threads: Understanding the multiple effects of diversity in organizational groups." *Academy of Management Review*, 21: 402–433.

Murnighan, J. Keith, and Donald E. Conlon 1991 "The dynamics of intense work groups: A study of British string quartets." *Administrative Science Quarterly*, 36: 165–186.

Murray, Alan I. 1989 "Top management group heterogeneity and firm performance." *Strategic Management Journal*, 10: 125–141.

Nelson, Reed E. 1989 "The strength of strong ties: Social networks and inter-group conflict in organizations." *Academy of Management Journal*, 32: 377–401.

O'Reilly, Charles A., David F. Caldwell, and William P. Barnett 1989 "Work group demography, social integration, and turn-over." *Administrative Science Quarterly*, 34: 21–37.

Parker, Thomas H., and Ira J. Dolich 1986 "Toward understanding retail bank strategy: Seemingly unrelated regression applied to cross-sectional data." *Journal of Retailing*, 62: 298–320.

Pedhazur, Elazar J., and Liora P. Schmelkin 1991 *Measurement, Design, and Analysis: An Integrated Approach*. Hillsdale, NJ: Erlbaum.

Pelled, Lisa H. 1996 "Demographic diversity, conflict, and work group outcomes: An intervening process theory." *Organization Science*, 7: 615–631.

Pelled, Lisa H., Thomas G. Cummings, and Mark A. Kizilos 1999 "The influence of organizational demography on customer-oriented prosocial behavior: An exploratory investigation." *Journal of Business Research*, in press.

Pelled, Lisa H., Gerald E. Ledford, and Susan A. Mohrman 1998 "Individual demographic dissimilarity and organizational inclusion: A field investigation." Academy of Management Best Paper Proceedings, OMT Division: A1–A17.

Reardon, Kathleen K. 1995 *They Don't Get It, Do They? Communication in the Workplace — Closing the Gap between Men and Women*. Boston: Little, Brown.

Ross, Raymond S. 1989 *Small Groups in Organizational Settings*. Englewood Cliffs, NJ: Prentice-Hall.

Ryan, Alan J. 1991 "Job rotations grease long-term IS wheels." *Computer-world*, 25 (24): 76.

Sarason, Irwin G. 1984 "Stress, anxiety, and cognitive interference: Reactions to tests." *Journal of Personality and Social Psychology*, 46: 929–938.

Schoonhoven, Claudia B. 1981 "Problems with contingency theory: Testing assumptions hidden within the language of contingency 'theory'." *Administrative Science Quarterly*, 26: 349–377.

Sessa, Valerie I., and Susan E. Jackson 1995 "Diversity in decision-making teams: All differences are not created equal." In Martin M. Chemers, Stuart Oskamp, and Mark A. Costanzo (eds.), *Diversity in Organizations: New Perspectives for a Changing Workplace*: 133–156. Thousand Oaks, CA: Sage.

Smith, Ken G., Ken A. Smith, Judy D. Olian, Henry P. Sims, Douglas P. O'Bannon, and Judith A. Scully 1994 "Top management team demography and process: The role of social integration and communication." *Administrative Science Quarterly*, 39: 412–438.

South, Scott J., Charles M. Bonjean, William T. Markham, and Judy Corder 1982 "Social structure and inter-group interaction: Men and women of the federal bureaucracy." *American Sociological Review*, 47: 587–599.

Staw, Barry M., Lloyd E. Sandelands, and Jane E. Dutton 1981 "Threat-rigidity effects in organizational behavior: A multi-level analysis." *Administrative Science Quarterly*, 26: 501–524.

Storms, Philip L., and Paul E. Spector 1987 "Relationships of organizational frustration with reported behavioural reactions: The moderating effect of locus of control." *Journal of Occupational Psychology*, 60: 227–234.

Tajfel, Henri 1972 "Social categorization." In Serge Moscovici (ed.), *Introduction à la Psychologie Sociale*, 1: 272–302. Paris: Larousse.

Tajfel, Henri 1978 *Differentiation between Social Groups*. New York: Academic Press.

Tajfel, Henri 1982 *Social Identity and Intergroup Relations*. Cambridge: Cambridge University Press.

Tajfel, Henri, M. G. Billig, R. P. Bundy, and C. Flament 1971 "Social categorization and intergroup behavior." *European Journal of Social Psychology*, 1: 149–177.

Teachman, Jay D. 1980 "Analysis of population diversity." *Sociological Methods and Research*, 8: 341–362.

Tjosvold, Dean 1986 "Constructive controversy: A key strategy for groups." *Personnel*, 63: 39–44.

Tjosvold, Dean 1991 *Team Organization: An Enduring Competitive Advantage*. New York: Wiley.

Tsui, Anne S., Susan J. Ashford, Lynda St. Clair, and Katherine R. Xin 1995 "Dealing with discrepant expectations: Response strategies and managerial effectiveness." *Academy of Management Journal*, 38: 1515–1543.

Turner, John C. 1975 "Social comparison and social identity: Some comparisons for intergroup behavior." *European Journal of Social Psychology*, 5: 5–34.

Van de Ven, Andrew H., Andre Delbecq, and R. Koenig 1976 "Determinants of coordination modes within organizations." *American Sociological Review*, 41: 322–338.

Vernon, J., and T. E. McGill 1960 "Utilization of visual stimulation during sensory deprivation." *Perceptual and Motor Skills*, 11: 214.

Wagner, W. Gary, Jeffrey Pfeffer, and Charles A. O'Reilly 1984 "Organizational demography and turnover in top management groups." *Administrative Science Quarterly*, 29: 74–92.

Waller, Mary J., George P. Huber, and William H. Glick 1995 "Functional background as a determinant of executives' selective perception." *Academy of Management Journal*, 38: 943–974.

Walsh, James P. 1988 "Selectivity and selective perception: An investigation of managers' belief structures and information processing." *Academy of Management Journal*, 31: 873–896.

Watson, W. E., K. Kumar, and L. K. Michaelson 1993 "Cultural diversity's impact on interaction process and performance: Comparing homogeneous and diverse task groups." *Academy of Management Journal*, 36: 590–602.

Weingart, Laurie 1992 "Impact of group goals, task component complexity, effort, and planning on group performance." *Journal of Applied Psychology*, 77: 682–693.

Wheaton, B., B. Muthen, D. Alwing, and G. Summers 1977 "Assessing reliability and stability in panel models." In D. Heise (ed.), *Sociological Methodology*: 84–136. San Francisco: Jossey-Bass.

Wickens, Christopher D. 1989 "Attention and skilled performance." In D. Holding (ed.), *Human Skills*: 71–105. New York: Wiley.

Wiersema, Margarethe F., and Karen Bantel 1992 "Top management team demography and corporate strategic change." *Academy of Management Journal*, 35: 91–121.

Wonnacott, Thomas H., and Ronald J. Wonnacott 1984 *Introductory Statistics for Business and Economics*. New York: Wiley.

Zellner, Arnold 1962 "An efficient method of estimating seemingly unrelated regressions and tests for aggregation bias." *Journal of the American Statistical Association*, 57: 348–368.

Zenger, Todd R., and Barbara S. Lawrence 1989 "Organizational demography: The differential effects of age and tenure distributions on technical communication." *Academy of Management Journal*, 32: 353–376.

Zimbardo, Philip G., and Michael R. Leippe 1991 *The Psychology of Attitude Change and Social Influence*. New York: McGraw-Hill.

Zuckerman, Marvin 1979 *Sensation Seeking*. Hillsdale, NJ: Erlbaum.

Zuckerman, Marvin 1984 "Sensation seeking: A comparative approach to a human trait." *Behavioral and Brain Sciences*, 7: 413–471.

54

Psychological Safety and Learning Behavior in Work Teams

Amy Edmondson

Source: *Administrative Science Quarterly* 44 (2) (1999): 350–383.

A growing reliance on teams in changing and uncertain organizational environments creates a managerial imperative to understand the factors that enable team learning. Although much has been written about teams and about learning in organizations, our understanding of learning in teams remains limited. A review of the team effectiveness and organizational learning literatures reveals markedly different approaches and a lack of cross-fertilization between them. An emerging literature on group learning, with theoretical papers on groups as information-processing systems and a number of empirical studies examining information exchange in laboratory groups, has not investigated the learning processes of real work teams (cf. Argote, Gruenfeld, and Naquin, 1999). Although most studies of organizational learning have been field-based, empirical research on group learning has primarily taken place in the laboratory, and little research has been done to understand the factors that influence learning behavior in ongoing teams in real organizations.

Studies of work teams in a variety of organizational settings have shown that team effectiveness is enabled by structural features such as a well-designed team task, appropriate team composition, and a context that ensures the availability of information, resources, and rewards (Hackman, 1987). Many researchers have concluded that structure and design, including equipment, materials, physical environment, and pay systems, are the most important variables for improving work-team performance (Goodman, Devadas, and Hughson, 1988; Campion, Medsker, and Higgs, 1993; Cohen and Ledford, 1994) and have argued against focusing on interpersonal factors (e.g., Goodman, Ravlin, and Schminke, 1987). According to this research, organization and team structures explain most of the variance in team effectiveness.

In contrast, organizational learning research has emphasized cognitive and interpersonal factors to explain effectiveness, showing, for example, that individuals' tacit beliefs about interpersonal interaction inhibit learning behavior and give rise to ineffectiveness in organizations (e.g., Argyris, 1993). This cognitive emphasis takes different forms. Organizational learning theorists have offered both descriptive theory explaining the failure of organizations to adapt rationally due to cognitive biases that favor existing routines over alternatives (e.g., Levitt and March, 1988) and prescriptive theory proposing interventions that alter individuals' "theories-in-use" to improve organization effectiveness (e.g., Argyris and Schön, 1978). The former theorists suggest that adaptive learning in social systems is fundamentally problematic and rare, and the latter, only slightly more sanguine, propose that expert intervention is necessary to bring it about (cf. Edmondson and Moingeon, 1998). This paper takes a different approach to understanding learning in organizations by examining to what extent and under what conditions learning occurs naturally in organizational work groups.

Much organizational learning research has relied on qualitative studies that provide rich detail about cognitive and inter-personal processes but do not allow explicit hypothesis testing (e.g., Senge, 1990; Argyris, 1993; Watkins and Marsick, 1993). Many team studies, in contrast, have used large samples and quantitative data but have not examined antecedents and consequences of learning behavior (e.g., Goodman, Devadas, and Hughson, 1988; Hackman, 1990; Cohen and Ledford, 1994). I propose that to understand learning behavior in teams, team structures and shared beliefs must be investigated jointly, using both quantitative and qualitative methods.

This paper presents a model of team learning and tests it in a multimethod field study. The results support an integrative perspective in which both team structures, such as context support and team leader coaching, and shared beliefs shape team outcomes. Organizational work teams are groups that exist within the context of a larger organization, have clearly defined membership, and share responsibility for a team product or service (Hackman, 1987; Alderfer, 1987). Their learning behavior consists of activities carried out by team members through which a team obtains and processes data that allow it to adapt and improve. Examples of learning behavior include seeking feedback, sharing information, asking for help, talking about errors, and experimenting. It is through these activities that teams can detect changes in the environment, learn about customers' requirements, improve members' collective understanding of a situation, or discover unexpected consequences of their previous actions.

These useful outcomes often go unrealized in organizations. Members of groups tend not to share the unique knowledge they hold, such that group discussions consist primarily of jointly held information (Stasser and Titus, 1987), posing a dilemma for learning in groups. More centrally, those in a position to initiate learning behavior may believe they are placing themselves at risk; for example, by admitting an error or asking for help, an individual may appear incompetent and thus suffer a blow to his or her image. In addition, such individuals may incur more tangible costs if their actions create unfavorable impressions on people who influence decisions about promotions, raises, or project assignments. Image costs have been explored in research on face saving, which has established that people value image and tacitly abide by social expectations to save their own and others' face (Goffman, 1955). Asking for help, admitting errors, and seeking feedback exemplify the kinds of behaviors that pose a threat to face (Brown, 1990), and thus people in organizations are often reluctant to disclose their errors (Michael, 1976) or are unwilling to ask for help (Lee, 1997), even when doing so would provide benefits for the team or organization. Similarly, research has shown that the sense of threat evoked in organizations by discussing problems limits individuals' willingness to engage in problem-solving activities (Dutton, 1993; MacDuffie, 1997). The phenomenon of threat rigidity has been explored at multiple levels of analysis, showing that threat has the effect of reducing cognitive and behavioral flexibility and responsiveness, despite the implicit need for these to address the source of threat (Staw, Sandelands, and Dutton, 1981). In sum, people tend to act in ways that inhibit learning when they face the potential for threat or embarrassment (Argyris, 1982).

Nonetheless, in some environments, people perceive the career and interpersonal threat as sufficiently low that they do ask for help, admit errors, and discuss problems. Some insight into this may be found in research showing that familiarity among group members can reduce the tendency to conform and suppress unusual information (Sanna and Shotland, 1990); however, this does not directly address the question of when group members will be comfortable with interpersonally threatening actions. More specifically, in a recent study of hospital patient-care teams, I found significant differences in members' beliefs about the social consequences of reporting medication errors; in some teams, members openly acknowledged them and discussed ways to avoid their recurrence; in others, members kept their knowledge of a drug error to themselves (Edmondson, 1996). Team members' beliefs about the interpersonal context in these teams could be characterized as tacit; they were automatic, taken-for-granted assessments of the "way things are around here." For example, a nurse in one team explained matter-of-factly, "Mistakes are serious, because of the toxicity of the drugs [we use] — so you're never afraid to tell the Nurse Manager"; in contrast, a nurse in another team in the same hospital reported, "You get put on trial! People get blamed for mistakes . . . you don't want to have made one." These quotes illustrate markedly different beliefs about the interpersonal context; in the first team, members saw it as self-evident that speaking up is natural and necessary, and in the other, speaking up was viewed as a last resort.

An aim of the present study was to investigate whether beliefs about the interpersonal context vary between teams in the same organization, as well as to examine their effects on team outcomes. Existing theories do not address the issue of how such beliefs may affect learning behavior in teams, instead focusing primarily on structural conditions associated with overall team effectiveness (e.g., Hackman, 1987) or on the skills that must be learned by individuals to enable learning in difficult interpersonal interactions (e.g., Argyris, 1982). Similarly, research on group training has focused primarily on task knowledge and has paid little attention to the role of social knowledge (Levine and Moreland, 1991). Thus, the role of beliefs about the interpersonal context in individuals' willingness to engage in otherwise-threatening learning behavior has been largely unexamined. This is the gap I seek to fill with a model and study of team learning.

A Model of Team Learning

Team Learning Behavior

Organizational learning is presented in the literature in two different ways: some discuss learning as an outcome; others focus on a process they define as learning. For example, Levitt and March (1988: 320) conceptualized organizational learning as the outcome of a process of organizations "encoding inferences from history into routines that guide behavior"; in contrast, Argyris and Schön (1978) defined learning as a process of detecting and correcting error. In this paper I join the latter tradition in treating learning as a process and attempt to articulate the behaviors through which such outcomes as adaptation to change, greater understanding, or improved performance in teams can be achieved. For clarity, I use the term "learning behavior" to avoid confusion with the notion of learning outcomes.

The conceptualization of learning as a process has roots in the work of educational philosopher John Dewey, whose writing on inquiry and reflection (e.g., Dewey, 1938) has had considerable influence on subsequent learning theories (e.g., Kolb, 1984; Schön, 1983). Dewey (1922) described learning as an iterative process of designing, carrying out, reflecting upon, and modifying actions, in contrast to what he saw as the human tendency to rely excessively on habitual or automatic behavior. Similarly, I conceptualize learning at the group level of analysis as an ongoing process of reflection and action, characterized by asking questions, seeking feedback, experimenting, reflecting on results, and discussing errors or unexpected outcomes of actions. For a team to discover gaps in its plans and make changes accordingly, team members must test assumptions and discuss differences of opinion openly rather than privately or outside the group. I refer to this set of activities as learning behavior, as it is through them that learning is enacted at the group level. This conceptualization is consistent with a definition of group learning proposed recently by Argote, Gruenfeld, and Naquin (1999) as both processes and outcomes of group interaction activities through which individuals acquire, share, and combine knowledge, but it focuses on the processes and leaves outcomes of these processes to be investigated separately.

The management literature encompasses related discussions of learning, for example, learning as dependent on attention to feedback (Schön, 1983), experimentation (Henderson and Clark, 1990), and discussion of failure (Sitkin, 1992; Leonard-Barton, 1995). Research has demonstrated performance benefits for feedback seeking by individual managers (Ashford and Tsui, 1991), for teams seeking information and feedback from outside the team (Ancona and Caldwell, 1992), and for research and development teams that experiment frequently (Henderson and Clark, 1990). Similarly, because errors provide a source of information about performance by revealing that something did not work as planned, the ability to discuss them productively has been associated with organizational effectiveness (Michael, 1976; Sitkin, 1992; Schein, 1993). On one hand, if feedback seeking, experimentation, and discussion of errors individually promote effective performance, learning behavior — which includes all of these activities — is also likely to facilitate performance, whether for individuals or teams. On the other hand, learning behavior consumes time without assurance of results, suggesting that there are conditions in which it may reduce efficiency and detract from performance, such as when teams are responsible for highly routine repetitive tasks with little need for improvement or modification. For teams facing change or uncertainty, however, the risk of wasting time may be small relative to the potential gain; in such settings, teams must engage in learning behavior to understand their environment and their customers and to coordinate members' actions effectively. Moreover, teams that perform routine production tasks may still require learning behavior for effective self-management as a team and for intermittent process improvement:

> Hypothesis 1 (H1): Learning behavior in teams is positively associated with team performance.

Team Psychological Safety

Team psychological safety is defined as a shared belief that the team is safe for interpersonal risk taking. For the most part, this belief tends to be tacit — taken for granted and not given direct attention either by individuals or by the team as a whole. Although tacit beliefs about interpersonal norms are sometimes explicitly discussed in a team, their being made explicit does not alter the essence of team psychological safety. The construct has roots in early research on organizational change, in which Schein and Bennis (1965) discussed the need to create psychological safety for individuals if they are to feel secure and capable of changing. Team psychological safety is not the same as group cohesiveness, as research has shown that cohesiveness can reduce willingness to disagree and challenge others' views, such as in the phenomenon of groupthink (Janis, 1982), implying a lack of interpersonal risk taking, The term is meant to suggest neither a careless sense of permissiveness, nor an unrelentingly positive affect but, rather, a sense of confidence that the team will not embarrass, reject, or punish someone for speaking up. This confidence stems from mutual respect and trust among team members.

The importance of trust in groups and organizations has long been noted by researchers (e.g., Golembiewski and McConkie, 1975; Kramer, 1999). Trust is defined as the expectation that others' future actions will be favorable to one's interests, such that one is willing to be vulnerable to those actions (Mayer, Davis, and Schoorman, 1995; Robinson, 1996). Team psychological safety involves but goes beyond interpersonal trust; it describes a team climate characterized by interpersonal trust and mutual respect in which people are comfortable being themselves.

For team psychological safety to be a group-level construct, it must characterize the team rather than individual members of the team, and team members must hold similar perceptions of it. Previous researchers have studied the similarity of beliefs in social systems such as organizations and work groups (for reviews, see Klimoski and Mohammed, 1994; Walsh, 1995). Perceptions of psychological safety, like other such beliefs, should converge in a team, both because team members are subject to the same set of structural influences and because these perceptions develop out of salient shared experiences. For example, most members of a team will conclude that making a mistake does not lead to rejection when they have had team experiences in which appreciation and interest are expressed in response to discussion of their own and others' mistakes.

Team psychological safety should facilitate learning behavior in work teams because it alleviates excessive concern about others' reactions to actions that have the potential for embarrassment or threat, which learning behaviors often have. For example, team members may be unwilling to bring up errors that could help the team make subsequent changes because they are concerned about being seen as incompetent, which allows them to ignore or discount the negative consequences of their silence for team performance. In contrast, if they respect and feel respected by other team members and feel confident that team members will not hold the error against them, the benefits of speaking up are likely to be given more weight. Support for the centrality of interpersonal inferences in groups is found in research on distributive justice, which shows that people are more focused on relational than instrumental considerations in their assessments of allocation decisions made by authority figures; people are very attentive to the tone and quality of social processes and are more willing to comply with these when they feel valued (Tyler and Lind, 1992). Argyris and Schön (1978) made a connection between interpersonal threat and learning when they posited that interpersonally threatening issues impede learning behavior, but they did not address the possibility that dyads or groups may differ in their tacit beliefs about interpersonal threat, thereby giving rise to different levels of learning. In contrast, I propose that psychological safety varies from team to team, such that otherwise interpersonally threatening learning behavior can occur if the team has a sufficiently safe environment:

> Hypothesis 2 (H2): Team psychological safety is positively associated with learning behavior in organizational work teams.

Psychological safety does not play a direct role in the team's satisfying customers' needs, the core element of performance; rather, it facilitates the team's taking appropriate actions to accomplish its work. Thus, learning behavior should mediate the effects of team psychological safety on performance outcomes:

> Hypothesis 3 (H3): Team learning behavior mediates between team psychological safety and team performance.

Team Efficacy and Team Learning

Building on earlier work on the role of self-efficacy in enhancing individual performance (Bandura, 1982), a body of research has established group efficacy as a group-level phenomenon (e.g., Guzzo et al., 1993) and also reported a relationship between group efficacy and performance (Lindsley, Brass, and Thomas, 1995; Gibson, 1996). This work has not specified mechanisms through which shared perceptions of efficacy lead to good performance, and one possibility is that efficacy fosters team members' confidence, which promotes learning behavior and helps accomplish desired team goals:

> Hypothesis 4 (H4): Team efficacy is positively associated with team learning behavior.

Team members deciding whether to reveal errors they have made are likely to be motivated to speak up if two conditions are satisfied: first, they believe they will not be rejected (team psychological safety) and, second, they believe that the team is capable of using this new information to generate useful results (team efficacy). Team psychological safety and team efficacy are thus complementary shared beliefs, one pertaining to interpersonal threat and the other characterizing the team's potential to perform. Team efficacy thus should supplement team psychological safety's positive effect on team learning:

> Hypothesis 5 (H5): Team efficacy is positively associated with team learning behavior, controlling for the effects of team psychological safety.

Team Leader Coaching and Context Support as Antecedents of Team Psychological Safety

A set of structural features — consisting of a clear compelling team goal, an enabling team design (including context support such as adequate resources, information, and rewards), along with team leader behaviors such as coaching and direction setting — have been shown to increase team effectiveness (Hackman, 1987; Wageman, 1998). These structural features provide a starting point for examining antecedents of team psychological safety. The extent of context support experienced by a team should be positively associated with team psychological safety because access to resources and information is likely to reduce insecurity and defensiveness in a team. Team leader coaching is also likely to be an important influence on team psychological safety. A team leader's behavior is particularly salient; team members are likely to attend to each other's actions and responses but to be particularly aware of the behavior of the leader (Tyler and Lind, 1992). If the leader is supportive, coaching-oriented, and has non-defensive responses to questions and challenges, members are likely to conclude that the team constitutes a safe environment. In contrast, if team leaders act in authoritarian or punitive ways, team members may be reluctant to engage in the interpersonal risk involved in learning behaviors such as discussing errors, as was the case in the study of hospital teams mentioned above (Edmondson, 1996). Furthermore, team leaders themselves can engage in learning behaviors, demonstrating the appropriateness of and lack of punishment for such risks.

> Hypothesis 6 (H6): Team leader coaching and context support are positively associated with team psychological safety.

Through enhancing psychological safety, team leader coaching and context support are likely to facilitate team learning. Team psychological safety thus serves as a mechanism translating structural features into behavioral outcomes:

> Hypothesis 7 (H7): Team psychological safety mediates between the antecedents of team leader coaching and context support and the outcome of team learning behavior.

Context support and team leader coaching should also affect team efficacy. Effective coaching is likely to contribute to members' confidence in the team's ability to do its job, as is a supportive context, which reduces obstacles to progress and allows team members to feel confident about their chances of success. If coaching and context support promote team efficacy and team efficacy promotes team learning, this suggests that team efficacy also functions as a mediator:

> Hypothesis 8 (H8): Team efficacy mediates between the antecedents of team leader coaching and context support and the outcome of team learning behavior.

Team Type

Organizations use a variety of types of teams. Team type varies across several dimensions, including cross-functional versus single-function, time-limited versus enduring, and manager-led versus self-led. These dimensions combine to form different types of teams, such as a time-limited new product development team or an ongoing self-directed production team. The team learning model should be applicable across multiple types of teams, because the social psychological mechanism at the core of the model concerns people taking action in the presence of others, and the salience of interpersonal threat should hold across settings. Therefore, although the utility of learning behavior may vary across types of teams, the association between team psychological safety and team learning behavior should apply across different team types. Thus, the effects of team type on learning behavior should be insignificant when assessed together with the other variables in the team learning model, shown in figure 1. For example, new product development teams might be expected to exhibit more learning behavior than production teams because of the nature of their task; nonetheless, mean differences in learning behavior that might be observed across types of teams should be explained by team psychological safety and team efficacy, as shown in figure 1, rather than by team type.

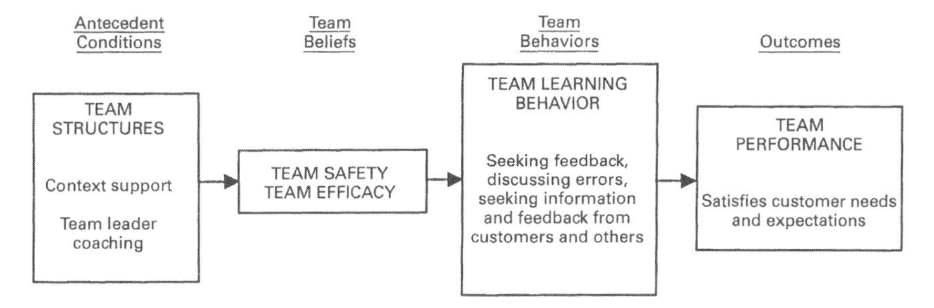

Figure 1: A model of work-team learning.

Methods

To test the hypotheses in the team learning model, I studied real work teams in an organization that has a variety of team types, using a combination of qualitative and quantitative methods to investigate and measure the constructs in the model. Preliminary observation and interviews in the organization suggested that there was considerable variation in the extent to which teams engaged in learning behavior, making it a good site in which to explore the phenomenon and to investigate factors associated with team learning.

Research Site and Sample

"Office Design Incorporated" (ODI), a manufacturer of office furniture with approximately 5,000 employees and a reputation for product and management innovation, provided the research site for this study. Teams in this company, implemented in 1979 to promote employee participation and cross-functional collaboration, consisted of four types. Most were functional teams, made up of managers or supervisors and direct reports, and these included sales teams, management teams, and manufacturing teams; this type of team existed within and supported the work of a single functional department. Although encompassing dyadic reporting relationships, functional teams had shared goals, and members were interdependent in reaching them. As with other teams at ODI, they also typically had some training in teamwork. Second, ODI had a growing number of self-managed teams in both manufacturing and sales; these teams consisted of peers from the same function. The third type was time-limited cross-functional product development teams, and the fourth was time-limited cross-functional project teams, convened to work on other projects that involved multiple departments. The company was willing to participate in this research to obtain feedback on how well its teams were working.

My primary contact at ODI was a manager in an internal organization development group who worked closely with me to facilitate data collection. She scheduled interviews and meetings, recruited teams to participate in the study, and identified recipients of the work of each of these teams. As ODI did not have a central roster of all work teams, she distributed a memo to managers throughout the company describing the goal of the study (to assess team effectiveness at ODI) and asking for lists of teams in their area. This yielded a list of 53 teams, encompassing differences in organization level, department, type, size, self- versus leader-managed, and tenure or team age. At the time of survey data collection, the oldest team had been together for about seven years, and the newest had been in place for four months; both the oldest and newest teams were production teams. These 53 teams included 34 functional teams (in sales, manufacturing, and staff services such as information technology and accounting), nine self-managed teams (in manufacturing and sales), five cross-functional product development teams, and three cross-functional project teams. As the purpose of the study was to test a theoretical model rather than to describe properties of this particular organization or to characterize teams of different types, this sample was not selected to ensure representativeness of the population of all teams at ODI, nor were the four subgroups of team types selected to ensure that they were representative of each type. The sample did satisfy the essential criterion to achieve the purposes of this study, however, which was to include sufficient variance on the variables in the model to test hypothesized relationships. Despite using a process characterized by voluntary participation in the research, the resulting sample was not a self-selected group of high-performing or highly satisfied teams; instead, there was substantial variance for all variables studied, including for such key measures as team psychological safety, learning behavior, and performance.

Procedure

The study involved three phases of data collection. First, I conducted interviews and observation that involved eight teams, selected from among those available during my two first visits to ODI, to ensure variance in team type. Second, I designed and administered two surveys and a structured interview instrument to obtain quantitative data for all teams in the sample. Third, I interviewed and observed seven teams, selected according to survey results as high or low in learning behavior.

Phase 1: Preliminary Qualitative Research

In two four-day visits to ODI, I observed eight team meetings, each of which lasted one to three hours, and conducted 17 interviews lasting from 45 minutes to an hour with members or observers of these eight teams. The eight teams included five product development teams, two management teams, and one self-managed production team. I interviewed at least one and as many as six members of each team, as well as one senior manager responsible for reviewing the work of one of the product development teams. The objectives of this phase of the study were to verify that the theoretical constructs of team psychological safety and team learning behavior could be operationalized at ODI and, if so, to develop survey items to assess these constructs in language that would be meaningful in this setting — a modified empathic strategy (Alderfer and Brown,

1972). In team meetings, I took notes and listened for examples of learning behavior, such as asking for feedback, asking for help, admitting errors, and proposing or describing instances of seeking help or information from others outside the team. In interviews, I asked team members to describe features of their team, such as the goal and the nature of its task, and to describe how the team organized its work and what challenges it faced. These general questions allowed me to listen for examples of learning behavior. I taped most interviews, except for some in the factory where noise levels made it difficult to do so, and reviewed tapes and notes to identify data that provided evidence of team psychological safety and learning behavior and to assess whether these constructs varied across teams. Examples of learning behavior and quotes that suggested the presence or absence of team psychological safety were transcribed, and these data suggested that both psychological safety and learning behavior varied across teams.

A set of related beliefs about the interpersonal context emerged as suggestive of the presence (or absence) of team psychological safety, including a belief that others won't reject people for being themselves, that team members care about and are interested in each other as people, that other members have positive intentions, and that team members respect each other's competence. Table 1 presents excerpts of these data to illustrate the constructs of team psychological safety and team learning behavior and to show the elements that made up each construct.

Phase 2: Survey Research

All members of the 53 teams in the sample (496 individuals) were administered a five-section survey developed for this study. Most teams were requested to complete surveys before or after a team meeting and to enclose them in sealed envelopes collected by ODI staff and mailed to me. In a few cases, surveys were mailed to team members with return envelopes attached and were then returned to me directly. In total, 427 team members from 51 teams completed the surveys, an 86-percent response rate; of these 51 teams, 90 percent of members responded. Two teams did not return any surveys; in both cases, the teams continued to express a desire to participate but ultimately failed to do so, attributing this to busy schedules. At the same time, for each team, two or three managers outside of each team were identified as recipients of the team's work and were given a short survey I developed to assess team learning behavior and performance; 135 of the 150 observers surveyed returned the survey, a 91-percent response rate. Three months after completing the survey, each team received an individual report, providing feedback about their team and department results compared with the overall ODI results, along with a brief explanation of how to interpret these data.

During this time, to obtain independent data that could help establish the construct validity of survey variables assessing team design, another researcher — blind to the survey results — interviewed 31 managers who were familiar with the design of one or more of the 51 teams and who had not served as team

Table 1: Construct Development from Preliminary Qualitative Data*

Constructs	Positive form	Negative form
Beliefs about the team interpersonal context (inferred from informant quotes)		
Members of this team respect each other's abilities.	''I trust the people here that they're making the right decision, for the function and for ODI. And they feel the same way about me.'' [Finance member, New Product Development Team 1 (NPD 1)] ''Each person is important. Everyone is respected.'' (Marketing member, NPD 1)	''The [other] team has a lot of trust in the expertise of other [member]s, unlike this one.'' (Engineering member, NPD 2)
Members of this team are interested in each other as people.	''There's much greater openness on this team — it's intangible. . . . We have a personal interest in each other. We're comfortable outside the realm of work, we've shared personal information . . . if you don't know anything about people, you don't know how to react to them.'' (Manufacturing member, NPD 1) ''Our efforts to get to know each other led to our mutual respect. . . . At the core, these are outstanding human beings.'' (Finance member, NPD 1)	''What gets in the way is guys who hold information close to their chests, so knowledge doesn't get filtered out to the team.'' (Management team 2)
In this team, you aren't rejected for being yourself or stating what you think.	''Sally and Sue both had been getting a hard time on the first shift for outperforming. . . . That's why they like being on this team.'' (Chair production team 2) ''[Members of this team are] willing to state what they believe . . . people, in other teams, if they don't get their way, they stay silent.'' (NPD 1)	''People try to figure out what [the team leader] wants to hear [before saying what they think].'' (Management team 2)
Members of this team believe that other members have positive intentions.	''They're not out to corrupt my success.'' (NPD 1 team member, referring to the other team members)	''. . . we struggled through the problem statement, because it [the project] was clearly for ODI's internal needs, not for customers. We had a lot of nay sayers who just wanted to do [the assignment from management] and not question it. They were worried about getting their hands slapped. . . .'' (NPD 2)

Team behaviors (observed by researcher or reported by team members or team observers)		
Seeking or giving feedback	"We talked to over a hundred customers; this changed the project goal slightly, to make it integrate more with the [other] product as a top priority." (Marketing team member, NPD 1) "We also bring in people from Advanced Applications to bounce ideas off of, to get a check on what we're doing." (Engineering team member, NPD 1) "[NPD 1 team leader] asks me to come to certain meetings; she wants my view, my industry experience, and how [this product] fits with ODI's systems strategy." (Senior manager, R&D) "Am I missing the mark with how to proceed? Is there anything you can add?" (Team leader, management team 1, in a team meeting)	"They were too methodical, too detailed in their wandering . . . they did not do enough checking with customers until too far along." (Senior manager, R&D, describing NPD 2) NPD 2 hired a vendor to conduct customer interviews, in contrast to NPD Team 1 members, who frequently spoke to customers themselves.
Making changes and improvements (vs. avoiding change or sticking with a course too long)	"Every three months we decide we need to improve how we get our information. *We look for better ways to do something and we make changes*." (NPD 1)	"We did make changes, but too slowly." (NPD 2) "They did learn, but not fast enough." (Senior manager describing NPD 2)
	". . . every six months, they take time out to look at what works . . . and a lot happens in those meetings." (NPD 1)	". . . [there were a lot of] blind alleys. . . . We had a preconceived notion of what was important that prevented us from seeing it. . . ." (NPD 2) "We found ourselves going around in circles a lot. Sometimes this took a lot of time." (NPD 2) "This team gets stuck. . . . It's hard to get a decision. The dynamics are that the conversation gets shut down." (Management team 2)
Obtaining or providing help or expertise	"[NPD 1] used the applications specialists [an ODI internal design group] more than any other team I know of." (Senior manager, R&D) "I've learned a lot about marketing a product – about how and why we make decisions." (Finance team member, NPD 1) "Are there any concerns right now on regional fleets?" (Team leader, management team 1)	

Table 1: *Continued*

Constructs	Positive form	Negative form
Experimenting	''There's a lot of testing of new ways to do stuff. We're doing design and engineering at the same time. It's wild. It's incredibly complex. We need to be constantly creative about the mechanisms. . . .'' (NPD 1) ''There have been a lot of iterations. It's like reducing a sauce by half. It's a more flavorful sauce, a more complex group of ingredients, but the end result is simpler. We made it easier to use . . . by continually challenging ourselves to find what is essential.'' (NPD 1) One team member called the other eight together at the beginning of the shift and asked who was interested in trying which new task. She listened carefully to responses and suggested a plan that she explained would allow several people to learn a new role. (Chair production team 1) Another team member raised the question of what goal to set for the shift; after discussion, the team settled on a new (ambitious) target of producing 83 chairs. (Chair production team 1)	
Engaging in constructive conflict or confrontation	''They bring conflict up directly; they don't let it fester. . . .'' (Team leader, NPD 1) ''People speak openly in team meetings, [whereas in other teams] they *wait until the meeting is over and 'speak privately in the hall* [about their frustrations]. (Finance team member, NPD 1)	

*NPD = new product development. Text in italics became the basis of a new survey item.

observers. The interview instrument included questions to elicit informants' descriptions of team design (goal, task, composition, and context support), probing for factual descriptions and examples rather than evaluations of the team. The interviewer reviewed the tapes, made notes and — using a five-point scale from very low to very high — assessed four variables: (1) presence of a clear goal, (2) team task interdependence, (3) appropriateness of team composition, and (4) context support.

Phase 3: Follow-Up Qualitative Research

From the team survey data, I identified teams with the six lowest and six highest means for team learning behavior; seven of these twelve (four high and three low) were available for follow-up observation and interviews. The set of seven teams consisted of three functional teams (one high- and two low-learning), two product development teams (high and low), one self-managed team (high), and one project team (high); none of these overlapped with the eight teams I studied in the first phase. I observed six of these teams, individually interviewed one or two members of each, and conducted interviews with every member of the seventh team. The objective of this phase was to explore differences between high- and low-learning teams and to learn more about how team learning behavior works. I reviewed these field notes and tapes to construct short cases describing each team, which were then used to suggest patterns related to team learning.

Measures

Antecedent Factors

I coded the team survey to identify respondents by team rather than by individual and to identify team type (functional, self-managed, product development, or project) and company department (operations, sales, staff services, or cross-functional). I included in the survey scales developed by Hackman (1990) to assess team design features, including context support and team leader coaching.

Team Shared Beliefs

I developed scales to measure team psychological safety and team efficacy, using items designed to assess several features of each theoretical construct. In doing this, I also drew from qualitative data obtained in phase-1 interviews. Sample items for psychological safety include "If you make a mistake on this team, it is often held against you" (reverse scored), "It is safe to take a risk on this team," and "No one on this team would deliberately act in a way that would undermine my efforts." Team efficacy was measured with items such as "With focus and effort, this team can do anything we set out to accomplish." As in other sections of the survey, a mix of negatively and positively worded items was used to mitigate response set bias. (See the Appendix for all items.) The survey also measured team tenure (the average number of years each member had worked in the team) and company tenure (respondents' years of employment at ODI). Between-scale correlations for variables in the model are shown in table 2, at the group level of analysis ($N=51$).

Team Behavior

I developed scales to assess the extent of learning behavior for both the team and observer surveys. Team learning behavior includes items such as "We regularly take time to figure out ways to improve our team's work process" and "Team members go out and get all the information they possibly can from others — such as from customers or other parts of the organization."

Performance

Hackman's team performance scale was used to obtain self-report measures of team performance, and I developed a similar scale for the observer survey, including "This team meets or exceeds its customers' expectations" and "This team does superb work."

Team Feedback Variables

Additional variables, not included in the team learning model, such as presence of a clear goal, adequacy of team composition, team task design, quality of team relationships, job satisfaction, job involvement, and internal motivation were included in the team survey for the purpose of providing supplementary feedback to the teams.

Adequacy of Measures

I conducted preparatory analyses to assess psychometric properties of the two new instruments, including internal consistency reliability and discriminant validity of the scales. The results supported the adequacy of most of the measures for substantive analysis, although Cronbach's alpha was low for both context support and team efficacy (see table 2). Discriminant validity was established through factor analysis.[1] As the team antecedent and outcome sections yielded, respectively, six and three distinct factors with eigenvalues greater than one, these results demonstrated that the team survey was not hampered by excessive common-method variance, according to Harman's one factor test for common-method bias.[2]

I computed two scales from the observer survey (team learning behavior and team performance), and both showed high internal consistency reliability (see table 2). Discriminant validity was lacking; many team learning behavior items were as correlated or more correlated with team performance items as with themselves. Some of this between-scale (multitrait) correlation can be attributed

Table 2: Chronbach's Alpha and Intercorrelations between Group-Level Survey Variables*

Variable	Mean	S.D.	1	2	3	4	5	6	7	8	9	10	11	12
1. Context support	4.78	.97	.65											
2. Team leader coaching	3.77	.81	.69	.80										
3. Team psychological safety	5.25	1.03	.70	.63	.82									
4. Team efficacy	5.07	1.07	.70	.50	.50	.63								
5. Team learning behavior	4.67	.93	.68	.63	.80	.50	.78							
6. Team performance	5.10	1.03	.60	.45	.72	.50	.71	.76						
7. Internal motivation	6.11	.68	(.03)	(−.06)	.15	(−.02)	.12	.33	.64					
8. Job involvement	3.30	1.69	−.16	−.22	(−.07)	−.26	−.09	(−.01)	.31	†				
9. Team tenure (in years)	2.40	1.70	(−.06)	.34	−.26	−.15	−.16	−.09	(.05)	(.01)	†			
10. Average company tenure (in years)	9.00	6.70	.33	−.31	.26	.15	.17	.14	(.06)	(−.01)	.16	†		
11. Team learning (observer rated)	3.48	.77	.49	−.48	.60	.52	.60	.34	−.16	(−.02)	−.21	.30	.84	
12. Team performance (observer rated)	4.95	1.29	.48	−.50	.47	.43	.52	.36	−.11	−.12	−.21	.22	.81	.87

*Chronbach's alpha coefficients are presented on the diagnonal. Correlations in parentheses not significant at $p < .05$; all other correlations are significant at $p < .05$.
†Only 1 survey item.

to a substantive relationship between team learning behavior and team performance; however, because of the lack of discriminant validity, I avoided analyses that tested relationships between the two variables in the observer survey. Because it is likely that the team observers or customers are in a better position to judge performance — defined in part as meeting recipients' needs — than to assess specific behaviors, which they may not always observe, substantive analyses reported below rely primarily on observers' ratings of performance and members' ratings of behavior. Observers' ratings of learning are used in certain analyses to illustrate consistency in results across different measures of the same construct. Pearson correlations between team members' and independent observers' responses about team learning ($r = .60$, $p < .001$) and team performance ($r = .36$, $p < .01$) provided one measure of construct validity for the team survey. A substantial degree of correspondence between analogous measures in the team survey and structured interview data also contributed to establishing the construct validity of the survey measures of teams' structural features; correlations between each team-structure scale in the survey and the corresponding variable in the structured interviews were positive and significant.[3]

Finally, a group-level variable must satisfy two criteria (Kenny and LaVoie, 1985). First, the construct must be conceptually meaningful at the group level; for example, team size is a meaningful group attribute, internal motivation is not. Second, data gathered from individual respondents to assess the group attribute must converge, such that the intraclass correlation (ICC) is greater than zero. Intraclass correlation coefficients, measuring the extent to which team members' responses agree with each other and differ from other teams, were calculated for all group-level variables in the team survey; all were significant at the $p < .0001$ level.[4] Table 3 shows the results. It is particularly noteworthy that

Table 3: Analysis of Variance and Intraclass Correlation Coefficients for Group-Level Scales

Team survey variables	$F_{(50,427)}$	p	ICC
Context support	4.80	<.001	.29
Team leader coaching	4.88	<.001	.30
Team psychological safety	6.98	<.001	.39
Team efficacy	5.70	<.001	.34
Team learning behavior	5.79	<.001	.27
Team performance	6.02	<.001	.35
Internal motivation*	1.13	.07	.03
Job involvement*	1.25	.06	.04
Observers' survey variables	$F_{(50,135)}$	p	ICC
Team learning behaviors	2.27	<.001	.19
Team performance	2.90	<.001	.21

*Two variables that are conceptually individual-level variables are included for purposes of comparison, to demonstrate the contrast between these results and those for the variables from the same survey that are conceptually group-level. One-way ANOVA shows these two variables are not significantly different across teams, in contrast to the group-level variables, which are significant to the $p < .0001$ level.

new measures such as team psychological safety and team learning behavior have high ICCs (.39 and .33, respectively), satisfying the methodological prerequisite for group-level variables. In contrast, ICCs were near zero for constructs that are conceptually meaningful at the individual rather than group level of analysis (*internal motivation*, with $r_{icc} = .03$ and *job involvement* with $r_{icc} = .04$); the data thus confirm that these individual-level constructs are less likely to be shared within and vary across teams. ICCs were calculated for observer variables as measures of interrater reliability for different observers of the same team; these were also positive and significant. These results allowed the creation of a group-level data set ($N = 51$) that merged group means for group-level variables from both surveys.

Results

Team Psychological Safety, Efficacy, Learning Behavior, and Performance

To test hypotheses relating team shared beliefs, learning behavior, and performance, I conducted a series of regression analyses, using customers' ratings of team performance as the dependent variable and measures obtained from team members as regressors. Because respondents belonging to the same team are not independent, I performed regression analyses on the group-level data set ($N = 51$) to avoid violating the regression assumption of independence. The results are shown in table 4. First, regressing team learning (self-reported) on team performance (observer-rated) reveals that learning behavior is a significant predictor of team performance, supporting H1 (model 1). This minimal test of two key variables in the model was utilized to increase power, given the small team *N*, and the same strategy was used to test other core relationships, such as between team psychological safety and team learning behavior. To explore alternative models, I then introduced additional regressors into the model — specifically, context support and team leader coaching, which in previous studies have been used to explain team performance — and these provided no additional explanatory value, nor did they, without learning behavior, account for more variance than learning behavior alone (table 4, models 2 and 3). Similarly, a series of alternative regressors (team psychological safety, team efficacy, context support, and team leader coaching) individually accounted for less of the variance in team performance than was accounted for by team learning behavior. Thus, of seven alternative models, team learning behavior accounts for the most variance in observer-rated team performance, providing support for H1.

I conducted four regressions to test hypotheses relating team psychological safety and team efficacy to team learning behavior. To assess the consistency of these predictions for differing data sources, I first used self-reported team learning behavior as the dependent variable and then repeated the same analyses using observers' ratings of team learning behavior. The results reveal a high degree of consistency across the two sets of equations using the two independent

Table 4: Regression Models of Observer-Assessed Team Performance ($N = 51$)

	Model						
Variable	(1)	(2)	(3)	(4)	(5)	(6)	(7)
Constant	1.31	.44	.87	1.41	2.04*	1.48	2.32*
Team learning behavior	.80**	.40					
Context support		.22	.43			.75**	
Team leader coaching		.45	.57				.93**
Team psychological safety				.67**			
Team efficacy					.60**		
Adjusted R-squared	.26	.27	.26	.21	.17	.22	.23

*$p < .05$; **$p < .01$.

measures of team learning, as shown in table 5. First, regressing psychological safety on self-reported team learning behavior shows a significant positive relationship, providing initial support for H2 (panel A, model 1). I then regressed team efficacy alone on team learning behavior to test H4, and although team efficacy accounts for substantially less variance than team psychological safety, the relationship was positive and significant (panel A, model 2). H5, that team efficacy is positively associated with team learning behavior when controlling for team psychological safety, was not supported for self-reported team learning (panel A, model 3). With observer-rated team learning as the dependent variable, the results for H2 and H4 were similar to those obtained using self-reported team learning (panel B). When team psychological safety and team efficacy were entered into the model together, however, team efficacy remained significant (model 3, panel B), providing some support for H5. Finally, to explore alternative models, I regressed other antecedent variables on team learning behavior, and, as shown in table 5 (models 4, 5, and 6), team psychological safety accounts for more variance in both self-reported and observer-assessed team learning behavior than context support or team leader coaching. Model 7 then shows that when all regressors are entered into the model — for either measure of team learning behavior — only team psychological safety is significant. Together, these results provide substantial support for H2, that team psychological safety is associated with team learning behavior; support for H4, that team efficacy is associated with team learning behavior; and mixed support for H5, that team efficacy predicts team learning behavior when controlling for team psychological safety.

To test H3, that team learning behavior mediates the effects of team psychological safety on team performance, I conducted a three-stage analysis to test whether three conditions for mediation were satisfied: (1) the proposed mediator significantly predicts the dependent variable, (2) the independent variable predicts the mediator, and (3) the contribution of the independent variables drops substantially for partial mediation and becomes insignificant for full mediation when entered into the model together with the mediator (Baron and Kenny, 1986). In these analyses, I used observers' ratings of performance as the dependent variable and self-reported team learning behavior as the

Table 5: Regression Models of Team Learning ($N=51$)

Variable	Model (1)	(2)	(3)	(4)	(5)	(6)	(7)
A. Team learning (self-report)							
Constant	.69	2.45**	.41	1.06	1.41**	1.94**	.42
Team psychological safety	.76**		.70				.51**
Team efficacy		.45**	.11				−.05
Context support				.52**	.69**		.28
Team leader coaching				.31		.75**	.14
Adjusted R-squared	.63	.23	.63	.52	.45	.38	.66
B. Team learning behavior (Observer-assessed)							
Constant	.99*	1.5**	.48	1.27*	1.53**	1.73**	.53
Team psychological safety	.46**		.35				.33*
Team efficacy		.38**	.22*				.21
Context support				.27*	.40**		−.07
Team leader coaching				.23		.46**	.12
Adjusted R-squared	.35	.26	.40	.26	.23	.21	.36

*$p<.05$; **$p<.01$.

mediating variable, because this created a higher hurdle for demonstrating a relationship between team learning and team performance. As shown above, team learning behavior is significantly positively associated with team performance, supporting the first of the three conditions. The second condition, that the independent variable (team psychological safety) significantly predicts the proposed mediator (team learning behavior) also was established above. Finally, the third condition for mediation is also satisfied: the contribution of team psychological safety becomes insignificant ($B=.25$, $p=.42$) when entered into the regression model together with team learning, which remains significant ($B=.60$, $p<.05$).

Context Support, Leader Behavior, Psychological Safety, and Learning Behavior

Next, I used regression to test H6, H7, and H8, followed by GLM analysis to further explore the relationships in H6, that team leader coaching and context support are positively associated with team psychological safety. Results are shown in table 6. As a first test, I regressed these two variables on team psychological safety. Both were positively related to the dependent variable; context support was significant and team leader coaching was close to significant (condition 2). In testing H7, that team psychological safety mediates between coaching, context support, and team learning behavior, the first two conditions of the three-step analysis were already satisfied — team psychological safety predicted team learning behavior, and context support and coaching predicted team psychological safety. The third condition was also satisfied; when all three predictors were entered into the model simultaneously, the effects of context support and coaching were insignificant, and team psychological safety remained significant. This result supports H7, as did repeating the three-step analysis for self-reported learning behavior.

In contrast, the results shown in table 7 do not support H8, that team efficacy functions as a mediator. Team efficacy predicted observer-rated team learning behavior, but of the two independent variables, only context support predicted the proposed mediator, team efficacy. Despite insufficient support for the second condition, I checked the third condition by regressing context support and team efficacy on observer-rated team learning and found that the effects of context support were insignificant, while team efficacy remained barely significant. Finally, using self-reported team learning, I found no support for mediation.

Next, I examined relationships between context support, team leader coaching, team psychological safety, and team learning behavior using GLM analyses on the individual-level data set ($N=427$). This allowed simultaneous testing of random effects of team membership and fixed effects of team type while exploring the relationship between predictor variables and either team psychological safety or team learning behavior. Despite mean differences across team types in both team psychological safety and team learning behavior, GLM analyses revealed that these differences could be explained by context support and team leader coaching. As shown in table 8, context support, team leader coaching, and team membership (random effects of belonging to the same team) were significant predictors of individuals' ratings of team psychological safety. In contrast, the effect of team type was insignificant. Controlling for team tenure revealed that its effects on team psychological safety were also insignificant. Similarly, team psychological safety, team efficacy, and team membership were significantly related to team learning behavior, while team type and team tenure again were insignificant. Although the GLM analyses allowed a more detailed apportioning of the variance in individuals' responses than the group-level data set, which uses team means as data points, the direction and magnitude of the results are consistent with those obtained using regression analysis.

Exploring Differences between High- and Low-learning Teams

I used the data from the seven teams identified in phase 3 as high- or low-learning to better understand the relationship between team psychological safety and learning behavior.

My goal in studying them was to learn more about how they functioned as teams rather than to confirm or disconfirm a model. Table 9 summarizes these qualitative data by comparing aspects of context support, leader behavior, team psychological safety, and team learning behavior. A few observations stand out. First, team psychological safety is associated with learning behavior across all seven teams. Second, in all cases but two, team leader coaching is associated with team psychological safety and team learning behavior. The exceptions, shown in table 9, include the troubled Help Desk team, for which the leader's reported efforts to help and coach the team are juxtaposed against persistent conflicts and difficulties reported by team members, and a new product development team (NPD 3), in which the leader functioned primarily as a coordinator and offered

Table 6: Tests of Team Psychological Safety as a Mediator between Coaching, Context Support, and Learning

Conditions to demonstrate mediation*	Independent variable	Observer-assessed learning behavior				Self-report learning behavior			
		B	*t*	*p*	R^2	*B*	*t*	*p*	R^2
1. Does psychological safety significantly predict *team learning?*	Team psychological safety	.46	5.26	$<.001$	.33	.76	9.16	$<.001$	.63
2. Do coaching and context support significantly predict *team psychological safety?*	Team leader coaching	.33	1.89	.06	.52	.33	1.89	.06	.52
	Context support	.56	3.82	$<.001$		.56	3.82	$<.001$	
3. Does the effect of the antecedents drop substantially or become insignificant? (*Team learning*)	Team psychological safety	.29	2.46	.02	.33	.51	4.56	$<.001$	.66
	Context support	.09	.66	.51		.24	1.83	.07	
	Team leader coaching	.12	.78	.21		.14	1.00	.32	

*Dependent variables are in italics.

Table 7: Tests of Team Efficacy as a Mediator between Coaching, Context Support, and Learning

Conditions to demonstrate mediation*	Independent variable	Observer-assessed learning behavior				Self-report learning behavior			
		B	*t*	*p*	R^2	*B*	*t*	*p*	R^2
1. Does team efficacy significantly predict *team learning?*	Team efficacy	.45	3.97	$<.001$	.23	.38	4.27	$<.001$	.26
2. Do coaching and context support significantly predict *team efficacy?*	Team leader coaching	.01	.06	.95		.01	.06	.95	
	Context support	.77	5.01	$<.001$	.49	.77	5.01	$<.001$	.49
3. Does the effect of the antecedents drop substantially or become insignificant? (*Team learning*)	Team efficacy	.26	2.05	.05		.02	.21	.83	
	Context support	.20	1.46	.15	.27	.67	4.44	$<.001$	.52

*Dependent variables are in italics.

Table 8: Results of GLM Analyses ($N = 427$)

Model	Independent variable	F-ratio	p
Team psychological safety	Team type	$F(3,51) = 2.02$	.12
$R^2 = .60$	Team membership*	$F(50,427) = 3.25$	<.001
	Context support	$F(1,427) = 26.83$	<.001
	Team leader coaching	$F(1,427) = 39.81$	<.001
	Team tenure	$F(1,427) = 0.10$	.74
Team learning behavior	Team type	$F(3,51) = 2.21$	.10
$R^2 = .53$	Team membership*	$F(50,427) = 2.64$	<.001
	Team psychological safety	$F(1,427) = 42.21$	<.001
	Team efficacy	$F(1,427) = 10.52$	<.001
	Team tenure	$F(1,427) = 0.22$	.49

*Team membership is the categorical variable identifying each team. The result that team membership accounts for significant variance in team psychological safety or in team learning behavior indicates that variance is attributable to unexplained effects of belonging to the same team.

an easy-going, passive style that did not match the team's energetic discussions, active brainstorming, and feedback seeking.

A third observation pertains to team design. The degree of context support varied across the teams in a way that was not tightly coupled with the high- or low-learning categorization. For example, a self-managed production team (the Stain Team) confronted persistent depletion of its members, who were frequently pulled off to work on other jobs in the plant, leaving two or three others to carry out the six-person team's work. Despite this obstacle, the team exhibited numerous examples of proactive learning behavior, illustrated below. The Publications Team had a similar initial design and degree of context support as two other publications teams in the larger sample, yet survey data suggested that this team was both lower in psychological safety (3.9 versus 5.0 and 5.9) and in learning behavior (3.9 versus 4.2 and 5.4) than the other two similarly appointed teams. Although their own survey responses show lower means for context support than the other two teams (3.3 versus 4.5 and 5.4), the structured interview data (capturing outside managers' views) place all three at roughly the same level (3 versus 3.5 and 3.5). This comparison shows that learning-oriented teams might be able to modify their work processes to be more interdependent, suggesting that team design features are not always unchangeable constraints. Implications of these observations are explored in the discussion section.

Examining in more detail the Stain Team, a self-managed production team (selected as a high-learning team from the survey data), and the Publications Team, a leader-led functional team (selected as low learning), provides some insight into how team members' experiences of psychological safety may enable learning behavior. Evidence of team psychological safety was evident in interviews with each member of the Stain Team. For example, Margie, a long-time team member, offered, "Two years ago in the Stain Team [under a different structure and leader], people were blaming each other and trying to make themselves look better, never taking responsibility ... but this team is

Table 9: Qualitative Data from High- and Low-learning Teams

	Team	Context support	Team leader behavior	Team psychological safety	Team learning behavior
High-Learning	Factory Support	Adequate support. Team has access to both resources and information to help structure and carry out the team's work.	Team leader is proactive, warm, coaching-oriented, and organized.	''We don't wear a mask.'' ''We don't have to have a 'workface'.'' ''They really care; that makes it easier to express yourself.''	Team member asks me right away, ''Were we OK as a team?'' Leader asks, ''Are you here to help us be better?'' The team distributed its own shifts around the clock, on its own initiative, so that someone is always available, and the team is able to have daily meetings in the early morning.
	NPD3	Well-supported new product development team. Skilled disciplinary members, important initiative for ODI.	Leader as (largely passive) coordinator, not boss. In team meeting, a new data book is introduced by member, and team agrees that no changes can be made without a team decision. ''Even Matt can't change it.'' Matt laughs, ''You wouldn't let me!''	Members express mutual respect. Team selects humorous team name to encourage irreverence and unconventional ideas.	''We learned [how to learn from customers]; don't [over]present what you're thinking in advance, because they won't disagree — so you end up not learning from them. Sometimes have a non-team member present, to help them respond openly. . . . We learned to listen to what they're saying.''
	Fusion	Formed to develop new protocols to provide a consistent interface for all ODI computer users to replace department-specific systems. Often short on time and resources as members were juggling other distracting responsibilities.	Leader is committed to being a good coach: ''I learned to ask myself why I might be tuning someone else out . . . what is it about me? I recognize that it's a reinforcing cycle. . . .''	''We all respect each other, which is critical for being open.'' ''Everyone on this team was very competent. . . . I respect them.''	''We had a problem defining the scope and roles at first. We didn't get a lot of momentum. . . . We tried to develop a statement of [the scope], but there was a lot of disagreement. So we brought it to our 'customers,' but then ended up defining the scope ourselves. . . ..''

Table 9: *Continued*

	Team	Context support	Team leader behavior	Team psychological safety	Team learning behavior
					''We experimented with different meeting lengths, from a whole day to an hour, and ended up with half-day meetings as the best length.''
	Stain	Lack of context support. Team members constantly pulled off the team to substitute for absent workers elsewhere.	Team is self-managed, but one member emerges as leader. ''Brian'' is approachable and interested in people, outside of work he acts as a director in community theater.	Member contrasts current stain team with former, which involved ''blaming each other. . . .''	''. . . if we have a quality issue – we're not sure about something we've just done–we'll bring others in without telling them what the issue is to ask them if they see a problem with this part. . . .''
Low-Learning	Help Desk	Adequate support, but team leader is not located adjacent to team. Team task lacks true interdependence.	Leader is located in office at a distance from the team. Team leader reports planning physical reorganization to relocate closer to team. Expresses concern about helping the team.	One team member ''monitors the others, judging their behavior . . . but she can't confront them. This makes the problem bigger. . . . She has the most seniority, and she questions their competence and makes decisions on her own.''	[One member) complains to others not on the team (''to vent'') but is ''not able to confront other team members openly.''
	Chair Production 2	Information and resources are available, but often not used by team, which values opportunity for overtime work. Large team (30+ members) responsible for producing chairs on first shift. Work divided according to specialized subtasks.	Leader acts as traditional supervisor, monitoring attend-ance, job assignments, and output.	According to internal team consultant, ''They don't want to look like brown-nosers.'' ''They have bad attitudes.''	''If there's a technical problem, they don't ask the engineers for help.'' ''They were having problems with the glue, but they didn't get help; they just sit and don't work, then they get to do overtime on Saturday.'' Team has actively resisted cross-training.
	Publications	Adequate context support; identical to two other publications teams.	Leader acts as boss, distributes work, monitors progress.	''People are put down for being different . . . [there is a] lack of trust.''	Apparent lack of learning behavior. See text.

different." Matt, another member, explained, "This team ... has more cooperation; we take more responsibility for helping each other. ... Right now, I think this is the best team I've been on ... [in other teams] it's people not carrying their share, or it's conflict. ..."

These descriptions allude to a shared belief that one will not be blamed by other team members, who can be counted on to help each other and who are not punitive. Interviews with members also generated numerous examples of learning behavior. For example, Margie described how the team listened to and acted on negative feedback from recipients of the team's work output: "This is going to sound very childish, but let's say I just did a part and I got drips on it. Now, if they [those next in the production process] told me I got drips on the edge, I say 'thanks' — and then I'm glad I can get these drips off. Where it used to be, when that happened, we'd just try to find something wrong that person did — we'd keep an eye out for it! It wasn't to be helpful, it was to bring them down to your level or something like that. ... Now, we don't think anything of it. We just fix it." In this quote, feedback about the team's work — and about mistakes, in particular — is seen as helpful. Margie offered a reason for this change: "I think that the reason we are now so open to that kind of thing is because we feel that the people who are telling us are not telling us because they want to put us down and say we're doing a bad job but because they want us to do a good job — do the product good — so they want to work together to make the product better."

Her explanation suggests that the team's interpretation of others' intentions plays an important role in its openness to feedback; by believing others' intentions to be helpful rather than critical, the team is more likely to interpret negative feedback as friendly rather than unfriendly data. Another member, Joe, described this phenomenon in similar terms, "Our group is very good; if something comes back to us, I think all of us will say, 'Yeah, I did that.' I don't think there is any of us who wouldn't — where before it was, 'I don't remember. ...' Now I think everyone takes responsibility." Finally, Matt provided an example of learning behavior that combines the construct of conducting an experiment with that of seeking feedback: "Sometimes, if we have a quality issue — we're not sure about something we've just done — we'll bring others in without telling them what the issue is to ask them if they see a problem with this part. Second opinion type stuff. We do a lot of second opinions [from others not on the team but in the stain area]." The device of keeping the others blind to the real concern is used to generate honest and useful feedback. Knowing they cannot provide an objective opinion themselves, they seek an objective eye elsewhere. The Stain Team illustrates how a shared sense of psychological safety can allow team members to take interpersonal risks and act in learning-oriented ways. Margie alluded to understanding others' intentions to be helpful as making it easy to accept constructive criticism; Matt described proactive feedback-seeking that is enabled by the belief that the team takes responsibility for doing good work and is not focused on placing blame.

In contrast, interviews with the Publications Team, a functional team responsible for preparing brochures and other publications for a group of ODI dealers, revealed very different perceptions of the team's interpersonal context.

The newest member (who joined the team two months earlier) volunteered, ". . . there are underlying tensions. I'm not sure where it comes from." A long-time member complained, "Amanda [the team leader] doesn't want to know if things aren't going well." Later he added, "I'm not being backed, not being supported." Another member said, "People are put down for being different . . . [there is a] lack of trust." These descriptions suggest a tangible lack of team psychological safety; the notion that someone does not "want to know if things aren't going well" exemplifies the construct.

There appeared to be little overt learning behavior in this team. The team's task was flexible enough to permit either a collaborative approach or division into relatively independent tasks, and the degree of interaction varied across the three publications teams included in the sample of 51 teams. All three teams had the same general task, which required integrating technical skills with an understanding of dealers' specific needs to create finished products for a certain group of dealers, but did not specify how team members were to work together. The leader of this team attempted to manage work allocation herself, by assigning tasks to individuals. As she explained, "Everyone has their own assignment, but they can help each other. . . . But are they pulling together to get it all done? No. . . . In the past, some were putting in more hours. I did not like that. It's not fair. . . . I try to take care of it by spreading out the volume, switching the dealers around." Other members explained that much of the interaction and information transfer is between team members and dealers rather than among team members, despite the fact that team members are co-located. Within the team, members revealed a lack of learning behavior such as asking questions. As one member reported, "If I have questions I ask others — but I'm pretty confident in what I do and I do it." Not surprisingly, team members also reported not receiving "honest feedback," "not feeling heard," and having "no opportunity to gain skills, no opportunity to grow." Another member complained, "People are leaving, but none of the problems get addressed." Overall, the data suggested that the team was stuck in a self-defeating pattern in which a lack of psychological safety discouraged reaching out to ask for or offer help or to discuss ways to improve the team's work process. Viewing the environment as unsafe, members developed their own coping strategies, such as planning to leave the team or planning to stay while remaining as insular as possible.

A similar contrast can be found in data collected in phase 1 of the study, in which the two new product development teams studied displayed very different experimentation behavior. NPD 1 demonstrated an eagerness to experiment, to try many things quickly and often simultaneously. One member, Bob, reported, "There's a lot of testing of new ways to do stuff. We're doing design and engineering at the same time. It's wild. It's incredibly complex. We need to be constantly creative about the mechanisms. . . ." Another member, Kim, said, "There have been a lot of iterations. It's like reducing a sauce by half. It's a more flavorful sauce, a more complex group of ingredients, but the end result is simpler. We made it easier to use . . . by continually challenging ourselves to find what is essential."

In contrast, members of NPD 2 described getting stuck in "blind alleys" in a process of perfecting one solution at a time before getting feedback: "We went down a lot of blind alleys. . . . We would go down a path for a while, develop details, then abandon it. Each path represented time wasted. . . ." The lack of experimentation behavior in NPD 2 appeared to be related to the team's concern that they had to produce a certain solution that "management" wanted. As one member explained, ". . . we had a lot of nay-sayers who just wanted to do [the project] and not question it. They were worried about getting their hands slapped by management." In contrast, NPD 1 members describe their team as able to make the right decisions: "I trust the people here that they're making the right decision for the function and for ODI. And they feel the same way about me" — a belief that may facilitate the rapid, low-stakes experimentation behavior the team exhibited.

Discussion

Overall, the study shows the usefulness of the construct of team psychological safety for understanding collective learning processes. The existence of team psychological safety, conceptualized as a shared belief about the consequences of interpersonal risk-taking, at the group level of analysis was supported by qualitative and quantitative data. A set of salient beliefs about the interpersonal context that were consistent with this construct emerged from qualitative data collected in phase 1 of the study, and survey items designed to capture the experience of team psychological safety showed high internal consistency reliability. The data also suggest that team psychological safety is something beyond interpersonal trust; there was evidence of a coherent interpersonal climate within each group characterized by the absence or presence of a blend of trust, respect for each other's competence, and caring about each other as people. But building trust may be an important ingredient in creating a climate of psychological safety. Team members interviewed often referred to others' intentions, such as the Stain Team's belief that others pointing out their "drips" don't "want to put us down" but, rather, "want to work together to make the product better"; such beliefs suggest that a team's proclivity to trust others' intentions plays a role in psychological safety and learning behavior. Although building trust may not necessarily create a climate of mutual respect and caring, trust may provide a foundation for further development of the interpersonal beliefs that constitute team psychological safety.

Support for the Team Learning Model

The relationship between team psychological safety and learning behavior received substantial empirical support. Team members' own descriptions, taken from different types of teams and settings, illustrated how a climate of safety and supportiveness enabled them to embrace error — for example, the Stain Team's drips — or to seek feedback from customers and make changes in a product design, as did NPD 1. Conversely, a lack of psychological safety contributed to reluctance to ask for help in preparing publications for dealers and, in NPD 2, to an unwillingness to question the team goal for fear of sanction by management. Their stories lend weight to the premise that learning behavior in social settings is risky but can be mitigated by a team's tolerance of imperfection and error. This appeared to be a tolerance (or lack of tolerance) that was understood by all team members.

The results of the study supported the proposition that team psychological safety affects learning behavior, which in turn affects team performance. Quantitative analyses provided consistent support for six of the eight hypotheses. This included support for two mediating relationships: learning behavior appears to mediate between team psychological safety and team performance, and team psychological safety appears to mediate the effects of context support and team leader coaching on learning behavior. Data from team observers on team performance, independent of other data sources, strengthen these results.

Two hypotheses — that team efficacy would be associated with learning behavior when controlling for team psychological safety and that team efficacy mediates the effect of context support and leader behavior on team learning — were not supported. This outcome may in fact strengthen the core argument in this paper — that engaging in learning behavior in a team is highly dependent on team psychological safety — by suggesting that team members' confidence that they will not be punished for a well-intentioned interpersonal risk enables learning behavior in a way that team efficacy, or confidence that the team is capable of doing its work, does not. In contrast to the uneven results for efficacy, one of the most striking results is the degree to which the proposed relationship between team psychological safety and team learning behavior received consistent empirical support across several analyses and independent measures. The implication of this result is that people's beliefs about how others will respond if they engage in behavior for which the outcome is uncertain affects their willingness to take interpersonal risks. Because beliefs about team efficacy are unrelated to this central interpersonal concern, it may be less important for learning behavior. Thus, the theoretical premise that lies at the core of the team learning model does not appear to require the supplementary effects of team efficacy. Moreover, the conclusion that team psychological safety fosters team learning behavior is both consistent with existing organizational learning theory and has a certain degree of face validity; that is, the juxtaposition of team members' descriptions of the interpersonal context in their team with their stories of learning behavior is not a surprising one.

Quantitative and qualitative results both suggested that context support accounts for variance in learning behavior but, also, that it provides an incomplete explanation. The quantitative data demonstrated a positive association between context support and psychological safety, and the qualitative data allowed isolation of specific cases from within this general trend that suggested different ways real teams handle the absence or presence of enabling design conditions. For example, the Stain Team lacked context support and yet was learning-oriented. The Publications Team, despite having a similar set-up and level of context support as two other publications teams, showed substantially less learning behavior than they did. Thus, a focus on just these two teams might suggest that context support and other features of team design account for little variance in learning behavior. In contrast, the seven high- and low-learning teams studied provided more data and do suggest an important role for team design in team learning.

First, the four high- and three low-learning teams differed markedly by team type. Functional teams were overrepresented in the low-learning teams (two of three), and product development, project, and self-managed teams made up three of the four high-learning teams. The two groups also differed somewhat on whether they exhibited key elements of a well-designed team (cf. Hackman, 1987). The low-learning teams' tasks at the time of the study tended to lack interdependence; for example, in the Publications Team, each member had his or her "own assignment"; other team members could be used as resources, but as a design for teamwork, the arrangement was suboptimal. But the fact that the two other publications teams in the survey sample had higher levels of learning demonstrates that the degree of task interdependence can be modified through learning behavior. Similarly, context support was adequate for the three low-learning teams and inadequate for two of the four high-learning teams, also illustrating that it is possible for teams to overcome limitations in their context through learning behavior. These few cases thus provided evidence that high-learning teams could overcome obstacles they faced in their initial set-up; a lack of structural support did not seal their fate. The Stain Team overcame personnel limits that repeatedly depleted their ranks, and the Fusion Team (described in table 9) overcame time and staffing constraints to push energetically forward on its shared project. In contrast, low-learning teams, such as the Publications and Help Desk teams, appeared vulnerable to a self-sealing pattern of members having private concerns about the team environment, which led to withholding relevant thoughts and actions and made it difficult to escape the low-learning condition. These cases suggest an asymmetry, in which high-learning teams can confront and work with design and other constraints to improve their situation, while low-learning teams are far more likely to get stuck and be unable to alter their situation without intervention.

An integrative perspective that mirrors and reinforces the results of the quantitative data can be articulated from the seven cases; in this, team psychological safety is a mechanism that mediates between effective team design and learning behavior. Effective team leader coaching and context support, such as access to information and resources, appear to contribute to, but not to fully shape, an environment in which team members can develop shared beliefs that well-intentioned interpersonal risks will not be punished, and these beliefs enable team members to take proactive learning-oriented action, which in turn fosters effective performance. Quantitative results also suggest that team psychological safety mediates between team structures (context support and coaching) and the behavioral outcome of team learning. These findings have important implications for theories of team effectiveness. They suggest an explanation for *how* effective team design and leadership enables effective team performance.

Study Limitations and Model Applicability

The results of this study represent a first step in establishing team psychological safety as a construct, but additional conceptual and empirical work is needed to refine and extend the implications of the construct before firm conclusions can be drawn. The qualitative data, although consistent with the proposed construct, did not map onto it precisely. Similarly, survey items used to capture the experience of team psychological safety also have conceptual relationships with other interpersonal constructs, especially trust. Thus, the empirical data were supportive of the existence of team psychological safety as a construct but could not conclusively differentiate it from related constructs. Further research is needed to establish construct validity.

The relationship between psychological safety and learning could be detected across the four types of teams in the study; for example, whether a team was a self-managed team producing chairs or a management team designing transportation strategy, team psychological safety generally was associated with team learning behavior. Although I cannot generalize from this study about the relationship between team learning and team performance for all types of teams, it is likely that under certain conditions, team learning behavior will not control much variance in team performance, such as for team tasks that are highly constrained with tightly specified criteria for success. For example, a team working to assemble a product on a machine-paced assembly line is less likely to benefit from learning behavior than a team with few inherent task constraints and uncertain criteria for success, such as a cross-functional product development team designing a new product. Highly constrained tasks leave little opportunity for information seeking to be helpful in improving team performance, and feedback tends to be built into the task, such that asking others for feedback becomes redundant. Unconstrained tasks such as designing a new product, in contrast, create ample opportunity for the team's output to benefit from new information and feedback. The utility of learning behavior across team types thus deserves further research.

This study provides a limited exploration of factors that managers can influence in their efforts to promote team learning. It focused on two antecedent conditions with clear conceptual relationships to team psychological safety but did not examine a wide range of managerial factors that might also affect team learning. For example, team leader coaching was included in the study, but the data do not specify leader behaviors precisely. Furthermore, analyses testing predictors of team psychological safety had to rely on variables from within the same survey. Although the team survey was not subject to excessive common-method variance, this is still a concern and suggests that findings on the antecedents of team psychological safety should be considered tentative. Thus, further research is needed to explore factors that promote team psychological safety.

The cross-sectional survey design prevented a demonstration of causality and also limited my ability to explore dynamic issues. The theoretical model also leaves out the dynamic interaction that is likely to take place in which team psychological safety facilitates taking the risks of learning behavior, which, when unpunished by the team, further reinforces team psychological safety. A team's history includes events that demonstrate to members that interpersonal risk is or is not worthwhile, and thus both psychological safety and learning may be influenced as much or more by the cumulative effects of interpreting these events as by initial design features, Some evidence of the effects of history could be seen in the Stain Team, where informants contrasted present conditions with those under an earlier leader. Nonetheless, how shared beliefs are created gradually in teams over time as a consequence of minor events and subtle interactions cannot be assessed in this study, nor can whether self-reinforcing cycles or spirals exist. Given the inherently dynamic nature of learning, this snapshot approach provides an incomplete picture. Issues of how team psychological safety develops over time and how team learning behavior might alter undesirable structural factors warrant careful consideration and future research.

Finally, conducting the study in a single company imposed limitations, suggesting caution in drawing conclusions for teams in other organizations. Although there was considerable diversity across teams in work context, organization level, education, and tenure, the sample may not be representative of the full spectrum of possible teams in work organizations. Moreover, with 51 teams, the sample size is small for multivariate analyses. The inclusion of four team types is both a strength and a weakness of the study. On one hand, unlike studies that include only one type of team, such as sales teams or production teams, the findings cannot be said to be merely a function of the nature of the team task. On the other hand, this inclusion also brought in more variables than could be thoroughly tested with only 51 teams. Larger studies could strengthen the validity of the findings.

Conclusion

Structural and interpersonal factors have been viewed in the literature as alternative explanations for team effectiveness. This study supported, instead, an integrative perspective, in which structural and interpersonal characteristics both influence learning and performance in teams. In particular, the results showed that psychological safety is a mechanism that helps explain how previously studied structural factors, such as context support and team leader coaching, influence behavioral and performance outcomes. Future team research has much to gain by investigating how structural and inter-personal factors are interrelated rather than which is more important. To do this, it is essential to study real work teams. There was some evidence in this study that a team's history matters in shaping psychological safety. Shared beliefs about how others will react are established over time; these cannot take shape in the laboratory in a meaningful way. Moreover, for the risks of learning to be salient, the interpersonal consequences must matter — as they do in on-going work relationships. Studying learning in laboratory groups is therefore likely to miss an essential source of variance. Beyond the need to study real groups, longitudinal research could help to develop an understanding of how psychological safety develops or erodes with changes in membership, leadership, or context.

In this study, my focus on learning behavior and its accompanying risk made the interpersonal context especially salient; however, the need for learning in work teams is likely to become increasingly critical as organizational change and complexity intensify. Fast-paced work environments require learning behavior to make sense of what is happening as well as to take action. With the promise of more uncertainty, more change, and less job security in future organizations, teams are in a position to provide an important source of psychological safety for individuals at work. The need to ask questions, seek help, and tolerate mistakes in the face of uncertainty — while team members and other colleagues watch — is probably more prevalent in companies today than in those in which earlier team studies were conducted. This may partially account for the empirical support I found for the role of psychological safety in promoting performance in these work teams; however, it also suggests that psychological safety and ways to promote it will be increasingly relevant for future research on, work teams.

Notes

1. Discriminant validity was also established by creating a multitrait multimethod (MTMM) matrix (Campbell and Fiske, 1959) for each group of variables, from which I confirmed that, for antecedent and outcome variables, correlations between items designed to measure the same construct were larger than correlations between these items and all other items in the section. For the antecedent variables, the average within-trait, between-method correlation was .35, and between-trait, between-method correlations (between each item of a given scale and all items in other scales) averaged .25. For the outcome variables, the average within-trait, between-method correlation was .36 and between-trait, between-method correlations averaged .25.

2. Factor analyses (principal components, varimax rotation), using a cut-off criterion of .40 for factor loadings and eigenvalues of 1.0 or above, yielded six factors for the antecedent variables, replicating most of the planned scales: items for team psychological safety, team efficacy, team task, and clear goal loaded onto four factors exactly as planned, while context support items loaded onto two factors, both conceptually related to context support, and team composition items loaded onto the first three factors. All items were retained in the planned scales because they made a positive contribution to Cronbach's alpha. For the team outcomes section, factor analysis replicated the planned scales — team learning behavior and team performance. To test whether team learning behavior and team psychological safety items were tapping into the same issues rather than into two distinct constructs, I ran a factor analysis on all items from both scales. Reassuringly, two clean factors resulted, replicating the planned scales precisely.

3. Three of the four interview variables were more correlated with the survey scale measuring the same construct than with any other scale in the survey. The interview measure of *adequacy of team composition* was more highly correlated with the survey measure of team composition ($r = .33$, $p < .01$) than with any of the other survey variables. The *degree of task interdependence and wholeness* was most correlated with the survey variable assessing task design ($r = .34$, $p < .01$), and *context support* was most correlated with context support ($r = .33$, $p < .01$). Although the differences between correlation values were in some cases small, the overall degree of

convergence between the two different instruments is striking. The fourth interview variable, *clear team direction*, is more correlated with context support ($r=.28, p<.05$) than with the survey measure of team direction ($r=.17$, $p=.12$); however, this result is, in fact, reassuring for measurement reliability, as the survey and interview "direction" variables measured two distinct constructs. The survey measured the extent to which time and effort had been spent on clarifying team goals, and the interviews asked to what extent the team had a clear shared goal; the low correlation between the two is thus not surprising.

4. To generate each intraclass correlation coefficient, one-way analysis of variance (ANOVA) was conducted on the full data set of 427 cases, with team membership as the independent variable and a team survey scale as the dependent variable. Intraclass correlations are significant when the one-way ANOVA from which the coefficients are derived are significant (Kenny and LaVoie, 1985).

References

Alderfer, Clayton P. 1987 "An intergroup perspective on organizational behavior." In J. W. Lorsch (ed.), *Handbook of Organizational Behavior*: 190–222. Englewood Cliffs, NJ: Prentice-Hall.

Alderfer, Clayton P., and D. L. Brown 1972 "Designing an empathic questionnaire for organizational research." *Journal of Applied Psychology*, 56: 456–468.

Ancona, Deborah G., and David F. Caldwell 1992 "Bridging the boundary: External activity and performance in organizatioeal teams." *Administrative Science Quarterly*, 37: 634–655.

Argote, Linda, Deborah Gruenfeld, and Charles Naquin 1999 "Group learning in organizations." In M. E. Turner (ed.), *Groups at Work: Advances in Theory and Research*. New York: Erlbaum (forthcoming).

Argyris, Chris 1982 *Reasoning, Learning and Action: Individual and Organizational*. San Francisco: Jossey-Bass.

Argyris, Chris 1993 *Knowledge for Action: A Guide to Overcoming Barriers to Organizational Change*. San Francisco: Jossey-Bass.

Argyris, Chris, and Donald Schön 1978 *Organizational Learning: A Theory of Action Perspective*. Reading, MA: Addison-Wesley.

Ashford, Susan J., and Anne S. Tsui 1991 "Self regulation for managerial effectiveness: The role of active feedback seeking." *Academy of Management Journal*, 34: 251–280.

Bandura, Albert 1982 "Self-efficacy mechanism in human agency." *American Psychologist*, 37: 22–147.

Baron, Reuben M., and David A. Kenny 1986 "The moderator-mediator variable distinction in social psychological research: Conceptual, strategic and statistical considerations." *Journal of Personality and Social Psychology*, 51: 1173–1183.

Brown, Roger 1990 "Politeness theory: Exemplar and exemplary." In I. Rock (ed.), The *Legacy of Solomon Asch: Essays in Cognition and Social Psychology*: 23–37. Hillsdale, NJ: Erlbaum.

Campbell, D. T., and D. W. Fiske 1959 "Convergent and discriminant validation by the multitrait-multimethod matrix." In D. N. Jackson and S. Messick (eds.), *Problems in Human Assessment*. New York: McGraw-Hill.

Campion, Michael A., Gina J. Medsker, and A. Catherine Higgs 1993 "Relations between work group characteristics and effectiveness: Implications for designing effective work groups." *Personnel Psychology*, 46: 823–850.

Cohen, Susan G., and Gerald E. Ledford 1994 "The effectiveness of self-managing teams: A quasi-experiment." *Human Relations*, 47: 13–43.

Dewey, John 1922 *Human Nature and Conduct*. New York: Holt.

Dewey, John 1938 *Logic: The Theory of Inquiry*. New York: Holt.

Dutton, Jane 1993 "The making of organizational opportunities: An interpretive pathway to organizational change." In L. L. Cummings and B. M. Staw (eds.), *Research in Organizational Behavior*, 15: 195–226. Greenwich, CT: JAI Press.

Edmondson, Amy C. 1996 "Learning from mistakes is easier said than done: Group and organizational influences on the detection and correction of human error." *Journal of Applied Behavioral Science*, 32: 5–32.

Edmondson, Amy C., and Bertrand Moingeon 1998 "From organizational learning to the learning organization." *Management Learning*, 29: 499–517.

Gibson, Christina B. 1996 "They do what they believe they can: Group-efficacy beliefs and group performance across tasks and cultures." *Academy of Management Journal*, 42: 138–152.

Goffman, Erving 1955 "On face-work: An analysis of ritual elements in social interaction." *Psychiatry*, 18: 213–231.

Golembiewski, Robert T., and Mark McConkie 1975 "The centrality of interpersonal trust in group process." In Cary L. Cooper (ed.), *Theories of Group Process*: 131–181. London: Wiley.

Goodman, Paul, S, Devadas, and T. L. Hughson 1988 "Groups and productivity: Analyzing the effectiveness of self-managing teams." In J. P. Campbell, R. J. Campbell, and Associates (eds.), *Designing Effective Work Groups*: 295–327. San Francisco: Jossey-Bass.

Goodman, Paul, E. Ravlin, and M. Schminke 1987 "Understanding groups in organizations." In L. L. Cummings and B. M. Staw (eds.), *Research in Organizational Behavior*, 9: 121–173. Greenwich, CT: JAI Press.

Guzzo, R. A., P. R. Yost, R. J. Campbell, and G. P. Shea 1993 "Potency in groups: Articulating a construct." *British Journal of Social Psychology*, 32: 87–106.

Hackman, J. Richard 1987 "The design of work teams." In J. Lorsch (ed.), *Handbook of Organizational Behavior*: 315–342. Englewood Cliffs, NJ: Prentice-Hall.

Hackman, J. Richard (ed.) 1990 *Groups that Work (and Those That Don't)*. San Francisco: Jossey-Bass.

Henderson, Rebecca H., and Kim B. Clark 1990 "Architectural innovation: The reconfiguration of existing product technology and the failure of existing firms." *Administrative Science Quarterly* 35: 9–30.

Janis, Irving L. 1982 *Groupthink*, 2d ed. Boston: Houghton-Mifflin.

Kenny, David A., and Lawrence La Voie 1985 "Separating individual and group effects." *Journal of Personality and Social Psychology*, 48: 339–348.

Klimoski, Richard, and S. Mohammed 1994 "Team mental model: Construct or metaphor?" *Journal of Management*, 20: 403–437.

Kolb, David A. 1984 *Experiential Learning*. Englewood Cliffs, NJ: Prentice-Hall.

Kramer, Roderick M. 1999 "Trust and distrust in organizations: Emerging perspectives, enduring questions." *Annual Review of Psychology*, 50: 569–598. Palo Alto, CA: Annual Reviews.

Lee, Fiona 1997 "When the going gets tough, do the tough ask for help? Help seeking and power motivation in organizations." *Organizational Behavior and Human Decision Processes*, 72: 336–363.

Leonard-Barton, Dorothy 1995 *Wellsprings of Knowledge: Building and Sustaining the Sources of Innovation*. Boston: Harvard Business School Press.

Levine, John M., and Richard L. Moreland 1991 "Culture and socialization inwork groups." In L. B. Resnick, J. M. Levine, and S. D. Teasley (eds.), *Perspectives on Socially Shared Cognition*: 257–279: Washington, DC: American Psychological Association.

Levitt, Barbara, and James G. March 1988 "Organizational learning." *Annual Review of Sociology*, 14: 319–340. Palo Alto, CA: Annual Reviews.

Lindsley, Dana H., Daniel J. Brass, and James B. Thomas 1995 "Efficacy-performance spirals: A multi-level perspective." *Academy of Management Review*, 20: 645–678.

MacDuffie, John Paul 1997 "The road to 'root cause': Shop-floor problem-solving at three auto assembly plants." *Management Science*, 43: 479–502.

Mayer, Roger C., James H. Davis, and F. David Schoorman 1995 "An integrative model of organizational trust." *Academy of Management Review*, 20: 709–734.

Michael, Donald N. 1976 *On Learning to Plan and Planning to Learn*. San Francisco: Jossey-Bass.

Robinson, Sandra L. 1996 "Trust and breach of the psychological contract." *Administrative Science Quarterly*, 41: 574–599.

Sanna, Lawrence J., and R. Lance Shotland 1990 "Valence of anticipated evaluation and social facilitation." *Journal of Experimental Social Psychology*, 26: 82–92.

Schein, Edgar H. 1993 "How can organizations learn faster? The challenge of entering the green room." *Sloan Management Review*, 34: 85–92.

Schein, Edgar H., and Warren Bennis 1965 *Personal and Organizational Change via Group Methods*. New York: Wiley.

Schön, Donald 1983 *The Reflective Practitioner*. New York: Basic Books.

Senge, Peter M. 1990 *The Fifth Discipline: The Art and Practice of the Learning Organization*. New York: Doubleday.

Sitkin, Sim B. 1992 "Learning through failure: The strategy of small losses." In L. L. Cummings and B. M. Staw (eds.), *Research in Organizational Behavior*, 14: 231–266. Greenwich, CT: JAI Press.

Stasser, Garold, and William Titus 1987 "Effects of information load and percentage of shared information on the dissemination of unshared information during group discussion." *Journal of Personality and Social Psychology*, 53: 81–93.

Staw, Barry M., Lance E. Sandelands, and Jane E. Dutton 1981 "Threat-rigidity effects in organizational behavior: A multi-level analysis." *Administrative Science Quarterly*, 26: 501–524.

Tyler, Tom R., and E. Allan Lind 1992 "A relational model of authority in groups." In *Advances in Experimental Psychology*, 25: 115–191. New York: Academic Press.

Wageman, Ruth 1998 "The effects of team design and leader behavior on self-managing teams: A field study." Working paper, School of Business, Columbia University.

Walsh, James 1995 "Managerial and organizational cognition: Notes from a trip down memory lane." *Organization Science*, 6: 280–321.

Watkins, Victoria J., and Karen E. Marsick 1993 *Sculpting the Learning Organization: Lessons in the Art of Systemic Change*. San Francisco: Jossey-Bass.

Appendix: Survey Scales and Item Correlations

Team survey item*	Pearson *R* correlations			
Supportiveness of organization context	2	3	4	5
1. This team gets all the information it needs to do our work and plan our schedule.	.43	.29	.26	.22
2. It is easy for this team to obtain expert assistance when something comes up that we don't know how to handle.		.26	.30	.24
3. This team is kept in the dark about current developments and future plans that may affect its work.			.26	.29
4. This team lacks access to useful training on the job.				.15
5. Excellent work pays off in this company.				
Task design			2	3
1. The work that this team does makes a difference for the people who receive or use it.			.23	.27
2. The work we do on this team itself provides us with plenty of feedback about how well the team is performing.				.25
3. Those who receive or use this team's output rarely give us feedback about how well our work meets their needs.				
Clear direction			2	3
1. It is clear what this team is supposed to accomplish.			.41	.38
2. This team spent time making sure every team member understands the team objectives.				.65
3. The team has invested plenty of time to clarify our goals.				

Appendix: *Continued*

Team survey item*	Pearson *R* correlations					
Team composition					2	3
1. Most people in this team have the ability to solve the problems that come up in our work.					.26	.34
2. All members of this team have more than enough training and experience for the kind of work they have to do.						.27
3. Certain individuals in this team lack the special skills needed for good team work.						
Team efficacy					2	3
1. Achieving this team's goals is well within our reach.					.37	.43
2. This team can achieve its task without requiring us to put in unreasonable time or effort.						.28
3. With focus and effort, this team can do anything we set out to accomplish.						
Team psychological safety*	2	3	4	5	6	7
1. If you make a mistake on this team, it is often held against you.	.36	.38	.49	.41	.34	.43
2. Members of this team are able to bring up problems and tough issues.		.28	.56	.35	.34	.37
3. People on this team sometimes reject others for being different.			.32	.45	.45	.33
4. It is safe to take a risk on this team.				.37	.37	.48
5. It is difficult to ask other members of this team for help.					.42	.41
6. No one on this team would deliberately act in a way that undermines my efforts.						.39
7. Working with members of this team, my unique skills and talents are valued and utilized.						
Team leader coaching†					2	3
The team leader . . .						
1. . . . initiates meetings to discuss the team's progress.					.38	.47
2. . . . is available for consultation on problems.						.70
3. . . . is an ongoing "presence" in this team-someone who is readily available.						
Team learning behavior*	2	3	4	5	6	7
1. We regularly take time to figure out ways to improve our team's work processes.	.23	.28	.35	.41	.30	.23
2. This team tends to handle differences of opinion privately or off-line, rather than addressing them directly as a group.		.26	.31	.27	.29	.22
3. Team members go out and get all the information they possibly can from others – such as customers, or other parts of the organization.			.38	.35	.37	.37
4. This team frequently seeks new information that leads us to make important changes. . . .				.41	.47	.37
5. In this team, someone always makes sure that we stop to reflect on the team's work process.					.47	.25
6. People in this team often speak up to test assumptions about issues under discussion.						.43
7. We invite people from outside the team to present information or have discussions with us.						

Appendix: *Continued*

Team survey item*	Pearson *R* correlations					
Team performance*			2	3	4	5
1. Recently, this team seems to be "slipping" a bit in its level of performance and accomplishments.			.38	.41	.44	.40
2. Those who receive or use the work this team does often have complaints about our work.				.26	.42	.47
3. The quality of work provided by this team is improving over time.					.36	.26
4. Critical quality errors occur frequently in this team.						.51
5. Others in the company who interact with this team often complain about how it functions.						
Internal motivation*					2	3
1. My opinion of myself goes up when I do my job well.					.33	.48
2. I feel bad and unhappy when I discover that I have performed less well than I should have in my job.						.30
3. I feel a great sense of personal satisfaction when I do my job well.						
Job Involvement*						
1. I live, eat, and breathe my job.						
Observer Survey Items						
Team learning behaviors†	2	3	4	5	6	7
This team . . .						
1. . . . asks its internal customers (those who receive or use its work) for feedback on its performance.	.24	.55	.39	.52	.37	.30
2. . . . relies on outdated information or ideas. (Reverse scored)		.46	.37	.36	.47	.36
3. . . . actively reviews its own progress and performance.			.47	.61	.38	.51
4. . . . does its work without stopping to consider all the information team members have. (Reverse scored)				.36	.44	.22
5. . . . regularly takes time to figure out ways to improve its work performance.					.49	.44
6. . . . ignores feedback from others in the company. (Reverse scored)						.30
7. . . . asks for help from others in the company when something comes up that team members don't know how to handle.						
Team performance*				2	3	4
1. This team meets or exceeds its customers' expectations.				.77	.62	.64
2. This team does superb work.					.57	.71
3. Critical quality errors occur frequently in this team's work. (Reverse scored)						.53
4. This team keeps getting better and better.						

*7-point scale from "very inaccurate" to "very accurate."
†5-point scale from "never" to "always."